# Molecular
# Approaches
# to
# Plant
# Physiology

*McGraw-Hill Series in Organismic Biology*

*Consulting Editors*

Professor Melvin S. Fuller
Department of Botany
University of Georgia, Athens

Dr. Paul Licht
Department of Zoology
University of California, Berkeley

# C. A. Price

Department of Biochemistry and Microbiology
Rutgers University

# Molecular Approaches to Plant Physiology

McGraw-Hill Book Company

*New York   St. Louis   San Francisco   Düsseldorf*
*London   Mexico   Panama   Sydney   Toronto*

This book was set in Garamond by The Maple Press Company, and printed on permanent paper and bound by The Maple Press Company. The designer was Marsha Cohen; the drawings were done by B. Handelman Associates, Inc. The editor was James R. Young. Sally R. Ellyson supervised the production.

MOLECULAR APPROACHES TO PLANT PHYSIOLOGY

*Library of Congress Catalog Card Number* 77-95822

50853

1 2 3 4 5 6 7 8 9 0   M A M M   7 9 8 7 6 5 4 3 2 1 0

Dedicated
to the
memory of
David Powell Hackett

# Preface

The purpose of this text is to help students understand plant physiology. Since it is clearly impossible for any of us to assimilate all that is known or thought in a field as hoary as this, I have aimed at the more attainable goal of learning the rationale of plant-physiological research. If I were to be completely successful, the student—upon reading a paper, review article, or treatise in the field of plant physiology—would impose his own intellectual gestalt on it. He would ask: What are the phenomena under discussion? What hypotheses are being tested? What are the kinds of evidence that are valid for testing these hypotheses? Do the data provide this evidence? Have the hypotheses been proved or disproved?

My approach has been to identify those properties of molecules which provide both the inspiration for our ideas about modern plant physiology and the means for testing them. Thus, our understanding of general biochemistry has stimulated hypotheses on the chemical transformations that occur in plants (Chap. 1). The laws of photochemistry provide the basis for photophysiology (Chap. 2). Similarly, certain areas of physical chemistry and chelate chemistry underlie notions about fluid movements (Chap. 3) and inorganic nutrition (Chap. 4). Finally, I have inquired how the theories of molecular genetics and the properties of informational macromolecules may provide a basis for testing some of the older and newer ideas about growth, differentiation, and development (Chap. 5).

Each chapter begins with a brief discussion of the kinds of phenomena which the hypotheses hope to explain. There follows a more detailed development of the relevant physical, chemical, or biological theory and the evidence required to formulate and evaluate hypotheses in the area. Finally, in the Case Studies the hypotheses themselves are presented, usually with some indication of parentage. The kinds of acceptable evidence by which they can be tested are then considered, and the evidence is evaluated as a basis for acceptance or rejection of the hypothesis.

The text has therefore been limited to a portion, hopefully representative, of the ideas in plant physiology. It has been assumed either that the phenomena are already known to the student or that information on them is readily accessible. The text is emphatically not encyclopedic, and the selection of topics is biased by the heady expectation that it is only a matter of time before those areas of plant physiology which are still largely descriptive will yield to molecular approaches.

The text is intended for advanced undergraduate and beginning graduate levels. I have assumed familiarity with general biochemistry, organic chemistry, physics, and those elements of physical chemistry which are included in current undergraduate courses in analytical chemistry or biochemistry. The text will be appropriate in those biology programs where the elements of biochemistry are taught as the basis for physiology, genetics, cell biology, and so on.

Several individuals kindly consented to read portions of this manuscript: I am most grateful to George Curry, Harold Evans, J. Woodland Hastings, A. Carl Leopold, Aubrey Naylor, and Conrad Yocum. My colleagues at Rutgers—Harold E. Clark and Cecil C. Still—deserve my deepest thanks, not only for correcting and criticizing the text but also for subjecting it to classroom tests.

Over the years which the text was evolving, the most trenchant criticism came from our students. I wish to thank those who were particularly generous in this respect: David Brown, Nikos Galvalas, Alice Hirvonen, George Knight, Leticia Mendiola, Willian K. Neal II, John Quigley, Alan Rutner, and Curtis Suerth. Despite the abundance of help, I have not always followed advice; so the final results are clearly my responsibility. Finally, I wish to thank Mary Lou Tobin, the most perfect of secretaries.

*C. A. Price*

# Contents

# Molecular Approaches to Plant Physiology

# 1.
# Chemical Transformations in Green Plants

*"O Lord, I beg upon my knees*
*That all my various syntheses*
*May not turn out to be inferior*
*To those conducted by bacteria."*
　　　　　*——Anon., discovered by K. V. Thimann*

## 1.1 GREEN PLANTS FROM THE PERSPECTIVE OF COMPARATIVE BIOCHEMISTRY

Let us imagine the set of all chemical reactions occurring in all living organisms—green plants, bacteria, fungi, vertebrate animals, invertebrates, etc.—and represent the set of these reactions by the large circle in Fig. 1-1. On the

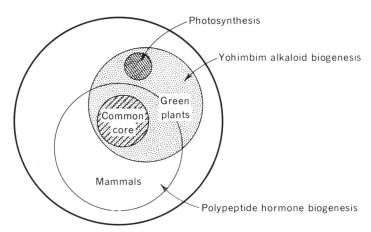

*Fig. 1-1* Distribution of chemical processes in living organisms.

basis of present knowledge we can imagine a subset consisting of reactions that are duplicated with only slight variations in *all* living organisms, such as the formation of activated amino acids preparatory to protein synthesis. We present this common subset by the area labeled "common core" in Fig. 1-1. The reactions occurring in any one kind of organism or group of organisms would be a large subset including the common core subset (e.g., circles labeled "green plants" and "mammals"). Since we do not know how many reactions occur in any one organism, much less in all living organisms, the relative sizes of the circles are arbitrary. The important point is that a large number of reactions occurring in, say, mammals also occur in green plants.

In addition to the reactions common to all living organisms there are processes characteristic of large groups of organisms, such as oxidative phosphorylation in aerobic organisms, photosynthesis in green plants, and hemoglobin formation in vertebrates. Finally, there are reactions that are specific to individual families, genera, and species, such as the reactions leading to the formation of certain alkaloids in plants and to the $\beta$-chains of hemoglobin in primates.

The chemical transformations in green plants are a series of subsets for which the "universe of discourse" is general biochemistry; therefore, we may search general biochemistry for deductions that may be relevant to green plants. Because of what Seymour Hutner called "mammalian chauvinism," to which we might add "*E. coli* chauvinism," we know much more about the biochemistry of rats and *E. coli* than the biochemistry of green plants.

There will be no attempt here to describe or even to outline all the chemical transformations that occur in plants. One has only to recognize that each of the thousands of chemical components of plants must arise through a chemical process (usually a chain of reactions) to understand that even a catalog of such processes would be encyclopedic (cf. Bernfeld, 1966). Instead we shall examine the *basis of our knowledge of the phenomena*. We shall inquire into the *epistemology,* the how-we-know-what-we-know of physiology.

## 1.2 KINDS OF EVIDENCE FOR THE IDENTIFICATION OF METABOLIC PATHWAYS

The identification of the grand reaction pathways in green plants is part of our classical scientific heritage. These include the overall reactions of photosynthesis, respiration, fat synthesis, protein and polysaccharide accumulations. As noted above, the chemical identification of aromatic compounds leads directly to the inference that reactions forming these compounds must exist. Further, we often have data from balance experiments which tell us about both reactants and products. What remains is the elucidation of reaction pathways.

One of the most exacting tasks in modern biology is the selection of criteria for the identification of reaction pathways in living systems. Serious errors have been made by a too-casual acceptance of evidence for the operation of a chemical transformation in the living cell. In physiology we employ several different kinds of evidence representing criteria of varying power and usefulness. We shall learn to recognize some of them and the extent to which they individually and collectively constitute proof.

### Demonstrations of Pathways in Cell-free[1] Systems

One of the first objectives of a biochemically oriented physiologist is the demonstration that a transformation can be made to occur in a cell-free system. Cell-free systems have the obvious advantage that one can hope to subject them to further analysis, involving the traditional scientific strategy of separation of variables. But the cell-free system is a biochemical one, no longer a physiological one. In order to understand what can and what cannot be deduced from biochemical systems, it is extremely important that biochemistry be placed in proper relation to physiology. In general, physiology is concerned with the functions of intact cells, tissues, and organisms, whereas biochemistry is concerned with the composition of these aggregations and with the chemical properties of the micromolecules, macromolecules, and multimolecules that compose the organism. This distinction should be especially clear when we consider the demonstration

[1] The term *in vitro* ("in glass") is used synonymously with "cell-free." The term *in vivo* ("in life"), on the other hand means in the intact or living organism. Tissue-culture artists often use *in vitro* in the more liberal sense to mean tissues or intact cells isolated from a plant and cultured *in glass*.

of metabolic processes in cell-free systems as evidence for the operation of these processes in physiological systems.

*The demonstration of enzyme activities in cell-free systems establishes only the possibility that the enzyme is active physiologically.* Information from cell-free systems is of great value; it tells what to look for physiologically; it may permit us to devise tests involving kinetics, inhibitors, or even direct observations of enzyme-substrate complexes. The demonstration of enzyme activity in a cell-free system in amounts adequate to catalyze the reaction at the observed physiological rate is therefore a necessary but not a sufficient proof for the operation of the enzyme in the intact cell, tissue, or organism.

Consider the process

$$A \rightarrow B$$

or more usually $A \rightarrow X_1 \rightarrow X_2 \rightarrow X_3 \rightarrow \cdots \rightarrow X_n \rightarrow B$. It is clear (Sec. 1.4) that the plant must have one or more enzymes to catalyze this reaction. If we discover an enzyme that catalyzes the process or a part of the process and the amount of enzyme activity is sufficient to account for the physiological process, we are naturally encouraged to consider the possibility that the enzyme lies along the pathway of the process in question. There are many instances, however, when a number of alternative enzyme activities can be demonstrated. In the area of starch transformations, for example, there are at least five quite independent enzyme systems in higher plants, any one of which can catalyze the conversion of amylose to glucose.

On some occasions a process is restricted to a certain stage in the life cycle or to certain tissues, or it is influenced by nutritional or environmental conditions. We may sometimes discover that the activity of an enzyme, putatively identified with a process, fluctuates with the rate of the physiological process. The Latin phrase *pari passu,* meaning "at an equal pace," is often used to designate this kind of evidence. It is clearly much weightier than the demonstration of the mere existence of the enzyme.

Even though evidences of a *pari passu* nature or the discovery of sequential enzyme systems do not constitute sufficient proof of the identity of an *in vivo* process, these kinds of evidence taken separately or together do reinforce the hypothesis that the enzyme or system is in fact operative. The notion of reinforcing or intersecting evidence is expanded in Sec. 1.3.

### Intracellular Localization

In a sense the difference between a cell and a bag full of enzymes is the essence of a good part of physiology, and part of this difference is the organization of enzymes into specific structures.

We suspect (and in some cases we have evidence) that the metabolic control and high efficiencies that distinguish *in vivo* from *in vitro* processes are associated with structural elements of the cell. It is clear, therefore, that the study of isolated organelles in which the structural elements are preserved represents an intermediate condition between purified enzymes and intact cells and tissues. This halfway point serves in analyzing a physiological process into its component enzymes, in reconstructing control mechanisms, and in elucidating the physical arrangements that permit high efficiencies and high reaction rates.

The discovery of *enzyme systems* as distinct from isolated enzymes has changed physiological thinking. These systems are sets of enzymes, typically contained in organelles, and capable of carrying out a sequence of reactions. For example, Hans Krebs (1943) proposed that carbohydrate is oxidized by a complex cycle of enzymatic transformations (the Krebs cycle, cf. Sec. 1.8 and Appendix A). If, at that time, he could have pointed to an organelle (the mitochondrion) in which all the requisite enzymes were nicely packaged, he would have had a far stronger case than the demonstration of the separate enzymes.

Enzyme systems may occur in smaller units. Lynen (1967) has described a 100-Å particle from yeast, containing seven enzymes and one coenzyme, which is capable of catalyzing the complete synthesis of fatty acids from malonyl coenzyme A, acetyl coenzyme A, and NADPH.

It seems unlikely that a catenary set of enzymes should occur in a single structure, unless such a set were functional. The discovery of enzyme systems therefore is a kind of intersecting evidence which we shall see is of great importance in our epistemology (cf. Sec. 1.3).

At the same time we should also recognize that we have suffered from severe methodological limitations in studying subcellular organelles. The term "organelle" itself is employed differently by different workers. Chloroplasts, as an example, have been identified at one time or another on the basis of their external appearance, composition, function, or ultrastructure.

It should be sufficient to characterize organelles on the basis of their *ultrastructure* (size, shape, membrane thickness, number of repeating units, etc.), but in order to understand their function, it is also necessary for us to study their metabolism and their chemical and enzymological composition. Such a study requires that we be able to isolate the particles from the cell and from other cellular constituents.

Anderson (1966; cf. also Price, 1970) has shown that a number of organelles can be separated cleanly from one another in the *zonal centrifuge*. In this kind of centrifuge particles can be separated on the basis of either their equilibrium density $\rho$ or their sedimentation coefficient $S$. Equilibrium density is the density of a medium in which the particles tend neither to rise nor to sink. $S$ is a corrected *rate* of movement of particles in a centrifugal field and is related to the factors of particle size and shape.

The twin parameters $\rho$ and $S$ can be used not only for the physical separation of organelles but for their identification and characterization. We can

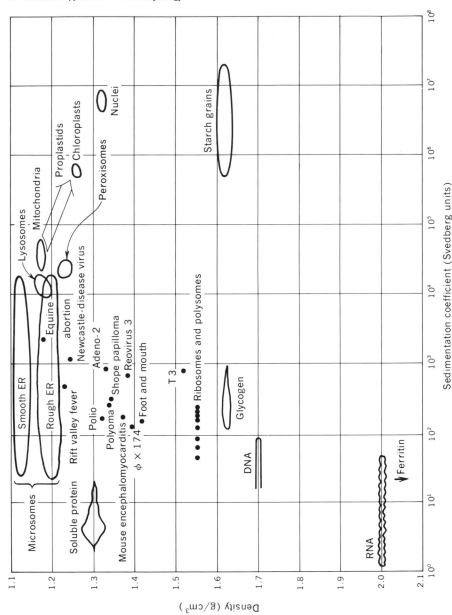

Fig. 1-2  S-ρ plot of subcellular particles (after Anderson, 1966). The organized structures and macromolecules of plant and animal cells can be characterized by their behavior in centrifugal fields. These particles have virtually unique combinations of sedimentation coefficients (S values) and equilibrium densities (ρ values). Several values in the figure are estimates based on the range of sizes of the particles.

treat $\rho$ and $S$ as cartesian coordinates. In the resulting $S$-$\rho$ plot, also known as an Anderson plot, the known classes of organelles occupy unique areas (Fig. 1-2). A convenient circumstance of nature, which has greatly aided virus isolation, is that most known viruses fall within a single area, the "virus window."

One can easily imagine extending the usefulness of the $S$-$\rho$ plot by determining the "migration" of a class of particles on the $S$-$\rho$ surface during biogenesis or some other transformation (Knight and Price, 1968; Neal, 1969).

Organelles can be separated by other procedures, such as differential centrifugation, but without repeated washings the fractions obtained are of uncertain purity.

Purity and the preservation of ultrastructure are the necessary preconditions for attributing physiological importance to findings with isolated organelles. Measurements of the enzyme activities of a suspension of mitochondria (cf. Sec. 1.8), for example, are meaningful only to the extent that the suspension contains mitochondria identifiable with organelles of intact cells. Plates 1 through 8 show something of the levels of purity and preservation of structure that can be achieved in the isolation of organelles.

*Note:* All the following lines of evidence were obtained with intact cells, tissues, or organisms.

### Detection of Enzyme-Substrate Complexes

The direct observation *in vivo* of an enzyme-substrate complex, together with knowledge of its substrate and product, is the only single method that provides *sufficient proof* for the operation of an enzyme *in vivo*. Unfortunately, the opportunities for direct observation of enzyme-substrate complexes are rare (cf. the detection of cytochromes in Sec. 1.8). Furthermore, the detection of an enzyme-substrate complex does not automatically identify the physiological process in which the reaction is occurring.

### Enzyme Inhibition

If one were to measure the amount of physiological research performed by the weight of the publications, the unquestioned heavyweight would be studies employing inhibitors.

The reasons are that inhibitors have proved very fruitful in characterizing enzymes and that the rich resources of organic chemistry were early applied to the design and synthesis of inhibitors.

Our concern is not with the undoubted usefulness of inhibitors in characterizing enzymes, but the use of inhibition as a criterion for the participation of an enzyme in a pathway of chemical transformation *in vivo*.

The overriding question in assessing physiological inhibitions is that of specificity. There are five broad classes of metabolic inhibitors: substances that

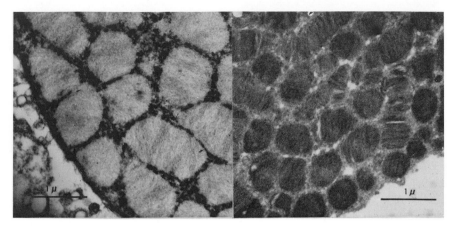

*Plate 1* Portion of the nucleus of a dyno-flagellate, *Gymnodinium nelsoni* (plate kindly prepared by Donna Kubai). This electron photomicrograph shows the condensed chromosome structure peculiar to the dynophyceae.

*Plate 2* Portion of an isolated nucleus from *Gymnodinium nelsoni* (Mendiola et al., 1966; plate kindly prepared by Donna Kubai). The ultrastructure of the nucleus is preserved although the limiting membrane may be largely removed or perforated. The nuclei were isolated by differential centrifugation.

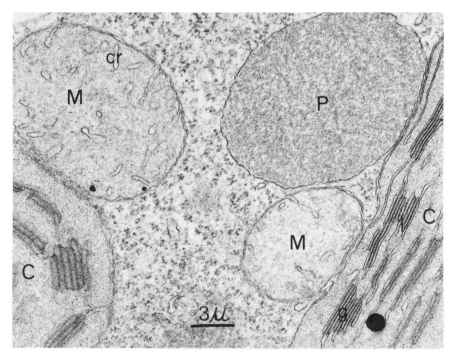

*Plate 3* Section of a tobacco leaf (cf. Frederick and Newcomb, 1969; plate kindly supplied by Eldon Newcomb). This remarkably clear photomicrograph shows two mitochondria (M), portions of chloroplasts (C), and a peroxisome (P) appressed to a chloroplast. Note the inner and outer membranes of the chloroplasts and mitochondria, the cristae of the mitochondria (cr), and the lamellar membranes of the chloroplast, which frequently form grana (g).

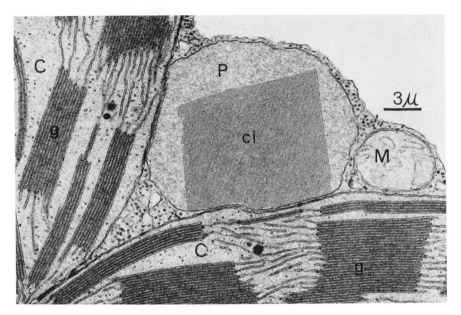

*Plate 4* Section of a tobacco leaf (cf. Frederick and Newcomb, 1969; plate kindly supplied by Eldon Newcomb). This photomicrograph is similar to Plate 3, except that the catalase of the peroxisome is organized into a crystal lattice (cl). The numerous small dots in the cytoplasm and chloroplasts are ribosomes.

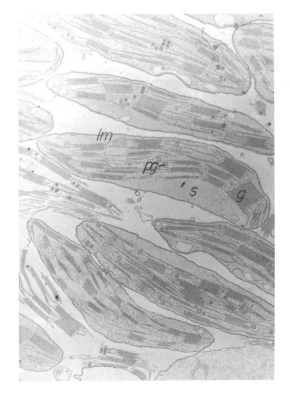

*Plate 5* Isolated spinach chloroplast (Eriksson et al., 1961; plate kindly supplied by Albert Kahn). The chloroplast shows intact limiting membranes (lm), and the lamellae are frequently aggregated as grana (g) and stroma (s). This section was fixed with permanganate, which heightens the contrast of membranes but destroys some features such as ribosomes.

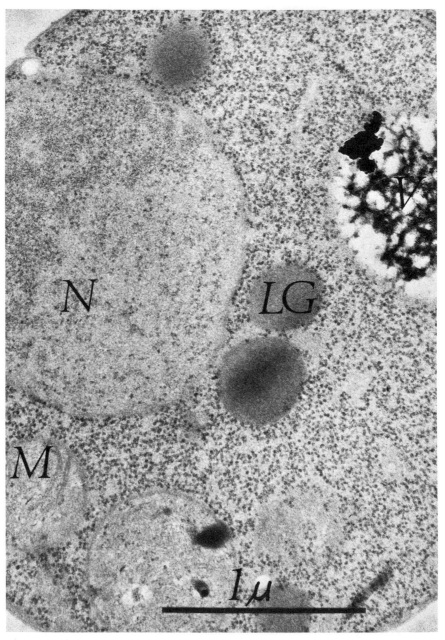

*Plate 6* Mitochondrion in a yeast cell (photomicrograph kindly supplied by H.-P. Hoffmann and W. K. Neal, II). Yeast (*Saccharomyces cerevisiae*) grown in high concentrations of fermentable sugars form a few large but inactive mitochondria (M). As in the mitochondria of higher plants (cf. Plates 3 and 4), there are cristae and ribosomes, and sometimes a DNA-containing nucleoid region is visible. Note also the nucleus (N), the vacuole (V), and the numerous ribosomes in the cytoplasm. The section was prepared from a spheroplast, that is, a cell from which the cell wall was stripped off by enzymatic digestion.

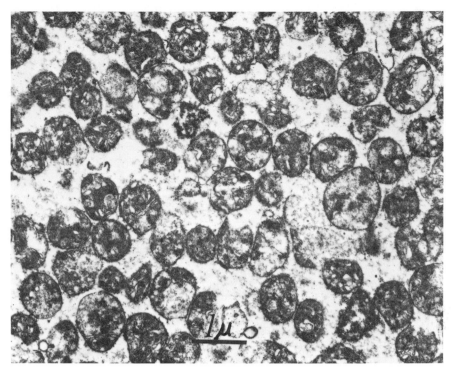

*Plate* 7 Isolated yeast mitochondria (photomicrograph kindly supplied by W. K. Neal. II, and H.-P. Hoffmann). The source of these mitochondria was yeast whose respiration was repressed with glucose; that is, the same kind of yeast as shown in Plate 6. The inner spaces of the cristae are swollen, but the membrane systems are intact and the ground substance remains dense. The mitochondria were isolated by zonal centrifugation.

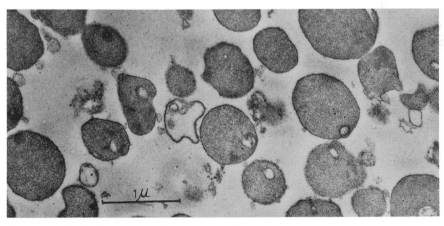

*Plate* 8 Isolated glyoxysomes (peroxisomes) from castor bean endosperm (cf. Breidenbach et al., 1968; electron photomicrograph kindly supplied by Harry Beevers). These single membrane-bound organelles are similar in ultrastructure and enzymatic composition to those in green leaves (Plates 3 and 4), but the particles additionally have certain key enzymes of the glyoxalate cycle (cf. 1.4).

react with functional groups of proteins; amino acid analogs and other substances that alter the activity of enzymes and coenzymes by incorporation as metabolic errors; substrate analogs, including most competitive inhibitors; substances that combine with or alter nucleic acids; and inhibitors whose sites of action are not known.

PROTEIN REAGENTS (TABLE 1-1).   The rational use of substances reacting with the functional groups of protein is almost completely restricted to isolated enzymes. The reason is fairly apparent: the number of enzymes with exposed sulfhydryl or mercaptide groups, essential disulfide bonds, amino or carboxyl groups are so large that little specific information can be obtained. A very incomplete list of physiological processes sensitive to —SH binding reagents would include respiration, photosynthesis, auxin-induced growth, and cell division.

Table 1-1   Inhibitors Which Act by Reacting with Functional Groups of Proteins*

Although there are many kinds of functional group reagents for proteins, only the following two groups have been of physiological importance.

| Functional Group | Inhibitor | Reaction |
|---|---|---|
| Sulfhydryl (RSH) | $p$-Chloromercuribenzoate, $p$-chloromercurisulfonate, etc. | $RSH + ClHg\langle\bigcirc\rangle COO^{\ominus} \rightarrow$ $RSHg$ $\langle\bigcirc\rangle COO^{\ominus} + H^+ + Cl^-$ |
|  | Iodoacetamide, iodoacetate, arsenite | $RSH + ICH_2CONH_2 \rightarrow$ $RSCH_2CONH_2 + H^+ + I^-$ $R_1SH \qquad\qquad R_1S$ $\mid \qquad + As^{3+} \rightarrow \quad \mid \rangle As^+ + 2H^+$ $R_2SH \qquad\qquad R_2S$ |
| Metals (Me) | Cyanide, azide, carbon monoxide, $o$-phenanthroline, etc. | $n\text{Me} + m\text{L} \rightleftharpoons [\text{Me}_n\text{L}_m]$ |

* After Hackett, 1960; Frankel-Conrat, 1959.

For the same reasons, substances that react with the prosthetic groups of enzymes—heme, metal ions, flavins, etc.—inhibit most physiological processes. Thus, if a physiological process is inhibited by a complexing agent such as azide or $o$-phenanthroline, it is reasonable to infer that a heavy metal or metal-enzyme is involved in the process; the unusual event, however, would be a process that is *not* inhibited by one or another complexing agent!

The obverse is not true: failure of inhibitions by a "functional group" inhibitor is not an unambiguous indication that the functional group is *not* involved. The reason stems directly from enzyme chemistry: (*a*) the sulfhydryl or mercaptide groups of certain enzymes may be masked, even when the group plays an essential role, e.g., bovine pancreatic carboxypeptidase; and (*b*) certain metalloproteins are resistant to the action of complexing agents (e.g., cytochrome *c*) or susceptible only under special conditions of pH, etc. (e.g., glutamate dehydrogenase).

The one apparent exception to the physiological nonspecificity of these functional group reagents in physiology is the strikingly specific light-reversible inhibition of cytochrome oxidase by carbon monoxide (cf. Sec. 1.8).

ANTIMETABOLITES (TABLE 1-2). There are numerous instances in which physiological processes have been selectively inhibited by the introduction of a substance that mimics a natural intermediate sufficiently well that it is incorporated into a metabolic product (e.g., a nucleic acid). The product may then fail to function normally.

Table 1-2 Inhibitors Which Act after Incorporation into Cell Constituents*

These substances are typically analogs of normally occurring molecules; the analogs are incorporated as "errors" into proteins, nucleic acids, or coenzymes, resulting in "defective" molecules.

| Analog | Normal Metabolite | Normal Product |
|---|---|---|
| Fluorophenylalanine | Phenylalanine | Proteins |
| Ethionine | Methionine | Proteins |
| Canavanine | Arginine or lysine | Proteins |
| 6-Mercaptopurine | Adenine | Ribonucleic acids |
| Benzimidazole | | Ribonucleic acids |
| 8-Azaguanine | Guanine | Ribonucleic acids |
| 8-Azaadenine | Adenine | Ribonucleic acids |
| 5-Fluorouracil | Uracil | Ribonucleic acids |

* From Hochster and Quastel, 1963.

The physiologist must determine that the antimetabolite is going into the expected product. Fluorodeoxyuridine, for example, has been employed as a means of generating defective DNA, but it may also interfere with RNA metabolism.

SUBSTRATE ANALOGS (TABLE 1-3). The classical example of a substrate analog is malonate, which can be mistaken for succinate by succinic dehydrogenase. Sometimes the normal product of an enzyme reaction may also act as an inhibitor for the substrate.

Table 1-3    Inhibitors Which Act as Substrate Analogs

Substrate analogs are molecules which resemble an enzyme substrate and are bound to an enzyme in the place of the normal substrate.

| Enzyme | Substrate | Substrate Analog |
| --- | --- | --- |
| ATP-phosphoglycerate transphosphorylase* | Phosphate | Arsenate |
| Aconitase | Citrate | Fluorocitrate |
| Succinate dehydrogenase | Succinate | Malonate |
| Mg-protoporphyrin-methionine transmethylase* | Methionine | Ethionine |

* And many others.

With isolated enzymes we may subject a substrate analog to a variety of tests from which we may determine the $K_I$ or dissociation constant for the enzyme-inhibitor system (Eq. 1-1; also see Appendix B).

$$E + I \underset{k_2}{\overset{k_1}{\rightleftharpoons}} EI$$

$$K_I \equiv \frac{k_2}{k_1} \qquad\qquad\qquad\qquad \text{Eq. 1-1}$$

Here E is the concentration of the free enzyme, I the concentration of the inhibitor, and EI the concentration of the enzyme-inhibitor complex.

Suppose now that we apply the same inhibitor to a physiological system. There are three possible kinds of results: inhibition identical to that with the intact enzyme, inhibition not identical, and no inhibition. If we obtain the same results as with the isolated enzyme, e.g., competitive kinetics or reversibility, we may take this as evidence that the enzyme is operative in the physiological system.

But just as with the protein reagents, we can draw no conclusion from negative results. As in the case of protein reagents, there are a number of factors that may cause an enzyme in a physiological system to react differently from in isolation.

- Because of permeability barriers, the enzyme may not be available to exogenously supplied substrates or inhibitors (cf. Sec. 1.8).
- The concentration of the substrate may be unknown. If the enzyme is localized in an organelle, it is possible for the natural substrate also to be concentrated there, so that a competitive inhibitor is less effective.
- The enzyme may be present in a form insensitive to the inhibitor. In the case of electron transport systems, an enzyme may exist in two forms with the steady-state proportions of the two forms dependent on physiological variables. In one form it is sensitive to the inhibitor and in the other insensitive; the properties *in vivo* may fluctuate and inhibition may deviate from that observed *in vitro*. Such an argument has been advanced for the carbon monoxide inhibition of cytochrome oxidase (cf. Sec. 1.8).

SUBSTANCES THAT COMBINE WITH OR ALTER NUCLEIC ACIDS (TABLE 1-4). Puromycin, actinomycin D, and mitomycin are extremely powerful tools for the physiologist in that they interfere, rather specifically, with the duplication, transcription, and translation of genetic information. Mitomycin C acts to link

Table 1-4   Inhibitors Which Combine with Nucleic Acids or Alter Nucleic Acid Functions*

| Function | Inhibitor |
| --- | --- |
| DNA duplication | Phleomycin, nalidixic acid, mitomycin C |
| DNA-directed RNA synthesis | Actinomycin D |
| Formation of ribosomal RNA | Levorphanol |
| Transfer of aminoacyl RNA to ribosomes | Tetracyclin |
| Recognition of acyl-tRNA by ribosomes | Streptomycin |
| Protein synthesis on ribosomes | Puromycin |
| Protein synthesis on bacterial-type ribosomes of chloroplasts and mitochondria | Chloramphenicol |
| Protein synthesis on cytoplasmic-type ribosomes | Cycloheximide |

* After Newton, 1965.

complementary DNA strands, thus preventing normal DNA replication. Actinomycin D reacts with guanine residues of DNA and prevents the formation of RNA. Puromycin strips uncompleted peptide chains off ribosomes, etc. From the standpoint of evolutionary advantage, it is probably no coincidence that most of the known inhibitors of nucleic acid metabolism are naturally occurring antibiotics.

UNRESOLVED INHIBITORS (TABLE 1-5). The molecular sites of action for 2,4-dinitrophenol (DNP) and certain related substances are unknown, but the general physiological effects are clearly defined and are correspondingly useful.

Table 1-5   Inhibitors of Enzyme Reactions Whose Role Is Not Established

| Function Inhibited | Inhibitor | Structure |
| --- | --- | --- |
| Oxidative phosphorylation | 2,4-Dinitrophenol | |
| Photosynthetic phosphorylation | CCP (also inhibits oxidative phosphorylation) | |
| Hill reaction | Substituted ureas and related molecules, e.g., monuron | |

The dinitrophenol family and the CCP family (Table 1-5) "uncouple" phosphorylation from oxidation and thus greatly lower the availability of ATP to the cell. There are few physiological processes that are insensitive to these reagents.

A general caveat for interpreting the effects of inhibitors in physiological systems is that the sensitivity of an enzyme in one organism may not be the same in another. For example, wide differences in sensitivity among organisms have even been observed for the classical inhibitor malonate acting on succinic dehydrogenases and for fluoroacetate acting on aconitases from different organisms (cf. Sec. 1.8). Joshi et al. (1965) measured the number of sulfhydryl groups which would react with silver ion in phosphoglucomutases from a variety of organisms. They observed a range of zero to eight. Thus, *the characteristics of an inhibitor on an enzyme must be evaluated in the test organism.*

Finally, it must be remembered that a reagent which may be specific *in vitro* for a functional group of proteins might attack membranes or other structural proteins *in vivo*.

### Promotion of Reaction by Putative Intermediates

Returning to the hypothetical reaction sequence

$$A \rightarrow X_1 \rightarrow X_2 \rightarrow \cdots \rightarrow X_n \rightarrow B$$

we may expect that the rate of formation of the product, $dB/dt = v_B$, will be increased by the addition of an intermediate $X_i$.

Let us perform an imaginary experiment. We feed a putative intermediate, $X_i$, to the system and test for increased $v_B$. There are three possible kinds of results: $v_B$ may increase, decrease, or remain the same. If $v_B$ increases, the results are consistent with the hypothesis. However there are alternative explanations: $X_i$ might merely stimulate $v_B$. Or $X_i$ might indeed be converted to B, but by a route other than that hypothesized. Such an alternative explanation becomes more likely the more generalized the reaction product. For example, ascorbic acid and polyphenols promote the rate of oxygen uptake in plants, although other evidence eliminates these substances as intermediates in respiration (Sec. 1.8). Thus even positive results are ambiguous.

The more usual experience with intact plant tissue is that feeding intermediates, even those firmly established by other lines of evidence, rarely results in increased formation of the end products of metabolic pathways. There may be problems of permeability and of substrate saturation of enzymes.

It is also possible that actual inhibition can occur, perhaps through substrate inhibition. Thus negative as well as positive evidence is ambiguous, so that reliance on evidence of promotion by intermediates is usually unsatisfactory. However, this procedure acquires new power when combined with isotope procedures described below.

## Accumulation of Intermediates

The complement of promotion of a reaction by an intermediate is the accumulation of intermediates. The two are about equally inadequate as sources of evidence for reaction pathways, but both can be greatly strengthened by combination with isotope procedures.

The principal problem in relying on evidence of accumulation is that reaction pathways in living systems are often in a steady state, whereas an accumulation of an intermediate can be observed only under non-steady-state conditions. Non-steady-state, or transient, conditions can be induced by either starting the process or stopping it. Let us take starch degradation as an example. We may start the oxidation of sugar by placing the tissue in the dark. The appearance of an intermediate under these conditions might be related to the *cessation* of photosynthesis rather than the *onset* of sugar oxidation. Second, the intermediate might be related to the formation of another sugar rather than to oxidation.

Alternatively, we may expect to observe the accumulation of intermediates when the reaction is stopped. For example, we may stop the oxidation of sugar by withholding oxygen. We shall indeed observe the accumulation of an intermediate: ethanol! Ethanol, of course, is not a true intermediate in the oxidation of sugar, but is rather the product of a side reaction.

We are on shifting sands, but so is any theory which is built on the accumulation of intermediates.

The use of cytochemical staining is a special case of accumulation of intermediates (Burstone, 1962).

## Method of Mutations

Although the method of mutations has been only an occasional source of evidence in green plants, it has contributed to our understanding of a variety of pathways in fungi and bacteria, and hence in general biochemistry. For this reason alone, it is important for us to understand the nature of this kind of evidence. In addition, there are numerous instances in which the method could be employed in the study of green plants.

Since each protein is specified by a segment of DNA (Sec. 5.1), which we can conveniently call a gene, it follows that each protein and therefore each enzyme is subject to alteration by mutation. Restricting ourselves to mutations which abolish enzyme activity (as opposed merely to altering it), we can expect to find organisms deficient in any specified gene. Referring again to our generalized reaction pathway

$$A \xrightarrow{E_A} X_1 \xrightarrow{E_1} X_2 \xrightarrow{E_2} \cdots X_n \xrightarrow{E_n} B$$

we can expect to find mutants in which any $E_i$ is absent. A mutant lacking $E_i$ would produce intermediates up to the lesion $X_i$, but none after. The mutants

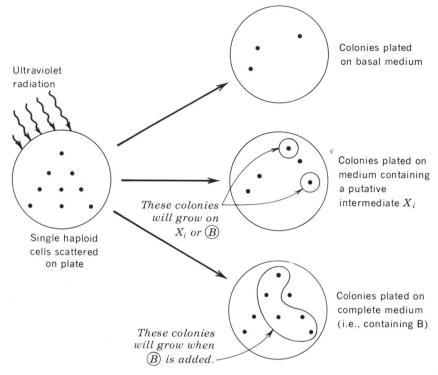

*Fig. 1-3* The method of mutations as evidence for reaction pathways.

are detected by demonstrating growth only when the organisms are grown on a "complete" medium (containing B) (Fig. 1-3 and Table 1-6) or by demonstrating the accumulation of intermediates before the lesion $X_i$. Obviously only

Table 1-6 Arrangement of Mutants to Correspond to Biosynthetic Pathway

| Designation of Mutant | Mutant Will Grow on Following Additions to Basal Medium | | | | Substance Accumulating* |
|---|---|---|---|---|---|
| | B only | B or $X_k$ | B, $X_k$, or $X_j$ | None | |
| B-less | + | | | | $X_k$ |
| $X_k$-less | + | + | | | $X_j$ |
| $X_j$-less | + | + | + | | $X_i$ |
| Wild type | + | + | + | + | B |
| | | | $X_j$-less | $X_k$-less | B-less |
| Model: | $A \longrightarrow$ | $\longrightarrow \cdots X_i \longrightarrow$ | $X_j \longrightarrow$ | $X_k \longrightarrow$ B | |

* Not infrequently, an intermediate $X_i$ whose concentration rises above some limiting value is converted by a side reaction into another substance, $X_i'$, not on the main pathway. In the case of a metabolic block after this kind of an intermediate, it would be the product of the side reaction which would accumulate.

such mutants that can survive on exogenous supplies of B or intermediates beyond $X_i$ will be detected.

For us the crucial feature of the method of mutations is that reaction pathways may be inferred from the identity of substances that will support growth in a series of mutants. This works as follows: a mutant that will grow only if supplied B is said to be "B-less." Another mutant will grow on either B or another substance $X_k$. We call it "$X_k$-less." Still another will grow on B, $X_j$, and $X_k$. If each of these is a single mutant, we can infer that each lacks a single enzyme. B-less lacks the enzyme for making B; $X_j$-less lacks the enzyme for making $X_j$; and $X_k$-less lacks the enzyme for making $X_k$. Since, however, the mutants can be expected to utilize an intermediate after the enzyme lesion or defect, the mutants can be arranged in a natural series

$$
\begin{array}{l}
\text{B-less} \\
X_k \dashrightarrow B \\
\quad X_k\text{-less} \\
\quad X_j \dashrightarrow X_k \longrightarrow B \\
\quad\quad X_j\text{-less} \\
X_i \dashrightarrow X_j \longrightarrow X_k \longrightarrow B
\end{array}
$$

In this way, at least in principle, one could identify all of the intermediates in a biosynthetic sequence (cf. Table 1-6). The method of mutations thus employs a form of the method of promotion by intermediates.

The method of mutations may give erroneous results if branched pathways exist.

It sometimes happens that a mutant will accumulate intermediates prior to a block; identification of accumulation products serves as additional evidence. In this form the method of mutations combines with that of promotion by intermediates. It not infrequently happens, however, that an intermediate prior to a genetic block is not itself accumulated, but the intermediate undergoes a secondary transformation (cf. Sec. 1.9). In this case the method would give misleading results.

The method of mutations reached its full flowering with the classic work of Tatum and Beadle, who employed *Neurospora.* It has also been employed in an exploratory manner with a "micro" higher plant, *Arabidopsis,* (Redei, 1967), and more extensively with the green alga *Chlamydomonas* (Levine, 1962). A wider use of the method of mutations can be employed with higher plants for the study of reaction pathways that are not essential to survival; in this case there is no necessity of maintaining the plants on organic media.

### Isotopic Labeling

Radioactive isotopes are perhaps the most powerful single tool in the arsenal of the physiologist. A molecule containing $C^{14}$ has almost exactly the same chemistry as a molecule containing $C^{12}$, but the molecule containing the radioactive isotope can be detected in vanishingly small (cf. Appendix D) amounts. Therefore $C^{14}$-"labeled" or "tagged" molecules can be introduced into a physiological system as a reactant or putative intermediate and followed through a reaction pathway without disturbing the existing equilibria or steady states.

The most primitive, but frequently effective, use of isotopic labeling is an adaptation of the method based on accumulation of intermediates. The limitations of the method are fairly obvious: the mere detection of a radioactive component following the introduction of a uniformly or randomly labeled reactant A proves that the component may be derived from A, but not necessarily as an intermediate on the pathway to B. For example, if A contains a carboxyl group and decarboxylation occurs, the resulting $C^{14}O_2$ may become incorporated into a variety of compounds unrelated to the pathway in question. In general, whenever there is a branched chain or a cyclic reaction, the mere detection of labeled components is apt to be more confusing than illuminating (cf. Sec. 1.6).

The method of promotion by intermediates when the intermediates are labeled becomes a method of great power and sensitivity. Limitations imposed by impermeability remain, but if a putative intermediate $X_i$ can penetrate to the reaction sites, we can hope to follow the course of conversion of $X_i$ along the reaction pathway to B. Moreover, specific activity provides us with a criterion of whether the added, labeled $X_i$ equilibrates with internal pools of $X_i$. In the simplest case the product B will have the same specific activity as the labeled intermediate $X_i$. Such data would provide evidence consistent with the hypothesis that $X_i$ is an intermediate. If the specific activity of B is less than that of $X_i$, we should infer that $X_i$ is transformed into B by an indirect process. Occasionally the specific activity of B is found to be *greater* than that of $X_i$; in this case we infer that labeled $X_i$ did not equilibrate with all of the endogenous $X_i$. In other words, separate pools of $X_i$ exist in the plant, some of which are not on the pathway to B. This is precisely the result when labeled amino acids or $C^{14}O_2$ are fed to plant tissues and the movement of radioactivity into protein is measured. The high specific activity of the proteins relative to the free amino acids may be reasonably interpreted to indicate the existence of separate pools of amino acids (Hellebust and Bidwell, 1964).

Finally, we may inquire into the efficiency of conversion: the fraction of labeled intermediate that becomes product.

SPECIFIC LABELING.   Let us return to the problem of error introduced from decarboxylation of a labeled carboxyl group, followed by incorporation of $C^{14}O_2$. Suppose we label the molecule in other than the carboxyl; better still, let us introduce radioactivity into the molecule in a specific position. By then following the movement of label through the system, we are observing the fate of this one carbon position. As a further refinement the location of radio-

boxylation, the direction being controlled by ADP/ATP and NAD/NADH ratios. In any event, PEP seems to be a likely intermediate since this is the only known gateway in going from organic acids to carbohydrates (Krebs and Kornberg, 1957 and Sec. 1.5).

## 1.8  CASE STUDY: RESPIRATION

Respiration, viewed most simply, is the oxidative production of ATP (Fig. 1-18). ATP may also be produced photochemically (cf. Sec. 2.5) and anaero-

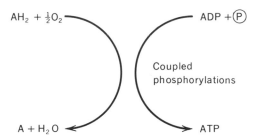

*Fig. 1-18*  Respiration viewed as the oxidative production of ATP.

bically (by fermentation), but the most general process, occurring in green and nongreen cells alike and at all stages of development, is the oxidative production of ATP. Indeed, the generality of this process, granting modifications here and there, appears to include all aerobic organisms.

Sir Hans Krebs (Krebs and Kornberg, 1957) has analyzed respiration conceptually into five phases (Fig. 1-19); depolymerization (hydrolysis or phosphorolysis), mobilization, the citric acid cycle, electron transport, and terminal oxidation (see Appendix A for detailed reactions). Each one of these phases presents challenging problems to the student of physiology, and in each area, substantial amounts of information concerning the occurrence of these processes in plants have been accumulated. We know, for example, that germinating seedlings are rich in $\alpha$- and $\beta$-amylases at times when vigorous hydrolysis of starch is occurring. These two enzymes are capable of "ganging up" on straight-chain starch (amylose) and branched-chain starch (amylopectin) as shown in Appendix A. We know also that there are at least two alternate pathways for the mobilization of the glucose residues, the EMP pathway and the pentose phosphate shunt (in Appendix A). Since the present evidences for pathways of depolymerization and mobilization in plants are of the same kinds that we encountered in Secs. 1.5 to 1.7, we shall leave them for the student to construct his own case studies (cf. Davies et al., 1964, pp. 112–122; Beevers, 1961b) and focus here on the remaining three phases of respiration.

It happens that the citric acid cycle, electron transport, and terminal oxida-

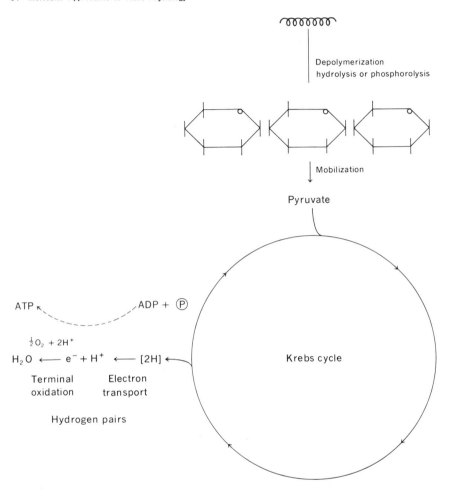

*Fig. 1-19*  Five phases of respiration. Starch is shown as the primary substrate of respiration. It may also be fat or, more rarely, organic acids or protein.

tion are joined by a set of five closely related hypotheses, all directly concerned with the oxidative production of ATP. We shall assume only that processes are available to accomplish phases 1 and 2 of Krebs' scheme so that we have a supply of pyruvate or acetyl CoA. The first hypothesis is that substrate transformations occur by means of the citric acid cycle (phase 3 of Fig. 1-17). The second is that electron transport occurs approximately as shown in Fig. 1-20. The third is that, in addition to the substrate-level phosphorylations of the citric acid cycle, phosphorylations are coupled to electron transport at the sites indicated in Fig. 1-20. The fourth is that terminal oxidation, i.e., the actual absorption of oxygen, is catalyzed by cytochrome oxidase. And finally there is a fifth hypothesis which says that the above four processes occur as an

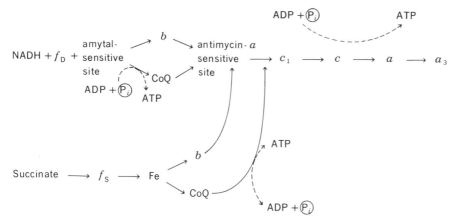

*Fig. 1-20* Electron transport in mitochondria (Lehninger, 1965). The exact sequence of electron transport has been more or less continuously disputed, especially the role of coenzyme Q. For recent arguments, see Lardy and Ferguson (1969).

integrated sequence in a single class of intracellular organelles, the mitochondria. Although we could seek evidence for the five hypotheses separately, it will be especially instructive if we organize our thinking around the most comprehensive and demanding one, the fifth hypothesis, which we might call the mitochondrial hypothesis.

### Respiratory Enzyme Activities in Plant Mitochondria

Most of the component enzymes required for the Krebs cycle (or, as Krebs prefers, the citric acid cycle) have been demonstrated not simply in extracts of plant cells and tissues, but "in association with the mitochondria" (Table 1-20).

Electron chain enzymes have also been found in plant mitochondria (cf. Hackett, 1964) and overall phosphorylation rates (P/O ratios) agree with predicted values (Table 1-21 and Fig. 1-21). There are several interesting generalizations and several inconsistencies in the data. The cytochromes are almost exclusively associated with the mitochondria, as is the phosphorylative machinery. Varying fractions of the Krebs cycle enzymes are nonmitochondrial. Whether, or to what extent, the nonmitochondrial enzymes can participate in respiration is unknown. Except for pyruvate dehydrogenation, the enzyme activities of the isolated mitochondria are the same order of magnitude as the observed rates of respiration. All of the cytochrome oxidase is associated with mitochondria, as are most of the other cytochromes. So far so good!

The comparison of the substrate-saturation curves of isolated enzymes with the corresponding curves of intact cells and tissues has been employed with decisive results in the case of terminal oxidation. This kind of evidence is particularly crucial in the case of higher plants for the following reason: a

Table 1-20  Intracellular Localization of Some Respiratory Enzymes and Carriers

Values in parentheses are percentages of enzyme activities in tissue brei. Theoretical rates are calculated from observed respiratory rates of intact tissue. Actual units employed vary from one set of measurements to another.

| Tissue Source | Enzyme | Nuclei, Plastids Debris | Mito-chondria | Micro-somes | Soluble | Activity in Mitochondria as Percent of Theoretical | Ref.* |
|---|---|---|---|---|---|---|---|
| Many | Pyruvate dehydrogenase complex | | Present | | | | 1 |
| Pea | Condensing enzyme | | Present | | | | 1 |
| stem | Aconitase | | Present | | | | 1 |
| | NAD-isocitrate dehydro-genase | | Present | | | | 1 |
| | α-Ketoglutarate oxidase | | 37 | | | 67 | 2 |
| | Succinate dehydrogenase | (6) | (74) 116 | | (10.5) | 210 | 2 |
| | Fumarase | | Present | | | | 1, 2 |
| | Malate dehydrogenase | (7) | (2) 126 | | (76) | 229 | 2 |
| Cauli- | NAD-cyto c reductase | 0.9 | 17.2 | 8.0 | 10.4 | | 3 |
| flower | Succinate dehydrogenase | 0.006 | 0.014 | 0.001 | | | 3 |
| buds | Cytochrome oxidase | 3 | 88 | 5 | 7 | | 3 |
| Aroid spadix | Succinate-cyto c reduc-tase | | 33 | 1 | | | 3 |
| | NAD-cyto c reductase | | 114 | 10 | | | 3 |
| | Cytochrome oxidase | | 127 | 15 | | | 3 |
| Potato | Cytochrome oxidase | | 1.22 | | | 1,000 | 3 |

* (1) Davies et al., 1964; (2) Price and Thimann, 1954; (3) Hackett, 1957.

number of oxidases other than cytochrome oxidase, such as ascorbic acid oxidase and polyphenol oxidases, have been isolated from many higher plants and the activities of these enzymes are often extremely high. On the other hand, cytochrome oxidase activities are relatively low, and it has been difficult even to detect activities in some mature tissue. However the substrate-saturation curves for $O_2$ with pea stems, potato tuber, and aroid spadix provided clear-cut data: only the curve of cytochrome oxidase corresponded even approximately to the oxygen-saturation curves for the intact tissues (Table 1-22).

In the case of mung bean mitochondria, Ikuma et al. (1964) found evidence for not one, but two, oxidases, with low $K_M$ values characteristic of cytochrome oxidase. The two oxidases also differed with respect to cyanide and carbon monoxide sensitivity (see below). We should not be surprised if cytochrome oxidases from different sources have somewhat different $K_M$, but it is surprising that cytochrome oxidases with different properties should coexist in the same organism (but cf. Sec. 1.2). Physical separation of the two oxidases has not been achieved.

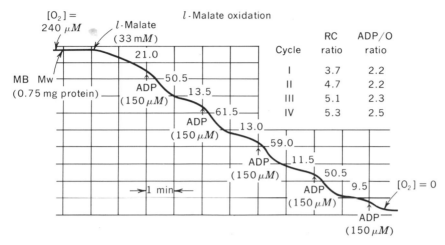

*Fig. 1-21* Polarographic measurement of oxidation, oxidative phosphorylation, and respiratory control in mitochondria (Ikuma and Bonner, 1967). Mitochondria were separated from mung bean hypocotyls and a suspension placed in the polarographic cell. Oxygen was slowly consumed as shown by the nearly level trace of the recorder. The rate increased when substrate was added. When a small amount of ADP was added, the rate increased much more for a time and then returned to nearly the original rate with substrate alone. The cycle was repeated by further additions of ADP.

The *rate of oxidation* is determined simply from the slope of the polarographic trace. The ADP/O ratio is a measure of oxidative phosphorylation and is the ratio of the moles of ADP added to the gram atoms of oxygen absorbed before the system returns to its initial rate. The *respiratory control* is a measure of the tightness of coupling between oxidation and phosphorylation and is the factor by which the rate of oxidation is stimulated by ADP.

One of the reasons why the mitochondrial hypothesis has engendered so much interest among physiologically oriented biologists has been that mitochondria, although derived from broken cells, retain a few physiological properties.

Table 1-21  Oxidative Phosphorylation by Plant Mitochondria*

| Source | Substrate | P/O ratio |
|---|---|---|
| Potato tuber | Succinate | 1.32 |
| | $\alpha$-Ketoglutarate | 2.30 |
| | Citrate | 1.95 |
| Castor bean endosperm | Succinate | 1.57 |
| | $\alpha$-Ketoglutarate | 1.98 |
| | Malate | 2.0 |
| Skunk cabbage | $\alpha$-Ketoglutarate | 3.7 |
| Tobacco leaves | Succinate | 1.3 |
| | Isocitrate | 2.1 |
| Cabbage heads | $\alpha$-Ketoglutarate | 3.8 |
| | Succinate | 2.3 |

* From Beevers, 1961b.

Table 1-22   Oxygen Saturation of Tissue Respiration and Various Oxidases*

|  | O$_2$-Absorbing System | $[O_2]^{50}$ or $K_M$, moles liter$^{-1}$ |
| --- | --- | --- |
| Tissue respiration | Aerobacter aerogenes | $3.1 \times 10^{-8}$ |
|  | Yeast | $10^{-6}$ |
|  | Potato tuber slices | $3 \times 10^{-6}$ |
|  | Aroid spadix | $3 \times 10^{-6}$ |
| Oxidase | Cytochrome $c$ | $2.5 \times 10^{-8}$ |
|  | Ascorbic acid | $3.3 \times 10^{-4}$ |
|  | Ascorbic acid | $1.7 \times 10^{-4}$ |
|  | Phenol | $3.3 \times 10^{-5}$ |

* After Yocum and Hackett, 1957.

For one thing, mitochondria have *structure;* indeed, the best present criteria for the identification of mitochondria is the presence of *cristae mitochondriales* observable with the electron microscope (Plates 6 and 7). The structure, moreover, is related to the metabolic state of the mitochondria (Lehninger, 1964): the particles are contracted under phosphorylative conditions and swollen under conditions when phosphorylations are uncoupled from oxidation. Mitochondria show permeability barriers, not only toward protein molecules, but selectively toward small molecules (cf. Sec. 4.4). Finally, mitochondria are capable of accumulating certain ions, although not nearly as efficiently as do intact cells (Sec. 4.4). In sum, mitochondria behave like subphysiological systems, hence the appropriate term, organelles.

Since nearly all studies concerning the localization in mitochondria of enzymes and enzyme systems have been made using differential centrifugation methods (cf. Sec. 1.2), we are obliged to consider the evidence in terms of localization within a centrifugal fraction which may include proplastids, lysosomes, and microbodies, as well as mitochondria. Because of the unresolved state of these fractions, we shall speak of enzymes "associated with" rather than "in" or "on" mitochondria.

DETECTION OF ENZYME-SUBSTRATE COMPLEXES.   If there is such a thing as sufficient proof in physiology, then surely the *in vivo* demonstration of an enzyme-substrate complex is sufficient proof for the functioning of that enzyme.

Through absorption and *difference spectrophotometry,* the oxidation and reduction of the electron carriers of mitochondria have been observed (Fig. 1-22). The same and similar spectral changes have been detected in intact tissues. Lundegårdh has observed difference spectra characteristic of cytochromes in wheat roots by making them anaerobic and recording their absorption spectrum at intervals (Fig. 1-23). Similar spectra have been observed with potato slices, pea stems, and *Arum* spadix. It is evident that, although the "classical"

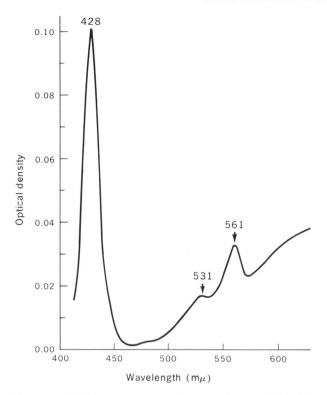

*Fig. 1-22* Difference spectra of some plant mitochondria (Shichi and Hackett, 1962). In the usual procedure one beam of a split beam spectrophotometer is passed through a suspension of reduced mitochondria and the other beam is passed through an oxidized suspension. The light absorption of one suspension is subtracted electronically from the other, and the difference is recorded as a function of wavelength.

electron transport components are observed with several plant preparations, which is to say they are similar to the components of heart mitochondria, others show exotic components or unusual concentrations of certain components, particularly cytochrome *b*. The respiration of high *b* tissues is typically resistant to poisoning by cyanide and carbon monoxide.

The amounts and turnover numbers[1] of putative intermediates can also be estimated (Table 1-23). The coupling of oxidation to phosphorylation may also be observed indirectly by optical methods. According to mitochondrial demonologists, the "best" preparations show large respiratory control ratios, i.e., electron transport is quantitatively dependent on addition of ADP (cf.

---

[1] "Turnover number" refers to the number of moles of substrate transformed per mole of enzyme in 1 sec.

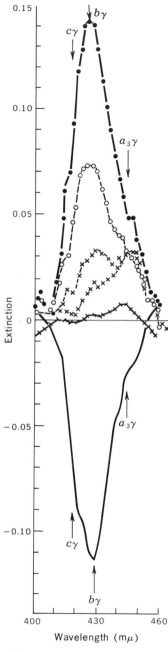

*Fig. 1-23* Difference spectra of wheat roots (after Lunde-gårdh, 1959). Spectra obtained through 14-mm bundle of wheat roots; aerated sample serves as reference. At zero time, roots were placed in O₂-free solution of succinate; curves obtained at various times afterwards. Soret peaks of different cytochromes shown by arrows.

Table 1-23   Amounts and Turnover Numbers of Respiratory Carriers as Determined by Difference Spectrometry

| | FP | $b_7$ | $c$ | $a_3$ | $a$ | | Ref.* |
|---|---|---|---|---|---|---|---|
| Potato tuber mitochondria: | | | | | | | |
| Fresh | 10.5 | 3.3 | 4.6 | 3.9 | 2.8 | $\mu$moles | 1 |
| Aged | 7.3 | 6.8 | 6.3 | 3.3 | 1.4 | | |
| | | $b$ | $c$ | $a_3 - a$ | | | |
| *Peltandra* spadix | | 1.1 | 1.7 | 0.7 | | $\mu$moles | 2 |
| | | 46 | 30 | 73 | | TN(sec$^{-1}$)† | |
| *Symplocarpus foetidus* spadix mitochondria | | 0.95 | 0.84 | 1.62 | | | 3 |

* (1) Hackett et al., 1960; (2) Yocum and Hackett, 1957; (3) Hackett and Haas, 1958.
† Assuming respiratory rate of 1,000 $\mu$l hr$^{-1}$ g FW$^{-1}$.

Fig. 1-21). In this way the absorption changes of cytochromes may actually be calibrated with the phosphorylation of ADP.

Garfinkel and Hess (1964) and others have constructed analog models for electron transport and phosphorylation by mitochondria in such systems as ascites tumor cells. These simplified models faithfully correspond to respiration kinetics and have some predictive capacity. Chance's (1967) assertion that the metabolic characteristics of a computer model of mitochondria were indistinguishable from those of real mitochondria led Lardy and Ferguson (1969) to lament, " *Sic transit gloria mitochondriorum!*"

INHIBITION.   With some precision, certain selective poisons can block the Krebs cycle and electron transport chain of mitochondria at various points (Table 1-24). These inhibitors are also active *in vivo* and their effects may be observed as altered difference spectra.

Table 1-24   Inhibitors of Respiratory Enzymes*

| Reaction Step | Inhibitor |
|---|---|
| Cytochrome *b*—cytochrome *c* | Antimycin A |
| Cytochrome *c*–oxygen | Cyanide |
| | CO and other complexing agents |
| Succinate dehydrogenase | Malonate |
| Aconitase | Fluoroacetate → fluorocitrate |

* After Hackett, 1964.

The inhibition of terminal oxidation presents some interesting difficulties: cytochrome oxidase behaves as if it contained a metal with one free coordination bond (cf. Sec. 4.3). In fact, the enzyme contains both iron and copper; but apparently only the iron is free for external complexation. In the normal course of enzyme action this metal complexes with oxygen. Carbon monoxide acts as a powerful inhibitor competitive with respect to oxygen (Eq. 1-8).

$$\frac{[\text{Cytochrome oxidase} \cdot O_2]}{[\text{Cytochrome oxidase} \cdot CO]} \cdot \frac{[CO]}{[O_2]} = K \simeq 10 \qquad \qquad \text{Eq. 1-8}$$

The enzyme is also inhibited by other metal-complexing agents: cyanide, azide, etc.

Many metal enzymes are inhibited by carbon monoxide and other complexing agents, but cytochrome oxidase is unique in that the enzyme–carbon monoxide complex is decomposed by light corresponding to the absorption spectrum of the complex (cf. Sec. 2.3). This property is extremely valuable as a criterion for identifying cytochrome oxidase activity, *in vivo*.

The respiration of certain tissues (e.g., washed potato slices, aroid spadix) are resistant to complexing agents. We might infer that their respiration was not mediated by cytochrome oxidase. Washed potato slices are especially interesting in that their respiration is actually promoted slightly by carbon monoxide. However the oxygen-saturation curve of respiration shows a tight affinity similar to that of cytochrome oxidases (Fig. 1-24). But, the *growth* of potato slices is *inhibited* by carbon monoxide and the inhibition is reversed by light, exactly as one would expect if cytochrome oxidase were the terminal oxidase (reviews of Hackett, 1959; Price, 1960). This confusing behavior is extremely difficult to rationalize. The problem of cyanide- and carbon monoxide-resistant respiration remains far from resolved.

PROMOTION BY INTERMEDIATES.   Given the citric acid cycle as a model, we may expect that each component of the cycle will act as a substrate in its own right and also as a catalyst in the oxidation of pyruvate. In the case of higher plant tissues, substantial increases in the rates of oxygen uptake by the addition of cycle intermediates can be demonstrated only after severe starvation. This is not too surprising; it may mean that endogenous levels of cycle intermediates are present in amounts that are normally not rate limiting.

Catalytic promotion of pyruvate oxidation has never been demonstrated with any intact tissue, plant or animal. Krebs originally (1943) obtained such responses to cycle intermediates with a brei of pigeon breast muscle. With the advent of mitochondria, breis, which never worked well with plants anyway, became unfashionable. First from animals and then from a wide variety of plants, isolated mitochondria were shown to oxidize cycle intermediates and to catalyze pyruvate oxidation as required by the hypothesis.

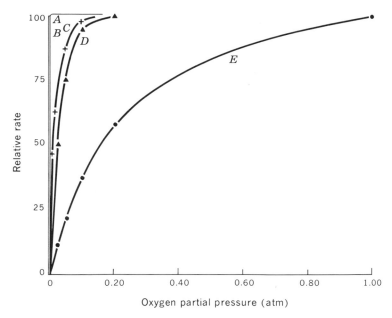

*Fig. 1-24* Comparison of oxygen affinity curves for cytochrome oxidase, ascorbic acid oxidase, and the respiration of potato slices (Thimann et al., 1954). (*A*) Yeast cytochrome oxidase; (*B*) potato discs, 0.5 mm thick, at 15° (curve is indistinguishable from *A*); (*C*) discs kept one day at 25°, measured at 15°; (*E*) ascorbic acid oxidase.

ACCUMULATION OF INTERMEDIATES.    It is commonly assumed that mitochondria carry out complete oxidation of cycle intermediates, that once a substrate molecule is taken in, it is completely converted to $CO_2$ and water (Fig. 1-19). In reality, mitochondria typically carry out sequential oxidations on pyruvate and the members of the citric acid cycle. Succinate is converted to fumarate and malate, but these intermediates are then returned to the medium and may accumulate. When mitochondria are exposed to oxalacetate, a dismutation probably to ketoglutarate and malate occurs; this phenomenon is predictable from a consideration of the malate-oxalacetate equilibrium (Krebs and Kornberg, 1957).

The names of most of the Krebs cycle acids derive from the plants in which they were discovered. The widespread and often very lopsided accumulations of organic acids in many tissues, especially fruits (Thimann and Bonner, 1950; Zelitch, 1964), is, paradoxically, an argument *against* the operation of the Krebs cycle. If these acids were functioning catalytically in the oxidation of food reserves, they should not accumulate. Moreover, from the study of the *kinetics of incorporation of labeled intermediate,* it appears that these organic acids typically equilibrate very slowly. It seems likely that these organic acids are secreted into the vacuole and are somehow kept out of the main stream of metabolism.

Citrate accumulation in fluoroacetate poisoning is a special case of accumulation of an intermediate. In this case of "lethal synthesis," fluoroacetate is mistaken for acetate, condensed with oxaloacetate, and then synthesized into fluorocitrate. This substance is a powerful competitive inhibitor of aconitase and prevents cycle operation beyond citrate.

HISTOCHEMICAL EVIDENCE.   When dyes such as neotetrazolium are introduced into tissues, they are reduced by electron transport components, probably flavoproteins. The reduced dyes are extremely insoluble and become visible as deposits of formazans, presumably at the site of reduction. The accumulation of reduced dyes and histochemical agents generally may be thought of as the accumulation of synthetic intermediates.

Formazan dyes accumulate in what appear to be mitochondria. Taken at face value the evidence is clearly consistent with mitochondria being the exclusive respiratory centers (or loci). A particularly interesting case is afforded in the epidermal cells of *Phleum* roots, which have special morphogenetic interest

Table 1-25   Histochemical Detection of Enzyme Activities Associated with Particles Putatively Identified as Mitochondria*

| Species and Treatment | Number of Stained Particles per Cell | Percent of Particles Stained with Janus Green B† |
|---|---|---|
| *Phleum pratense:* | | |
| Janus green B | $90.6 \pm 1.7$ | 100 |
| Cytochrome oxidase | $94.5 \pm 1.8$ | 104 |
| Acid phosphatase | $108.1 \pm 1.6$ | 119 |
| Succinic dehydrogenase | $46.8 \pm 1.2$ | 52 |
| *Festuca arundinacea:* | | |
| Janus green B | $84.7 \pm 1.4$ | 100 |
| Cytochrome oxidase | $89.7 \pm 3.0$ | 106 |
| Acid phosphatase | $105.0 \pm 1.8$ | 124 |
| Succinic dehydrogenase | $42.4 \pm 0.6$ | 50 |
| *Panicum virgatum:* | | |
| Janus green B | $90.8 \pm 1.7$ | 100 |
| Cytochrome oxidase | $94.6 \pm 2.1$ | 104 |
| Acid phosphatase | $98.1 \pm 2.0$ | 108 |
| Succinic dehydrogenase | $43.6 \pm 0.8$ | 48 |
| *Chloris gayana:* | | |
| Janus green B | $95.4 \pm 1.6$ | 100 |
| Cytochrome oxidase | $92.8 \pm 1.7$ | 97 |
| Acid phosphatase | $101.4 \pm 1.7$ | 106 |
| Succinic dehydrogenase | $57.6 \pm 2.0$ | 60 |

\* After Avers and King, 1960.
† Janus green B is a stain that is thought to be specific for mitochondria.

because of the peculiar alternating patterns of hair cells. In this tissue the number of particles that stain in the presence of citric acid cycle intermediates is surprisingly constant (Table 1-25), but the number is only one-half of those which stain with a cytochrome oxidase substrate and is a still smaller fraction of the number of mitochondria observable by electron microscopy. Thus, either formazan deposited with cycle intermediates fails to stain all the mitochondria, or there is a division of function among mitochondria.

In summary, there appears to be sufficient enzyme activity associated with mitochondria to account for the respiration of intact plant tissues. Isolated mitochondria can carry out all the reactions normally associated with respirations: the complete oxidation of substrates to $CO_2$ and water. The $-\Delta G$ from this oxidation may be employed to drive the phosphorylation of ADP to ATP.

The similarity of the $K_M$ $(O_2)$ of cytochrome oxidase to oxygen-saturation curves of respiration together with the specific release of CO inhibition by light is strong evidence that cytochrome oxidase is the terminal oxidase of respiration.

The most commanding evidence for the mitochondrial hypothesis is the spectrophotometric observation of cytochrome changes which proceed with respiratory changes in both mitochondria and intact tissues.

## 1.9 CASE STUDY: CHLOROPHYLL SYNTHESIS

> ". . . Chlorophyll . . . is the real Prometheus, stealing fire from the heavens."
>
> ———Timiryazev

The three most abundant porphyrins in living organisms are chlorophyll *a* and *b* and heme.

From the obvious similarity of these and other porphyrins we can well imagine that these porphyrins might have a common pathway of formation. The general hypothesis (Bogorad, 1963) for chlorophyll *a* synthesis is shown in Fig. 1-25. Quite as interesting as the evidence for this pathway are the kinds of reasoning that led to it. This case study is another example of the beauty and economy in reasearch when scientists are able to construct generalities with predictive capacity.

In the early 1950s Granick (Granick and Mauzerall, 1961) adapted the method of mutations to chlorophyll synthesis. Mutants of *Chlorella vulgaris,* obtained by x-irradiation, were selected for pigment defects: failure to synthesize chlorophyll, and accumulation of other porphyrins. (No method was then available for obtaining recombinations in algae, although it could be done now with *Chlamydomonas.*) Granick perceived close relations among the porphyrins of plants, animals, and microorganisms and, reasoning by analogy, was able to place the mutants in an order based on the chemical structure of the porphyrins they excreted or accumulated (Fig. 1-25). Even though there were

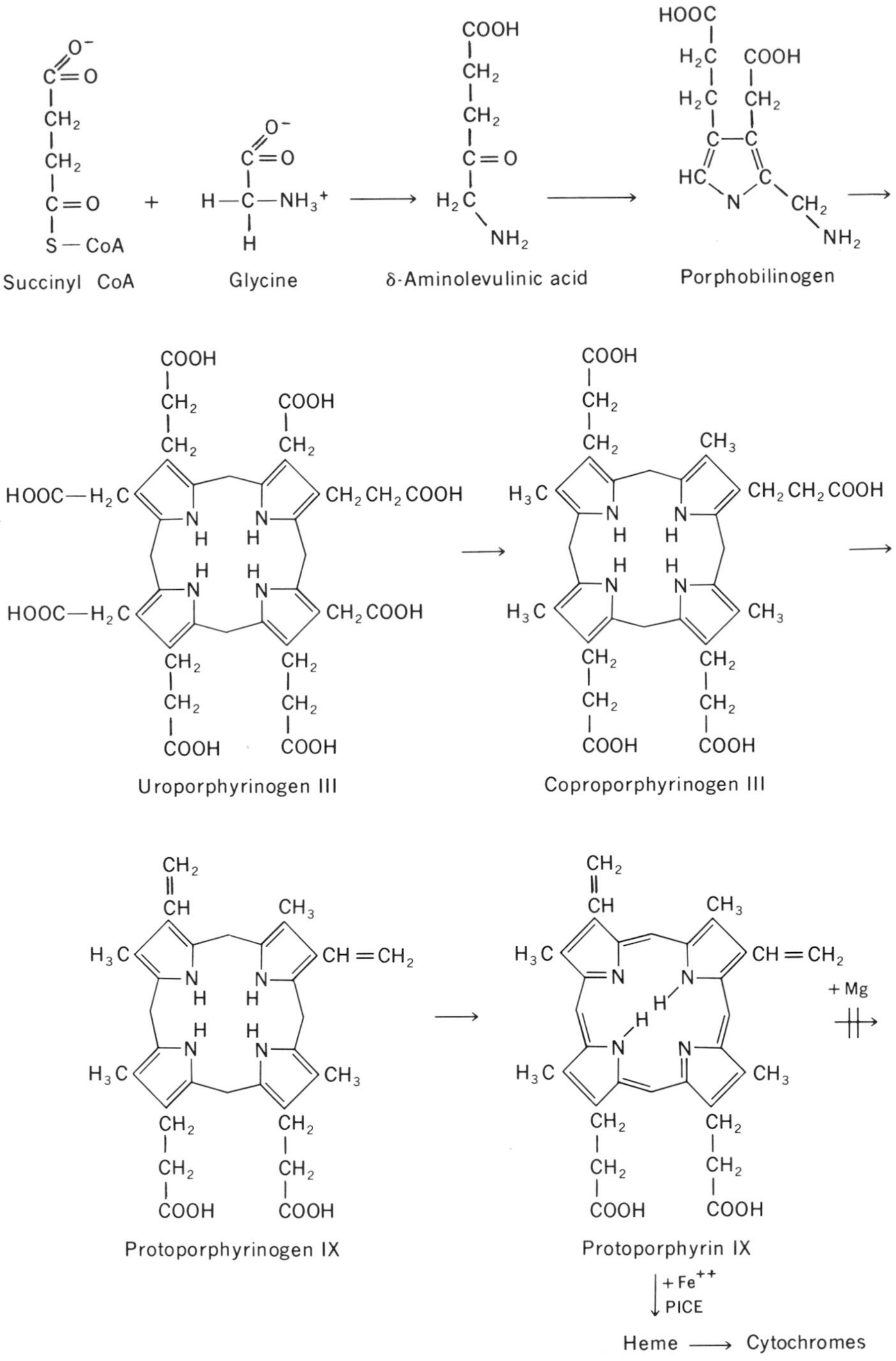

*Fig. 1-25* General hypothesis for chlorophyll synthesis. Genetic blocks observed in *Chlorella* mutants are shown as ―‖→.

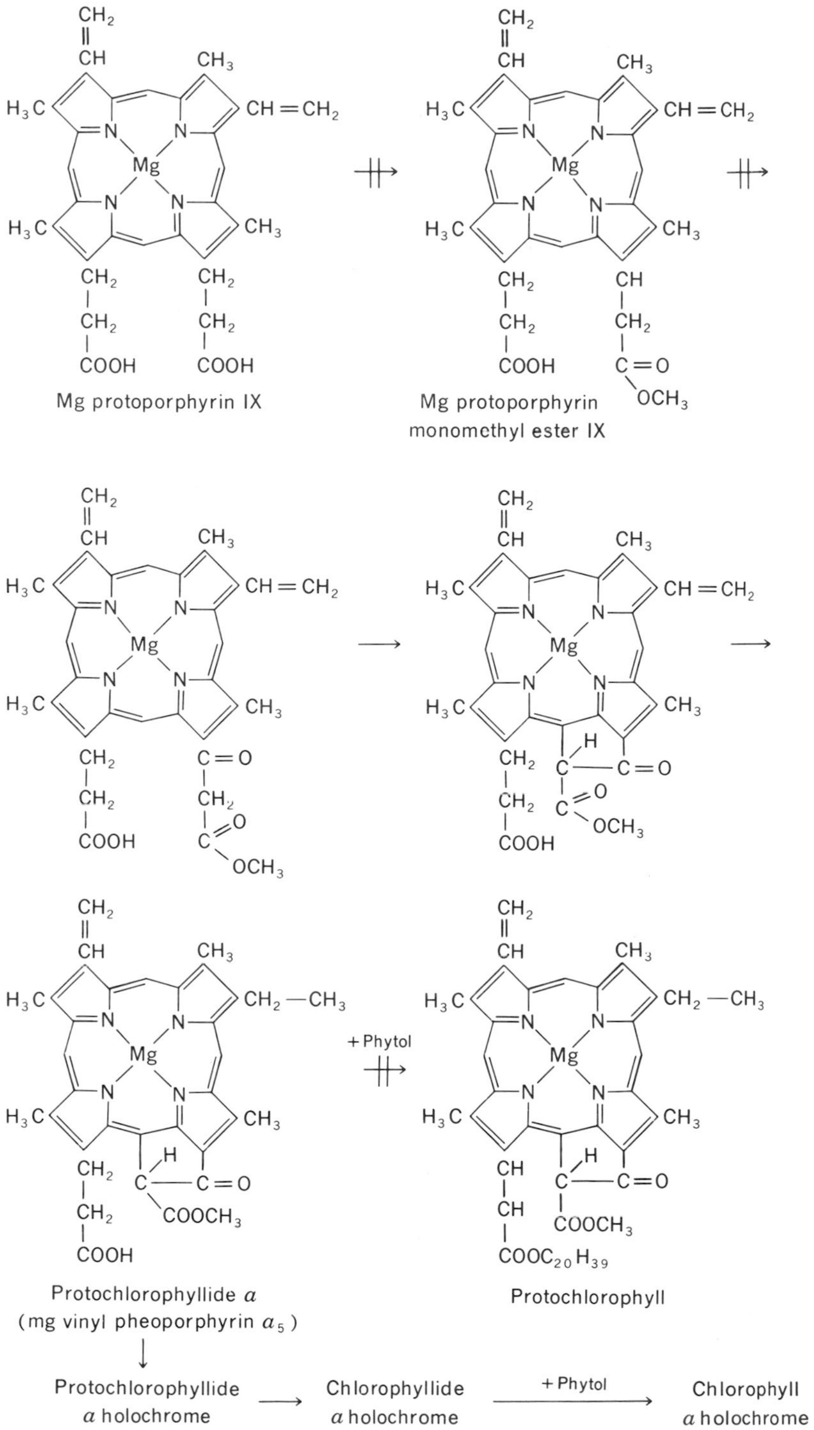

Mg protoporphyrin IX

Mg protoporphyrin monomethyl ester IX

Protochlorophyllide $a$
(mg vinyl pheoporphyrin $a_5$)

Protochlorophyll

Protochlorophyllide
$a$ holochrome $\longrightarrow$ Chlorophyllide
$a$ holochrome $\xrightarrow{\text{+ Phytol}}$ Chlorophyll
$a$ holochrome

*Fig. 1-25* (Continued).

many obvious gaps in the sequence, they provided a logical sequence from protoporphyrin to chlorophyll *a*.

Later Shemin and Rittenberg in a brilliant feat of chemical degradation employed the specific labeling method in avian erythrocytes to deduce the origins of heme carbons and nitrogen in succinyl CoA and glycine. Several investigators predicted and then found δ-aminolevulinate (ALA) as an intermediate. The corresponding enzymes were subsequently discovered (see Granick and Mauzerall review, 1961).

Let us now look at the other evidence for the general hypothesis.

ACTIVITIES IN CELL-FREE REACTIONS. With the exception of the initial or "branch point" enzyme, most of the enzyme activities required by the hypothesis have been found in plants (Table 1-26). The important point for us is that the existence of these enzymes was predicted by the hypothesis; their discovery is therefore intersecting evidence.

Table 1-26  Occurrence in Plants Putatively Involved in Chlorophyll Synthesis

| Enzyme | Reaction | Occurrence in Plants |
|---|---|---|
| δ-ALA synthetase | Succinyl CoA + glycine → ALA | Sought, but not reported in green plants |
| δ-ALA dehydratase | δ-ALA → porphobilinogen (PBG) | Widespread: *Chlorella*, *Euglena*, spinach |
| Urogen I synthetase (PBG deaminase) | 4 PBG → Urogen I | Spinach |
| Urogen III cosynthetase (acts with Urogen I synthetase) | 4 PBG → Urogen III | Wheat germ, *Chlorella* |
| Urogen decarboxylase | Urogen → Coprogen + 4CO₂ | *Euglena*, *Chlorella* |
| Coprogen oxidase | Coprogen → Proto + 2CO₂ | *Euglena*, *Chlorella* |
| Proto-magnesium chelating enzyme (PMCE) | Proto + Mg → Mg-Proto | Not detected |
| Mg-Proto-methylpherase | Mg-Proto + 5-adenosylmethionine → Mg-Proto monomethyl-ester | Not detected |

*Note:* No enzymes catalyzing steps after Proto have been detected in higher plants.

INTRACELLULAR LOCALIZATION. Granick (see Granick and Mauzerall, 1961) has found that various steps along the pathway of Proto biosynthesis in animal tissues are carried out in different cell compartments; for example, ALA dehydratase is soluble where ALA synthetase and Coprogen oxidase are mitochondrial. In plants, the same division may exist, but in addition there appears to be an independent pathway contained wholly within the chloroplasts (Carell and Kahn, 1964).

We may imagine that Proto for cytochromes, etc., is formed from enzymes that are physically separate and perhaps different from those producing Proto for chlorophyll synthesis. In this way separate feedback control mechanisms could operate (cf. Sec. 5.2).

PROMOTION BY INTERMEDIATES AND ACCUMULATION OF INTERMEDIATES. When δ-ALA or PBG are fed to a variety of organisms, including *Phaseolus* and *Chlorella,* a number of free porphyrins accumulate. In the case of etiolated barley, magnesium Proto has been detected (Granick and Mauzerall, 1961).

The part of the hypothesis up to Mg-Proto is supported by this type of evidence. Between Mg-Proto and protochlorophyllide we have only the evidence from mutations and some chemical guess work.

Both protochlorophyllide and protochlorophyll accumulate in etiolated tissues. Since protochlorophyll is inactive in the photochemical formation of chlorophyll (Sec. 2.4), and chlorophyllide accumulates in plants that are caused suddenly to produce chlorophyll (Price and Carell, 1964), it seems plausible that in the normal course of events, the phytol side chain is added after the formation of chlorophyll *a*. Phytol may add to chlorophyll in the presence of chlorophyllase as a simple reversal of hydrolysis.

We shall deal later (Sec. 2.3) with the photochemical formation of chlorophyllide from protochlorophyllide.

In conclusion, evidence first from the method of mutations and subsequently from comparative biochemistry led to a fairly complete scheme for the synthesis of chlorophyll. Further evidence on the occurrence of enzymes, together with promotion by and accumulation of intermediates, is fully consistent with the hypothesis. Many details must still be worked out.

The synthesis of chlorophyll is tied by one or more control mechanisms to the morphogenesis of chloroplasts. As we shall discuss in detail later (Sec. 5.8), a number of environmental factors (for example, the iron supply) influence chlorophyll indirectly by controlling morphogenesis.

# SUGGESTED CASE STUDIES

## The Formation of Starch

The only plausible pathway for sucrose synthesis in plants includes UDPglucose as an intermediate, but for starch synthesis UDPglucose and ADPglucose serve about equally well *in vitro* (see Appendix A). Efforts have been made to deduce which of these sugar nucleotides actually functions in starch synthesis. Another problem concerning starch synthesis is the means of control of amylopectin content, i.e., the proportion of $1 \to 6$ glucan branching in otherwise linear $1 \to 4$ glucans.

The available information includes *in vitro* properties of the sugar-transferring enzymes, the chemical and enzymatic composition of various mutants, especially in maize and rice, and intracellular localization in plastids of some of the starch-synthesizing enzymes.

Is the evidence sufficient to identify the role of one of these nucleotides and exclude the other? What additional evidence is required?

*References:* Akatsura and Nelson, 1966; de Fekete and Cardini, 1964; Ghosh and Preiss, 1966; Leloir, 1964; Nomura et al., 1967.

## The Biosynthesis of Phenylpropanoids

Plants are given to the incontinent production of substances with the phenyl-propane carbon skeleton

These include phenylalanine, lignins, and anthocyanins. There is genetic evidence and evidence of isotopic incorporation that the pathway involves shikimic acid.

*References:* Neish, 1965; Yoshida, 1969.

## Photorespiration

In many higher plants (but not in tropical grasses with the Hatch-Slack type of photosynthesis), respiration is two to three times higher in the light than in the dark. An indirect consequence of photorespiration is that plants that do not photorespire can continue to photosynthesize at very low partial pressures of $CO_2$. Such a capability could be of ecological importance.

A current hypothesis of photorespiration is that the substrate is glycolic acid and that it arises as a side reaction from the Calvin cycle. Some of the glycollate-oxidizing enzymes are thought to be in peroxisomes. One proposal calls for a close collaboration between chloroplasts and peroxisomes.

*References:* Ellyard and Gibbs, 1969; Kisaki and Tolbert, 1969.

## Lysine Biosynthesis

The living world appears to be divided between two kinds of organisms in the manner in which they synthesize lysine. In one case synthesis occurs through diaminopimelic acid; in the other, through $\alpha$-aminoadipic acid. By feeding aspartate-3-$C^{14}$, aspartate-4-$C^{14}$, and alanine-1-$C^{14}$, one obtains diametrically opposite labeling into lysine with organisms utilizing one or the other pathway. The plant world appears to be sharply divided along systematic lines in the distribution of the two pathways.

*Reference:* Vogel, 1965.

### Lysosomes in Plants

A single membrane-bound organelle occurs in many animal tissues and contains a variety of hydrolases with similar acid pH optima. It has been proposed that these organelles, called lysosomes, are involved in a variety of physiological processes. While it has been more difficult to identify lysosomes unambiguously in plants, there is some evidence from ultrastructure, differential centrifugation, and histochemistry that the vacuoles perform an analogous function.

*References:* de Duve, 1963; Matile, 1966, 1968.

## REFERENCES

Akatsura, T., and O. E. Nelson, 1966. Starch granule-bound adenosine diphosphate glucose-starch glucosyltransferases of maize seeds, *J. Biol. Chem.*, **241**:2280–2285.

Anderson, N. G. (ed.), 1966a. Zonal centrifugation, *J. Natl. Cancer Inst. supplement.*

———, 1966b. Zonal centrifuges and other separation systems, *Science,* **154**:103–112.

Avers, C. J., and E. E. King, 1960. Histochemical evidence of intracellular enzymatic heterogeneity of plant mitochondria, *Am. J. Botany,* **47**:220–225.

Baldry, C. W., D. A. Walker, and C. Burke, 1966. Calvin cycle intermediates in relation to inductive phenomena in photosynthetic carbon dioxide fixation by isolated chloroplasts, *Biochem. J.,* **101**:642–646.

Bamberger, E. S., and M. Gibbs, 1965. Effect of phosphorylated compounds and inhibitors on $CO_2$ fixation by intact spinach chloroplasts, *Plant Physiol.,* **40**:919–926.

Bassham, J. A., 1963. Energy capture and conversion by photosynthesis, *J. Theoret. Biol.,* **4**:52–72.

———, 1964. Kinetic studies of the photosynthetic carbon reduction cycle, *Ann. Rev. Plant Physiol.,* **15**:101–120.

——— and M. Calvin, 1957. "The Path of Carbon in Photosynthesis," Prentice-Hall, Inc., Englewood Cliffs, N.J.

——— and Martha Kirk, 1960. Dynamics of the photosynthesis of carbon compounds: I. Carboxylation reactions, *Biochem. Biophys. Acta,* **43**:447–464.

Beevers, H., 1956. Utilization of glycerol in the tissues of the castor bean seedling, *Plant Physiol.,* **31**:440–445.

———, 1957. Incorporation of acetate-carbon into sucrose in castor bean tissues, *Biochem. J.,* **66**:23P.

———, 1961a. Metabolic production of sucrose from fat, *Nature,* **191**:433–436.

———, 1961b. "Respiratory Metabolism in Plants," Row, Peterson, and Co., New York.

——— and D. A. Walker, 1956. The oxidative activity of particulate fractions from germinating castor beans, *Biochem. J.,* **62**:114–120.

Bendall, D. S., and R. Hill, 1956. Cytochrome components in the spadix of *Arum maculatum, New Phytol.,* **55**:206–212.

Benedict, C. R., and H. Beevers, 1961. Formation of sucrose from malate in germinating castor beans: I. Conversion of malate to phosphoenolpyruvate, *Plant Physiol.,* **36**:540–544.

——— and ———, 1962. Formation of sucrose from malate in germinating castor beans: II. Reaction sequence from phosphoenolpyruvate to sucrose, *Plant Physiol.,* **37**:176–178.

Bernfeld, P. (ed.), 1966. "Biogenesis of Natural Compounds," 1,204 pp., Pergamon Press, New York.

Bieleski, R. L., and G. G. Laties, 1963. Turnover rates of phosphate esters in fresh and aged slices of potato tuber tissue, *Plant Physiol.*, **38**:586–593.

Boatman, S. G., and W. M. L. Crombie, 1958. Fat metabolism in the West African oil palm (*Elaeis guineensis*): II. Fatty acid metabolism in the developing seedling, *J. Exp. Botany*, **9**:52–74.

Bogorad, L., 1963. The biogenesis of heme, chlorophylls, and bile pigments, in P. Bernfeld (ed.), "Biogenesis of Natural Compounds," pp. 183–231, Macmillan, New York.

Bonner, J., 1950. "Plant Biochemistry," pp. 1–537, Academic Press, New York.

Boyer, P. D., 1959. Sulfhydryl and disulfide groups of enzymes, in P. D. Boyer, H. Lardy, and K. Myrbäck (eds.), "The Enzymes," 2d ed., vol. 1, chap. 11, pp. 511–588.

Bradbeer, C., and P. K. Stumpf, 1959. Fat metabolisms in higher plants: XI. The conversion of fat into carbohydrate in peanut and sunflower seedlings, *J. Biol. Chem.*, **234**:498–501.

Bradbeer, J. W., S. L. Ranson, and Mary Stiller, 1958. Malate synthesis in crassulacean leaves: I. The distribution of $C^{14}$ in malate of leaves exposed to $C^{14}O_2$ in the dark, *Plant Physiol.*, **33**:66–70.

Breidenbach, R. W., A. Kahn, and H. Beevers, 1968. Characterization of glyoxysomes from castor bean endosperm, *Plant Physiol.*, **43**:705–713.

Bruinsma, J., 1958. Studies on the crassulacean acid metabolism, *Acta Bot. Neerl.*, **7**:531–590.

Bucke, C., D. A. Walker, and C. W. Baldry, 1966. Some effects of sugars and sugar phosphates on carbon dioxide fixation by isolated chloroplasts, *Biochem. J.*, **101**:636–641.

Bull, H. B., 1951. "Physical Biochemistry," 2d ed., John Wiley & Sons, New York.

Burstone, M. S., 1962. "Enzyme Histochemistry," Academic Press, New York.

Calvin, M., and J. A. Bassham, 1962. "The Photosynthesis of Carbon Compounds," pp. 1–127, W. A. Benjamin, Inc., New York.

Canvin, D. T., and H. Beevers, 1961. Sucrose synthesis from acetate in the germinating castor bean: Kinetics and pathway, *J. Biol. Chem.*, **4**:988–995.

Carell, E. F., and J. S. Kahn, 1964. Synthesis of porphyrins by isolated chloroplasts of *Euglena*, *Arch. Biochem. Biophys.*, **108**:1–6.

Carpenter, W. D., and H. Beevers, 1959. Distribution and properties of isocitritase in plants, *Plant Physiol.*, **34**:402–409.

Chance, E. M., 1967. A computer simulation of oxidative phosphorylation, in "Computers in Biomedical Research," vol. 1, pp. 251–264, Academic Press, New York.

Cooper, T. G., and H. Beevers, 1969a. Mitochondria and glyoxysomes from castor bean endosperm: Enzyme constituents and catalytic capacity, *J. Biol. Chem.*, **244**:3507–3514.

————— and —————, 1969b. β-Oxidation in glyoxysomes from castor bean endosperm, *J. Biol. Chem.*, **244**:3515–3520.

—————, D. Filmer, M. Wishnick, and M. D. Lane, 1969. The active species of "$CO_2$" utilized by ribulose diphosphate carboxylase, *J. Biol. Chem.*, **244**:1081–1083.

Crombie, W. M. L., and R. Comber, 1956. Fat metabolism in germinating *Citrullus vulgaris, J. Exp. Botany,* **7**:166–180.

Crozier, W. J., 1924. On biological oxidation as function of temperature, *J. Gen. Physiol.,* **7**:189–216.

Davies, D. D., J. Giovanelli, and T. ap Rees, 1964. "Plant Biochemistry," pp. 1–454, Blackwell, Oxford, England.

Davis, B. D., and D. S. Feingold, 1962. Antimicrobial agents: mechanism of action and use in metabolic studies, in I. G. Gunsalus and R. Y. Stanier (eds.), "The Bacteria," vol. IV, Academic Press, New York.

de Duve C. 1963. The lysosome concept, in de Reuck and Cameron (eds.), "Lysosomes," pp. 1–33, Little, Brown and Company, Boston.

de Fekete, Maria A. R., and C. E. Cardini, 1964. Mechanism of glucose transfer from sucrose into the starch granule of sweet corn, *Arch. Biochem. Biophys.*, **104**:173–184.

Desveaux, R., and M. Kogane-Charles, 1952. Germination of oleaginous seeds, *Ann. inst. recherches agron., Ser. A., Ann. agron.,* **3**:385–416.

Dixon, M., and E. C. Webb, 1964. "Enzymes," 2d ed., 950 pp., Longmans Green & Co., London.

Ellyard, P., and M. Gibbs, 1969. Inhibition of photosynthesis by oxygen in isolated spinach chloroplasts, *Plant Physiol.,* **44**: in press.

Eriksson, G., A. Kahn, B. Walles, and D. V. Wettstein, 1961. *Ber. deut. Botan. Ges.,* **74**:221.

Fewson, C. A., M. Al-Hafidh, and M. Gibbs, 1962. Role of aldolase in photosynthesis: I. Enzyme studies with photosynthetic organisms with special reference to blue-green algae, *Plant Physiol.,* **37**:402–406.

Fraenkel-Conrat, H., 1959. Other reactive groups of enzymes, in P. D. Boyer, H. Lardy, and K. Myrbäck (eds.), "The Enzymes," 2d ed., vol. 1, chap. 12, pp. 589–618, Academic Press, New York.

Frederick, S. E., and E. H. Newcomb, 1969. Cytochemical localization of catalase in leaf microbodies (peroxisomes), *J. Cell Biol.,* in press.

Fuller, R. C., and M. Gibbs, 1959. Intracellular and phylogenetic distribution of ribulose-1,5-diphosphate carboxylase and D-glyceraldehyde-3-phosphate dehydrogenases, *Plant Physiol.,* **34**:324–329.

Garfinkel, D., and B. Hess, 1964. Metabolic control mechanisms: VII. A detailed computer model of the glycolytic pathway in ascites cells, *J. Biol. Chem.,* **239**:971–983.

Gassman, M., J. Plusec, and L. Bogorad, 1968. δ-Aminolevulinic acid transaminase in *Chlorella vulgaris, Plant Physiol.,* **43**:1411–1414.

Ghosh, H. P., and J. Preiss, 1966. Adenosine diphosphate glucose pyrophos-phorylase: A regulatory enzyme in the biosynthesis of starch in spinach leaf chloroplasts, *J. Biol. Chem.,* **241**:4491–4505.

Gibbs, M., E. Latzko, R. G. Everson, and W. Cockburn, 1967. Carbon mobilization by the green plant, in A. San Pietro, F. Greer, and T. J. Army (eds.), "Harvesting the Sun," Academic Press, New York.

Gibson, R. E., 1964. Our heritage from Galileo Galilei, *Science,* **145**:1271–1276.

Goddard, D. R., and W. D. Bonner, 1960. Cellular respiration, in F. C. Steward (ed.), "Plant Physiology," vol. 1A, pp. 209–312, Academic Press, New York.

Granick, S., 1961. Magnesium protoporphyrin monoester and protoporphyrin monomethyl ester in chlorophyll biosynthesis, *J. Biol. Chem.,* **236**:1168–1172.

———, and D. Mauzerall, 1961. The metabolism of heme and chlorophyll, in D. Greenberg (ed.), "Metabolic Pathways," vol. II, chap. 20, pp. 525–616, Academic Press, New York.

Gregory, F. J., I. Spear, and K. V. Thimann, 1954. The interrelation between $CO_2$ metabolism and photoperiodism in *Kalanchöe, Plant Physiol.,* **29**:220–229.

Hackett, D. P., 1957. Respiratory mechanisms in the aroid spadix, *J. Exp. Botany,* **8**:151–171.

———, 1959. Respiratory mechanisms in higher plants, *Ann. Rev. Plant Physiol.,* **10**:113–146.

———, 1960. Respiratory inhibitors, in W. Ruhland (ed.), "Encyclopedia of Plant Physiology," XII/2, pp. 23–41, Springer-Verlag, Berlin.

———, 1964. Enzymes of terminal respiration, in H. F. Linskens, B. D. Sanwal, and M. V. Tracey (eds.), "Modern Methods of Plant Analysis," vol. VII, pp. 647–694, Springer-Verlag, Berlin.

———, and D. Haas, 1958. Oxidative phosphorylation and functional cytochromes in skunk cabbage mitochondria, *Plant Physiol.,* **33**:27–32.

———, ———, S. K. Griffiths, and D. J. Niederpruem, 1960. Studies on development of cyanide-resistant respiration in potato tuber slices, *Plant Physiol.,* **35**:8–19.

Havir, E. A., and M. Gibbs, 1963. Studies on the reductive pentose phosphate cycle in intact and reconstituted chloroplast systems, *J. Biol. Chem.,* **238**:3183–3187.

Heber, U., N. G. Pon, and M. Hever, 1963. Localization of carboxydismutase and triosephosphate dehydrogenases in chloroplasts, *Plant Physiol.,* **38**:355–360.

Hellebust, J. A., and R. G. S. Bidwell, 1964. Protein turnover in attached wheat and tobacco leaves, *Can. J. Botany,* **42**:1–12.

Hochster, R. M., and J. H. Quastel (eds.), 1963. "Metabolic Inhibitors: A Comprehensive Treatise," vol. I, 669 pp., and vol. II, 753 pp., Academic Press, New York.

Hock, B., and H. Beevers, 1966. Development and decline of the glyoxalate-cycle enzymes in watermelon seedlings (*Citrullus vulgaris Schrad.*), *Z. Pflanzenphysiologie,* **55**:405–414.

Hutton, D., and P. K. Stumpf, 1969. Fat metabolism in higher plants: XXXVII. Characterization of the $\beta$-oxidation systems from maturing and germinating castor bean seeds, *Plant Physiol.,* **44**:508–516.

Ikuma, H., F. J. Schindler, and W. D. Bonner, Jr., 1964. Kinetic analysis of oxidases in tightly coupled plant mitochondria, *Plant Physiol.,* **39**:1x.

——— and W. D. Bonner, Jr., 1967. Properties of higher plant mitochondria: I. Isolation and some characteristics of tightly-coupled mitochondria from dark-grown mung bean hypocotyls, *Plant Physiol.,* **42**:67–75.

Jensen, R. G., and J. A. Bassham, 1966. Photosynthesis by isolated chloroplasts, *Proc. Natl. Acad. Sci.,* **56**:1095–1101.

Joshi, J. G., T Hashimoto, K. Hanabusa, H. W. Dougherty, and P. Handler, 1965. Comparative aspects of the structure and function of phosphogluco-mutase, in V. Bryson and H. Vogel (eds.), "Evolving Genes and Proteins," Academic Press, New York.

Kindel, P., and M. Gibbs, 1963. Distribution of carbon-14 in polysaccharide after photosynthesis in carbon dioxide labelled with carbon-14 by *Anacystis nidulans, Nature,* **200**:260–261.

Kisaki, T., and N. E. Tolbert, 1969. Glycolate and glyoxylate metabolism by isolated peroxisomes and chloroplasts, *Plant Physiol.,* **44**:242–250.

Knight, G. J., and C. A. Price, 1968. Measurements of $S$-$\rho$ coordinates during the development of *Euglena* chloroplasts, *Biochim. Biophys. Acta,* **158**:283–285.

Kornberg, H. L., and H. Beevers, 1957. Glyoxalate cycle as a stage in the conversion of fat to carbohydrate in castor beans, *Biochim. Biophys. Acta,* **26**:531–537.

——— and H. A. Krebs, 1957. Synthesis of cell constituents from $C_2$ units by a modified tricarboxylic acid cycle, *Nature,* **179**:988–991.

Kortschak, H. P., C. E. Hartt, and G. O. Burr, 1965. Carbon dioxide fixation in sugarcane leaves, *Plant Physiol.,* **40**:209–213.

Krebs, H. A., 1943. The intermediary stages in the biological oxidation of carbohydrate, *Adv. Enzymol.,* **111**:191–252.

———, 1953. Some aspects of the energy transformation in living matter, *Brit. Med. Bull.,* **9**:97–104.

——— and H. L. Kornberg, 1957. Energy transformations in living matter, *Ergeb. Physiol.,* **49**:212–298.

Kunitake, G., C. Stitt, and P. Saltman, 1959. Dark fixation of $CO_2$ by tobacco leaves, *Plant Physiol.,* **34**:123–127.

Lardy, H., and S. M. Ferguson, 1969. Oxidative phosphorylation in mitochondria, *Ann Rev. Biochem.,* **38**:991–1034.

Lehninger, A. L., 1964. "The Mitochondrion," 263 pp., W. A. Benjamin, Inc., New York.

———, 1965. "Bioenergetics," 258 pp., W. A. Benjamin, Inc., New York.

Leloir, L. F., 1964. Nucleoside diphosphate sugars and saccharide synthesis, The Fourth Hopkins Memorial Lecture, *Biochem. J.*, **19**:1–8.

Levine, R. P., 1962 "Genetics," pp. 1–180, Holt, Reinhart and Winston, Inc., New York.

Lundegårdh, H., 1959. Spectrophotometric technique for studies of respiratory enzymes in living material, *Endeavour*, **18**:191–199.

Lynen, F., 1967. Multienzyme complex of fatty acid synthetase, in H. J. Vogel, J. O. Lampen, and V. Bryson (eds.), "Organizational Biosynthesis," Academic Press, New York.

MacLennan, D. H., H. Beevers, and J. L. Harley, 1963. "Compartmentation" of acids in plant tissues, *Biochem. J.*, **89**:316–327.

Marcus, A., and J. Velasco, 1960. Enzymes of the glyoxylate cycle in germinating peanuts and castor beans, *J. Biol. Chem.*, **235**:563–567.

Margoliash, E., and E. L. Smith, 1965. Structural and functional aspects of cytochrome *c* in relation to evolution, in V. Bryson and H. Vogel (eds.), "Evolving Genes and Proteins," Academic Press, New York.

Matile, Ph., 1966. Enzyme der Vakuolen aus Wurzelzellen von Maiskeimlinger: Ein Beitrag zur funktionellen Bedeutung der Vakuole bei der intrazellulären Verdauung, *Z. Naturforsch.*, **21b**:871–878.

———, 1968. Lysosomes of root tip cells in corn seedlings, *Planta*, **79**:181–196.

Mendiola, L. R., C. A. Price, and R. R. L. Guillard, 1966. Isolation of nuclei from a marine dinoflagellate, *Science* **153**:1661–1663.

Nash, L. K., 1952. "Plants and The Atmosphere," 122 pp., Harvard University Press, Cambridge.

Neal, W. K., II, 1969. "Biophysical and Morphological Changes That Occur in Yeast Mitochondria under Derepressing Conditions," Ph.D. thesis, Rutgers University.

Neish, A. C., 1965. Coumarins, phenylpropanes, and lignin, in J. Bonner and J. E. Varner (eds.), "Plant Biochemistry," pp. 581–617, Academic Press, New York.

Newton, B. A., 1965. Mechanism of antibiotic action, *Ann. Rev. Microbiol.*, **19**:209–240.

Nomura, T., N. Nakayama, T. Murata, and T. Akazawa, 1967. Biosynthesis of starch in chloroplasts, *Plant Physiol.*, **42**:327–332.

Palade, G. E., 1964. The organization of living matter, *Proc. Nat. Acad. Sci.*, **52**:613–634.

Pauling, L., 1955. The stochastic method and the structure of proteins, *Am. Scientist*, **43**:285–297.

Peterkofsky, A., and E. Racker, 1961. Reductive pentose phosphate cycle: III. Enzyme activities in cell-free extracts of photosynthetic organisms, *Plant Physiol.*, **36**:409–414.

Price, C. A., 1960. Respiration and development of vegetative plant organs and tissues, in W. Ruhland (ed.), "Encyclopedia of Plant Physiology," XII/2, pp. 493–520, Springer-Verlag, Berlin.

———, 1970. Zonal centrifugation, in W. W. Umbreit et al. (eds.), "Manometric Techniques," 5th ed., Burgess Press, Madison, Wisconsin.

——— and K. V. Thimann, 1954. Dehydrogenase activity and respiration; a quantitative comparison, *Plant Physiol.*, **29**:495–500.

——— and E. F. Carell, 1964. Control by iron of chlorophyll formation and growth in *Euglena gracilis, Plant Physiol.*, **39**:862–868.

——— and A. P. Hirvonen, 1967. Sedimentation rates of plastids in an analytical zonal rotor, *Biochim. Biophys. Acta,* **148**:531–538.

Ranson, S. L., and M. Thomas, 1960. Crassulacean acid metabolism, *Ann. Rev. Plant Physiol.,* **11**:81–110.

Redei, G. P., 1967. Biochemical aspects of a genetically determined variegation in *Arabidopsis, Genetics,* **56**:431–443.

Rutter, W. J., 1965. Enzymatic homology and analogy in phylogeny, in V. Bryson and H. J. Vogel (eds.), "Evolving Genes and Proteins," Academic Press, New York.

Shichi, H., and D. P. Hackett, 1962. Studies on the *b*-type cytochromes from mung bean seedlings, *J. Biol. Chem.*, **237**:2959–2964.

Slack, C. R., and M. D. Hatch, 1967. Comparative studies on the activity of carboxylases and other enzymes in relation to the new pathway of photosynthetic carbon dioxide fixation in tropical grasses, *Biochem. J.,* **103**:660–665.

Spanner, D. C., 1964. "Introduction to Thermodynamics," 278 pp., Academic Press, New York.

St. Angelo, A. J., and A. M. Altschul, 1964. Lipolysis and the free fatty acid pool in seedlings, *Plant Physiol.*, **39**:880–883.

Stiller, Mary, 1962. The path of carbon in photosynthesis, *Ann. Rev. Plant Physiol.,* **13**:151–170.

Stumpf, P. K., 1955. Fat metabolism in higher plants: III. Enzymic oxidation of glycerol, *Plant Physiol.*, **30**:56–58.

——— and G. A. Barber, 1956. Fat metabolism in higher plants: VII. $\beta$-oxidation of fatty acids by peanut mitochondria, *Plant Physiol.*, **31**:304–308.

——— and C. Bradbeer, 1959. Fat metabolism in higher plants, *Ann. Rev. Plant Physiol.*, **10**:197–222.

Sugiyama, T., N. Nakayama, and T. Akazawa, 1968a. Structure and function of chloroplast proteins: V. Homotropic effect of bicarbonate in RuDP carboxylase reaction and the mechanism of activation by magnesium ions, *Arch. Biochem. Biophys.*, **126**:737–745.

————, Y. Goto, and T. Akazawa, 1968b. Pyruvate kinase activity of wheat plants grown under potassium deficient conditions, *Plant Physiol.*, **43**:730–734.

Thimann, K. V., and W. D. Bonner, 1950. Organic acid metabolism, *Ann. Rev. Plant Physiol.*, **1**:75–108.

————, C. S. Yocum, and D. P. Hackett, 1954. Terminal oxidases and growth in plant tissues: III. Terminal oxidation in potato tuber tissue, *Arch. Biochem.* **53**:239–257.

Vennesland, B., 1966. Involvement of $CO_2$ in the Hill reaction, *Fed. Proc.,* **25**:893–898.

Vickery, H. B., 1952. The behavior of isocitric acid in excised leaves of *Bryophyllum calcinium* during culture in alternating light and darkness, *Plant Physiol.*, **27**:9–17.

Vogel, H., 1965. Lysine biosynthesis and evolution, in V. Bryson and H. Vogel (eds.), "Evolving Genes and Proteins," pp. 25–40, Academic Press, New York.

Walker, D. A., 1960. Physiological studies on acid metabolism: 7. Malic enzyme from *Kalanchöe crenata:* Effects of carbon dioxide concentration on phosphoenolpyruvic carboxylase activity, *Biochem. J.,* **74**:216–223.

————, 1967. Carboxylation in plants, *Endeavour,* **25**:21–26.

Warburg, O., 1964. Prefatory chapter, *Ann. Rev. Biochem.,* **33**:1–14.

Wilson, E. B., 1952. "Introduction to Scientific Research," 375 pp., McGraw-Hill Book Company, New York.

Wolf, J., 1960. Der diurnale Säurerhythmus, in "Handbuch der Pflanzenphysiol.," XII/2, pp. 809–889, Springer-Verlag, Berlin.

Yamamoto, Y., and H. Beevers, 1960. Malate synthetase in higher plants, *Plant Physiol.*, **35**:102–108.

Yocum, C. S., and D. P. Hackett, 1957. Participation of cytochromes in the respiration of the aroid spadix, *Plant Physiol.,* **32**:186–191.

Yoshida, S., 1969. Biosynthesis and conversion of aromatic amino acids in plants, *Ann. Rev. Plant Physiol.,* **20**:41–62.

Zelitch, I., 1964. Respiratory mechanisms in plants including organic acid metabolism, *Ann. Rev. Plant Physiol.,* **15**:121–142.

Zuckerkandl, E., and L. Pauling, 1965. Evolutionary divergence and emergence in proteins, in V. Bryson and H. Vogel (eds.), "Evolving Genes and Proteins," pp. 97–166, Academic Press, New York.

# 2
# Photophysiology: The Responses of Plants to Light

## 2.1  CENTRAL IMPORTANCE OF LIGHT TO THE PHYSIOLOGY OF PLANTS

Although plants do not have a total monopoly on photophysiology, no other group of organisms has the same range of responses to light: photosynthesis and other light-stimulated chemical transformations, photoperiodism, and phototropism, to name the most prominent. Moreover, studies with plants have laid the groundwork for photophysiological research generally and, in some cases, have contributed to or stimulated understanding in photochemistry and spectroscopy.

The analysis of a photophysiological phenomenon consists essentially in obtaining answers to two photochemical questions together with some attendant ones: What is the absorbing pigment? And what are the reactions of the excited pigment?

The theories of photochemistry and spectroscopy can become quite complicated, especially if we begin to worry about energy levels in excited states and the mechanisms of photochemical reactions. But we shall see that the limited questions of identity and reaction products may be answerable with the evidence of action spectra, absorption or difference spectra, spectra of light emission, and electron spin resonance.

## 2.2  SOME PHOTOCHEMISTRY[1]

The study of known photochemical processes permits us to anticipate the kinds of processes we shall encounter in photosynthesis and other biological responses to light.

### Electronic Transitions in Atoms, Molecules, and Crystals

Light can be thought of as traveling in packets called photons. Photons are often designated as $h\nu$, which is at once a symbol and a representation of their energy (see below). Every time an ultraviolet or visible photon is absorbed, the absorbing material undergoes an *electronic transition*. The simplest case of an electronic transition occurs in a hydrogen atom (Fig. 2-1). The single hydrogen electron is excited to a new energy level; the *entire* energy of the photon is converted into additional kinetic energy of the electron. Just as a violin string may vibrate only in a fundamental tone and higher harmonics, the hydrogen atom is stable only in certain definite energy states, so photons of certain definite energies, and only these energies, are absorbed by hydrogen atoms (Fig. 2-2). It is precisely as if the hydrogen atoms resonated at certain critical frequencies (or wavelengths) of light. If, with a spectograph, we were to observe white light that had passed through hydrogen atoms, we

[1] References: Clayton, 1965; Claesson, 1964; Kamen, 1963; Butler and Norris, 1962.

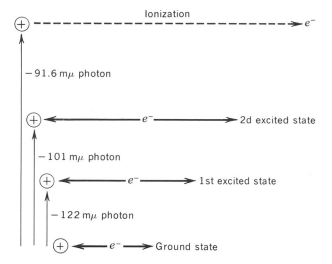

*Fig. 2-1* Excited states of hydrogen atoms. The single electron of a hydrogen atom oscillates between the nucleus and some outer limit. In excited states the oscillation occurs over a longer radius. If a sufficiently energetic photon is absorbed, the electron escapes completely and we say the hydrogen atom is *ionized* (Pauling, 1960).

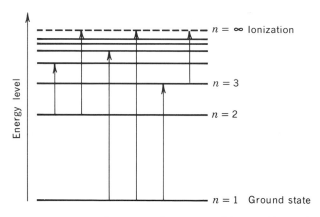

*Fig. 2-2* Energy diagram and some possible transitions of hydrogen atoms. The energy levels of excited states in hydrogen atoms are shown as increasing levels of the ordinate. Transitions between states are shown by arrows. The length of the arrows are proportional to the energy changes in transitions and would be proportional to the energies of exciting photons.

should see absorption at the wavelengths corresponding to transitions of the hydrogen atoms. The result is a series of absorption lines (Fraunhofer lines) (Fig. 2-3); atoms such as hydrogen are thus said to exhibit "line spectra."

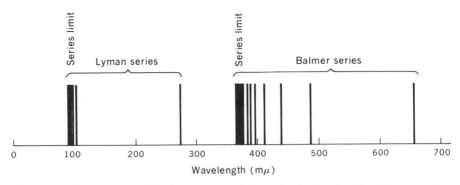

*Fig. 2-3*  Line spectrum of hydrogen atoms. The series in the far ultraviolet was discovered by Lyman and represents transitions from the ground state. The Balmer series represents transitions fom the first excited state.

The general method of wave mechanics, which treats orbital electrons as simple harmonic oscillators, is fairly successful even with complicated atoms such as sodium and iron. In contrast, the transitions of even the simplest molecules are enormously more difficult to dissect, but approximations are possible by starting from the premise that the most stable states for chemical bonds correspond to a minimum energy mode in which an electron wave can oscillate between two nuclei. The same kind of electronic transitions occurs in molecules as in atoms and is responsible for the absorption of visible and ultraviolet light by molecules.

Wave mechanics also provides a rational basis for the phenomena of "allowedness" and "forbiddenness," i.e., the complete or partial exclusion of certain types of transitions. Similar considerations apply to the energy levels of crystals, in which there can exist free electrons, i.e., electron waves oscillating within conduction bands or whole regions of the crystal.

### Other Transitions

We have been talking about electronic transitions. There are other kinds of transitions in polyatomic molecules and crystals: (1) vibrational transitions, where the energies are contained in elastic vibration among the nuclei of the molecule or crystal, and (2) rotational transitions, where the energies are associated with the twisting of nuclei along the axis of the bond joining them (Fig. 2-4).

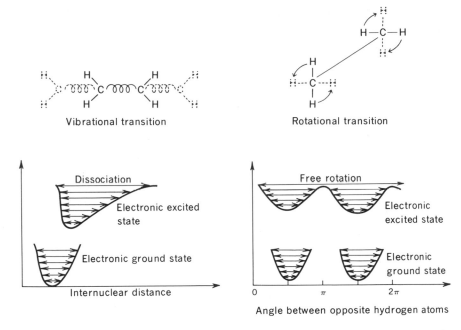

*Fig. 2-4* Vibrational and rotational transitions. Some transitions that can occur in the double bond of ethylene are shown as examples of vibrational and rotational transitions.

The energies of pure vibrational and rotational spectra are weak, and so the photons that can induce such transitions are restricted to the infrared. (Indeed, infrared absorption spectra are specific indicators of molecular structure for the simple reason that they correspond to the vibrational and rotational modes possible for the molecule.)

The photon energies for these different kinds of transitions are easily calculated. The energy of a photon is equal to the product of Planck's constant and the frequency of light

$$E = h\nu \quad \text{or} \quad E = \frac{hc}{\lambda}$$

where $c$ is the velocity of light and $\lambda$ is wavelength. The energy of a photon thus may be determined from the value of Planck's constant and the wavelength or frequency of light. When the wavelength is expressed in $m\mu$ (or nm), the energy in electron volts is

$$E_{ev} = \frac{1,235}{\lambda_{m\mu}}$$

Eq. 2-1

The approximate energy relations among the different kinds of transitions are:

| Transition | Effective Wavelengths, mμ | Energy | |
|---|---|---|---|
| | | electron volts | calories |
| Electronic | 100–1,000 | 1–10 | 300,000–30,000 |
| Vibrational | 1,000–10,000 | 0.1–1 | 30,000–3,000 |
| Rotational | 10,000–100,000 | 0.01–0.1 | 3,000–300 |

In the case of molecules, we usually observe mixed spectra. In these cases, the molecule exists in one of several possible vibrational states associated with the ground electronic state. The molecule may be raised to any of several possible vibrational states associated with an excited electron state (Fig. 2-5).

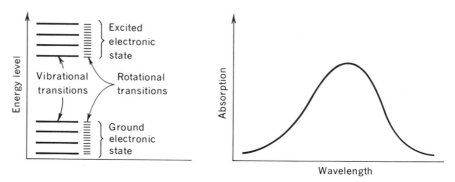

*Fig. 2-5*   Energy diagram for mixed transitions.

Instead of exhibiting the relatively simple line spectra of atoms, molecules show extremely complex spectra. In the gaseous state, the spectra consist of bands of relatively diffuse lines. In the condensed phase, the multitude of vibrational and rotational states smears into a single blurred peak for each electronic transition.

The ultraviolet and visible absorption spectra which have become such a commonplace in chemistry and biochemistry therefore have their physical basis in electronic transitions, greatly modified by vibrational and rotational transitions. Transitions may occur to the first, second, or higher excited states. For example, with chlorophyll *a*, light of 663 mμ which corresponds to 41 kcal is sufficient to excite the molecule to the first excited state. Light of 410 mμ (65 kcal) excites chlorophyll *a* to the second excited state.

Once in the excited state, a molecule may be excited still further. Again in the case of chlorophyll *a*, it has been possible with tremendous light intensities to convert most of a given sample to another excited state (in this case the *triplet state*). It remains in this state long enough that one may irradiate the solution again, this time with monochromatic light, and this time observe transitions to the second triplet state.

Unlike electronic transitions, vibrational and rotational states are functions of the temperature. We may therefore expect that different absorption spectra might be observed at different temperatures. In fact, absorption spectra at extremely low temperatures show greatly sharpened peaks. An example of low-temperature spectra of cytochromes was presented in Fig. 1-21.

### Extinction Coefficient and Oscillator Strength

If we are to characterize a particular electronic transition for a polyatomic molecule in the condensed phase, we have two quantities that we can measure: (1) the wavelength of the absorption peak and (2) the intensity of the aborption peak. The most familiar measure of absorption intensity is the extinction coefficient, simply defined in the Lambert-Beer equation (Appendix B)

$$\log_{10} \frac{I_0}{I} = \epsilon c d \qquad \text{Eq. 2-2}$$

where $I_0$ and $I$ are initial and final intensities of light passing through a sample, $c$ is the concentration of the sample in moles per liter, and $d$ is the light path in centimeters; the extinction coefficient, $\epsilon$, then has the units of liter mole$^{-1}$ centimeter$^{-1}$.

A more fundamental quantity is *oscillator strength,* which is the efficiency with which an oscillator (electron) resonates with the field (light). An ideal oscillator (e.g., a hydrogen atom) absorbs light in an allowed transition with 100 percent efficiency. We say it has an oscillator strength $f_A = 1$. Real molecular oscillators absorb light over a band of wavelengths, so that we must integrate (see Eq. 2–3). The integrated oscillator strength $f_A$ is

$$\bar{f}_A = \int_{\lambda_1}^{\lambda_2} f_A(\lambda)\, d\lambda \qquad \text{Eq. 2-3}$$

There is a convenient relation between oscillator strength, absorption coefficient (Eq. 2-2), and the index of refraction of the medium, $n$.

$$\bar{f}_A = 4.319 \times 10^{-9} n \int_{\lambda_1}^{\lambda_2} f(\lambda)\, d\lambda \qquad \text{Eq. 2-4}$$

Oscillator strength sums the electronic transition over all of the vibrational and rotational substates. This quantity represents, therefore, the total probability or allowedness of the electronic transition occurring. Since the probability of the transition occurring is the same in both directions, the oscillator strength is also related (as is the absorption coefficient) to the reciprocal of the mean lifetime of the excited state. Nearly ideal organic oscillators have strengths of 1 corresponding to absorption coefficients of about $10^6$ liters mole$^{-1}$ cm$^{-1}$ and mean lifetimes of about $19^{-9}$ sec.

Fates of Excited Substances

When a substance absorbs a photon in the visible or ultraviolet, it undergoes an electronic transition and becomes excited.

$$A + h\nu \rightarrow A^* \qquad\qquad \text{Eq. 2-5}$$

The excited substance may then undergo the following, either singly or in some cases in sequence:

1. Fluorescence.
2. A long-lived or metastable state, such as the triplet state.
3. A chemical reaction.
4. Ionization.
5. Dissociation (if a molecule).
6. Isomerization (if a molecule).
7. Conversion of all or part of its energy of excitation to kinetic energy through what is called internal quenching or a radiationless transition.
8. Transference of the energy of excitation to another atom or molecule at a small distance; this is called energy transfer, exciton migration, or resonance transfer.

Since all of these processes may possibly occur in intact plants, we shall examine them more closely.

FLUORESCENCE.   Atoms in the gas phase may emit the simplest kind of fluorescence, namely reemission of the absorbed photon. This is called resonance fluorescence. In polyatomic molecules and in condensed phases, fluorescence occurs when the species is excited to a higher singlet state. Part of the energy is lost by internal quenching (alternative 7 above), bringing the species down to the first singlet state. The remaining energy is then reemitted as a photon (Fig. 2-6). Since the probability for a transition is the same in either direction and the average lifetime of a single state (assuming an allowed transition to the ground state) is not much longer than $10^{-9}$ sec, fluorescence usually occurs within $10^{-9}$ sec. Longer intervals (up to $10^{-6}$ sec) have been observed when exciton migration intervenes between excitation and fluorescence.

LONG-LIVED AND METASTABLE STATES.   Although transition from the ground state to forbidden states is highly improbable, it may be quite probable for a molecule to pass from an excited singlet state to a forbidden one (Fig. 2-7). Photons may then leak slowly from the excited species, resulting in *phosphorescence*. A typical forbidden state is one with two unpaired electrons. From the observation that, under the influence of an external magnetic field, atoms with unpaired electrons can exist in three excited states with closely similar energy levels, spectroscopists named such little constellations "triplet states."

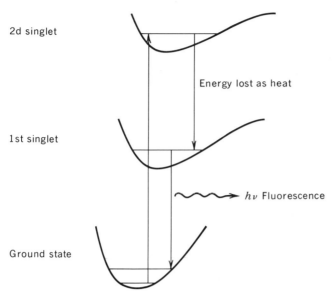

2d singlet

1st singlet

Energy lost as heat

$h\nu$ Fluorescence

Ground state

*Fig. 2-6*   Energy diagram for molecular fluorescence.

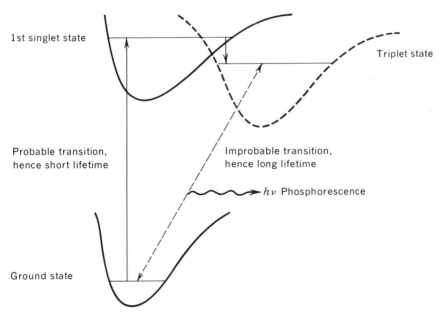

1st singlet state

Triplet state

Probable transition, hence short lifetime

Improbable transition, hence long lifetime

$h\nu$ Phosphorescence

Ground state

*Fig. 2-7*   Energy diagram for the metastable state and phosphorescence.

The term has been used loosely for any excited species which emits light over a long interval, even when the triplet energy splitting has not been demonstrated. This imprecision is unfortunate, especially in biological systems where highly ordered, near-crystalline arrays may occur in which other long- lived or metastable states are possible.

CHEMICAL REACTIONS.   Excited species may enter directly into chemical reactions. The reaction may occur simply because the energy of excitation supplies the energy of activation of the reaction; that is, the energy of excitation "drives" the reaction, producing a net decrease in free energy. A reaction may also occur because the orbitals of the excited state labilize a bond, greatly increasing the possibility of orbital overlap with another reactant. The last-named mechanism is in effect a decreased energy of activation for the reaction. Because of the extremely short lifetimes of singlet states, the probability is low for their experiencing a favorable collision. Efficient bimolecular reactions involving singlet excitation would require extremely favorable steric factors. Reactions with molecules in forbidden states are quite another matter for the very reason that they are long lived. There have been frequent suggestions that triplet states mediate a wide variety of reactions.

IONIZATION.   From the energy diagram of the hydrogen atom (Fig. 2-1), we saw that a photon of 13.60 ev (or 91.1 m$\mu$) was sufficient to move the hydrogen electron completely free of the nucleus. This is of course an ionization. Organic molecules may lose electrons at much lower energy levels: radiation of wavelengths shorter than about 300 m$\mu$, frequently called "ionizing radiation," is famous for producing genetic mutations. Ionized species may subsequently enter into chemical reactions. Moreover, the acid dissociation constants of molecules may be enormously altered by excitation. The $p_k$ of $\beta$-naphthol is 9.5, whereas the $p_k^*$ is 2.5. Thus $\beta$-naphthol in the excited state is a moderately strong acid!

DISSOCIATION.   If the energy of excitation is transformed into vibrational modes, it is quite possible for the new vibrational level to be sufficiently great to break the chemical bond. As can be seen in Fig. 2-4, the "envelope" for vibrational energies is apt to be much more shallow in the excited than in the ground state and displaced toward larger distances between the nuclei, so that after an electronic excitation from a stable vibrational state, a molecule can find itself in an unstable vibrational state and the molecule dissociates. The resulting fragments are free radicals, which are typically extremely reactive. Many light-catalyzed polymerizations and other chain reactions are initiated by photochemically generated free radicals.

A photodissociation is also thought to be involved in the release by light of cytochrome oxidase from carbon monoxide inhibition (cf. Sec. 1.8).

ISOMERIZATION.   An extremely interesting example of photochemical isomeriza-

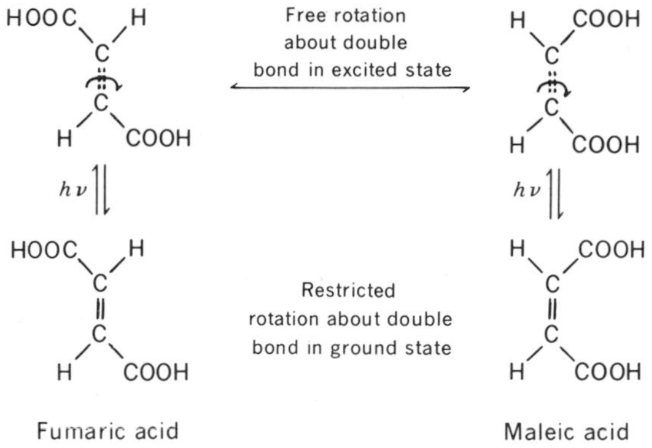

HOOC   H
      \  /
       C
       ⫴
       C
      /  \
    H    COOH

Free rotation
about double
bond in excited state

H    COOH
 \  /
  C
  ⫴
  C
 /  \
H    COOH

$h\nu$ ‖

$h\nu$ ‖

HOOC   H
      \  /
       C
       ‖
       C
      /  \
    H    COOH

Restricted
rotation about double
bond in ground state

H    COOH
 \  /
  C
  ‖
  C
 /  \
H    COOH

Fumaric acid

Maleic acid

*Fig. 2-8* Photoisomerization of maleic and fumaric acids.

tion is the interconversion of maleic to fumaric acid (Fig. 2-8). Although these two acids have quite distinct structures, corresponding to the cis and trans configurations around a carbon-carbon double bond, the excited states of the two are nearly identical, so it is possible for an excited fumaric molecule to return to the maleic ground state and vice versa. Photoisomerizations may be of great importance in biology.

QUENCHING.   Even though increased electronic energy resulting from excitation may labilize a bond sufficiently to permit a chemical reaction, the more usual event is that the excitation energy is converted to vibrational energy among all the bonds of the molecule. Through the principle of equal partition of energy, ultimately all of the excitation energy is converted to heat. The initial conversion of electronic to vibrational energy is called internal conversion or quenching. In the liquid phase and at ordinary temperatures, collisions frequently enhance quenching, so that most photochemical reactions, including fluorescence and phosphorescence, are strongly quenched in liquid phases at ordinary temperatures.

Considering the enormous variety of ways in which molecules can lose their excitation energies without entering into a chemical reaction, energy-conserving photochemical reactions such as photosynthesis seem highly improbable!

ENERGY TRANSFER.   Exactly as a plucked violin string will induce resonant vibrations in an adjacent string tuned to the same frequency, an excited electron in a molecule will induce an electronic transition in a nearby molecule. More than that, if the resonance energy of the neighboring molecule is somewhat less than that of the excited molecule but still sufficiently near that their absorption bands overlap, a transference of all of the excitation energy *must* occur. In this way excitation energy can migrate across many molecules in a closely packed array. The phenomenon is called energy transfer, exciton migration, and resonance transfer.

According to one model, the efficiency of energy transfer should decrease with the sixth power of the distance separating the molecules (cf. Hoch and Knox, 1968).

## 2.3   PHOTOCHEMICAL EVIDENCE[1]

*"Come, listen my men, while I tell you again*
*The five unmistakeable marks*
*By which you may know, wheresoever you go,*
*The warranted genuine Snarks"*

——*Lewis Carroll*

Let us suppose that we come across a biological process in which light appears to be involved. The process might be oxygen evolution, movement, change in pigmentation, flowering, luminescence, release from CO inhibition, etc. What questions might we appropriately ask about such a system? And by what criteria should we judge answers to these questions? Questions and the kinds of answers to them are the subjects of this section.

The questions we ask are simple enough. What pigment or pigments absorb light? Where are the pigments? What is the excited state. What is the fate of the excited pigment molecules. That is, what transformations or reactions does the pigment experience? How many quanta are required for a unit step in the process? Is there transfer of radiant energy in the system and, if so, how does transfer occur? After the strictly photochemical questions, there are questions that grade rapidly into biochemistry. For example, is the process strictly photochemical, or are there accompanying "dark" reactions, which might be required to regenerate the pigment? How does the product of the light reaction influence the physiological process?

The kinds of evidence required to answer these questions are all based squarely on the principles of photochemistry.

### The Bunsen-Roscoe Law

Our first step in appraising a light-influenced process is to determine if the response is proportional to the amount of light absorbed.

$$R = kIt \qquad\qquad\qquad \text{Eq. 2-6}$$

where $R$ may be taken as the extent of a process, $I$ is the light intensity, and $t$ is time. This statement is known as the Bunsen-Roscoe law. It follows directly from the notion of excitation

$$A + h\nu \rightarrow A^* \rightarrow P$$

[1] References: Clayton, 1965; Giese, 1964; Kamen, 1963; Gaffron, 1960; Allen, 1964.

We can expect that, since every A* has an equal probability of resulting in P, the rate or extent of the reaction should be proportional to A*; the light absorbed is proportional to the concentration of pigment A and the amount of light, which is clearly the product of $I$ and $t$.

Nonconformity with the Bunsen-Roscoe law means that either A varies over the time intervals chosen, A varies with $I$, or the extent of the reaction is not proportional to the immediate photochemical product P.

Variation of A with $I$ can occur in the case of optical pumping of lasers, but is rarely encountered in biological systems. In the scheme

$$x \rightarrow A + h\nu \rightarrow A^* \rightarrow P$$

variation of A with time can mean that the rate of formation and/or regeneration of A is slower than (or of the same order of magnitude as) the rate of excitation (absorption).

$$\frac{d(x \rightarrow A)}{dt} + \frac{d(A^* \rightarrow A)}{dt} \leq \frac{d(A \rightarrow A^*)}{dt}$$

We may use flashing light tests to tell us if a "dark" reaction is required for the light-influenced process. By varying the dark interval between flashes, we may learn something about limiting rates of the dark reaction. The effects of variations in the intensity of duration of the flashes give corresponding information about the relative amount of absorbing pigment.

## Action Spectra

The formation of the immediate photochemical product P is proportional to A* and therefore also to the absorption of light by A. From this rather obvious statement, it follows that if a chemical or physiological response to light, R, is proportional to P, then R *is proportional to* A (Eq. 2-7) *for all competent transitions.*

$$R(\lambda) = kA(\lambda) \qquad\qquad \text{Eq. 2-7}$$

A graph of $R(\lambda)$ is said to be an *action spectrum.* Equation 2-7 tells us that the *action spectrum should have the same shape as the absorption spectrum of the absorbing pigment.* Equation 2-7 is subject to certain obvious restrictions: that R is in fact proportional to P, that other pigments do not shade or mask A, that energy transfer does not occur, etc. Stated differently, when the action spectrum does *not* have the same shape as the absorption spectrum of a pigment, determined on other grounds to be *the absorbing pigment,* then one or more of the above restrictive conditions must obtain. For further discussion of some of the pitfalls of action spectra, see Clayton (1965) and Fork and Amesz (1969).

In general, the experimenter must ensure that the Bunsen-Roscoe law is obeyed when measuring action spectra.

### Absorption Spectra

The simple requirements that (except in lasers) there must be an absorbing pigment in order for light to produce an effect and that $A \rightarrow A^*$, mean that photochemical processes may result in changes in the pigments of the system; e.g., disappearance of A. This is obviously not a necessity; $A^*$ may be vanishingly small compared with A, and A may be rapidly regenerated. If absorption changes are observed, we may imagine them to be due to either disappearance of A or appearance of either $A^*$ or P. The occurrence of absorption changes due to the appearance of products may be further tested by studying the sequence of absorption changes following excitation. The changes observed *in vivo* may correspond to absorption spectra or difference spectra of the participating pigments.

### Light Emission

If the reader has gained the impression that measurements of light absorption are powerful and versatile instruments in the physiological armory, he should immediately set down measurements of light emission as an even more sensitive and almost as versatile means of inspecting physiological processes. The reason is purely technical: light absorption requires the comparison of the intensity of one light beam with another and is dependent on the stability of detectors and amplifiers. In contrast, the emission of light may be detected at the highest sensitivities of photocells and photomultipliers where the "control" is simply the dark current of the detector. The sensitivity for measuring fluorescence can be enhanced even further by modulating the exciting radiation and locking the detector to the same frequency.

When an excited molecule emits light, we may measure the rate of decay of light following removal of the exciting radiation. From the lifetime of the excited state, we can determine whether the light is fluorescence or phosphorescence. From measurements of depolarization we can also infer whether exciton migration intervened between excitation and emission.

We may also determine the *emission spectrum,* from which the identity of the excited species may be deduced. And, finally, we may determine the *excitation spectrum* or *action spectrum* of fluorescence, which has the same shape as the absorption spectrum.

A fluorescence emission spectrum resembles a long-wavelength mirror image of the singlet absorption band (Fig. 2-9). This occurs because internuclear distances are greater in the excited state than in the ground state (cf. Fig. 2-4). Consequently an excited molecule is apt to return to a higher vibrational state than the one it occupied in the ground state. The energies of the photons

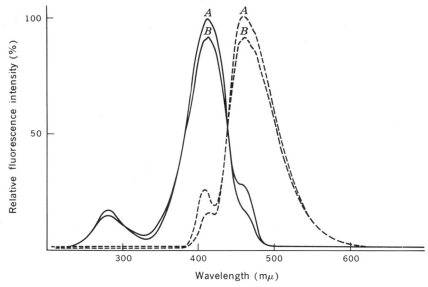

*Fig. 2-9* Fluorescence excitation and emission spectra of gibberellins. Because of the Franck-Condon principle—the tendency for the vibrating nuclei of molecules to spend most of their time at the ends of their oscillatory excursions—and because the internuclear distance of excited species is greater than those in the ground state, fluorescence emission spectra tend to form mirror images of absorption (i.e., excitation) spectra and to be shifted to longer wavelengths. Shown above are excitation (solid line) and emission (broken line) spectra of (*A*) gibberellic acid and (*B*) barley gibberellin (cf. pp. 310–316, Jones et al., 1963).

emitted will be less energetic, and hence of longer wavelengths, than those absorbed by the molecule in the ground state.

In the case of exciton migration, the absorbing species will very likely be different from the emitting species. This is typically the case in chlorophyll fluorescence in green plants.

A caveat: emitted light energy of fluorescence and phosphorescence is of course not available for photochemistry. The molecule emitting fluorescence or phosphorescence *may* be the same as those involved in a photobiological process, but it may also represent a diversion from the photochemistry of the physiological response. We should not conclude, therefore, that the emitting species are necessarily in the sequence of intermediates leading to the physiological reaction.

## Temperature Effects

Since electronic transitions can proceed regardless of the vibrational or rotational state of the absorbing molecules, pure photochemical reactions proceed equally rapidly at all temperatures. Zero temperature dependence becomes then an identifying feature of pure photochemical reactions or the photochemical part of mixed photochemical and dark reactions. Low temperatures can of course modify absorption spectra and therefore action spectra.

## Electron Spin Resonance[1]

Metastable excited states result when an excited singlet passes to a state that can be reached from the ground state only by a forbidden transition. The most familiar cause of forbiddenness is the rule that the number of unpaired electrons must be unchanged by a transition. In most molecules ($O_2$ is an exception), electron spins are scrupulously paired in the ground state. We may write such a spin configuration as $A_{\uparrow\downarrow}$. Allowed transitions preserve spin pairing; forbidden transitions change it. Thus

$$A_{\uparrow\downarrow} \rightarrow A^*_{\uparrow}{}^{\downarrow} \qquad \text{or} \qquad A_{\uparrow\downarrow} \rightarrow A^*_{\downarrow}{}^{\uparrow}$$

would be allowed, but

$$A_{\uparrow\downarrow} \rightarrow A^*_{\uparrow}{}^{\uparrow}$$

is forbidden.

Now unpaired spins interact with magnetic fields of the appropriate strength and may be detected as perturbations of an imposed field as each electron takes a quantum jump to a paired configuration.

This method, known as electron spin resonance (ESR) or electron paramagnetic resonance, can be employed to detect free unpaired electrons. We can infer from an ESR signal that a sample contains free radicals or triplet states and can also infer something about the abundance, the rates of decay, and the environment of the unpaired electrons.

ESR spectrometers display their signals as derivatives (cf. Fig. 2-32) with the applied magnetic field strength ($H_0$) as the abscissa. The *g-value*, the field position at which maximum absorption occurs, is related to the magnetic field strength by the relation

$$g = \frac{h\nu}{\beta H_0}$$

where $h$ is Planck's constant, $\nu$ is the frequency of the microwave radiation (typically 9.5 kMHz), $\beta$ is the Bohr magneton, and $H_0$ is the field strength. Most absorption occurs near $g = 2$ where $H_0$ is about 3,400 gauss. Other identifying features of ESR spectra are the width of the signal, $\Delta H$, measured in gauss between points of maximum slope, the amplitude, and the total signal strength, which is proportional to the number of spins.

[1] References: Blois and Weaver, 1964; Weaver, 1968.

## 2.4   CASE STUDY: CONVERSION OF PROTOCHLOROPHYLLIDE *a* TO CHLOROPHYLLIDE[1] *a*[2]

When higher plants are grown in the dark. they fail to develop chlorophyll. Instead of appearing green, they are pale yellow or yellow-green. Upon exposure to light there is a sudden formation of a small amount of chlorophyllide *a,* followed by a slowly accelerating accumulation of chlorophylls and chlorophyllides of *a* and finally *b* (Fig. 2-10).

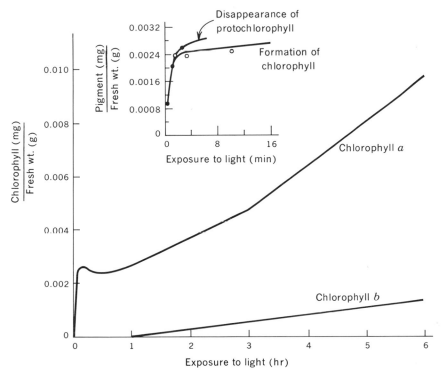

*Fig. 2-10*   Time course of formation of chlorophyll upon illumination of etiolated plants (after Koski, 1950; cf. Egle, 1960).

We shall confine our attention to the first, rapid phase of chlorophyll formation. As we shall see, this process behaves like a relatively simple photochemical system.

The existence in etiolated plants of a pale green form of chlorophyll,

---

[1] The chlorophyllides are chlorophylls without the phytol side chain (Fig. 1-25). Chlorophyllides differ from the analogous chlorophylls by greater solubility in water and by small differences in light absorption in the ultraviolet.

[2] Cf. reviews of Egle, 1960; Smith and Young, 1956; Smith, 1960, 1963; Boardman, 1966; cf. Sec. 1.9.

protochlorophyll, has been known since 1893. It differs from chlorophyll *a* by having two less H in the porphyrin ring (Fig. 1-25).

We shall consider the hypothesis that chlorophyllide *a* (chl *a*) is formed from protochlorophyllide (protochl) by a relatively simple photochemical reaction

$$\text{Protochl} + [2H] + h\nu \rightarrow \text{chl } a \qquad \text{Eq. 2-8}$$

### Stoichiometry

That protochl is transformed quantitatively into chl *a* is clearly evident from the stoichiometry (Table 2-1 and Fig. 2-10). During the short illumination of 5 min when 80 percent of the protochl has disappeared, the sum of protochl plus chl *a* remains constant.

Table 2-1 Stoichiometry of the Transformation of Protochlorophyll to Chlorophyll *a* by Etiolated Corn Leaves[*]

| Time, sec | Transformation, % | Protochlorophyll plus Chlorophyll *a*, mg $g^{-1} \times 10^3$ |
|---|---|---|
| 0 | 0.76 | 8.29 |
| 10 | 12.0 | 7.89 |
| 20 | 20.8 | 8.68 |
| 40 | 34.7 | 8.64 |
| 80 | 50.2 | 8.93 |
| 160 | 69.8 | 8.55 |
| 320 | 80.0 | 9.27 |
| | Average | 8.61 |

[*] Smith and Young, 1956.

### Bunsen-Roscoe Law

The photochemical yield of chl *a* in etiolated barley leaves remains constant when the duration of light flashes is varied down to 13 msec (Virgin, 1955). Therefore, if there are any dark reactions, they must go to completion in less than 13 msec.

### Action Spectrum

The shape of the action spectum should be the same as the shape of the absorption spectrum of the absorbing pigment. In the case of ordinary etiolated corn seedlings, the action spectrum for the transformation of protochl to chl *a* shows a red peak at 650 m$\mu$ and a blue peak at 445 m$\mu$ (Fig. 2-11). The red peak corresponds to the absorption decrease observed *in vivo* and might reasonably be due to some form of protochl. The blue peak is also somewhat shifted to a shorter wavelength. Thus far the data appear consistent with the notion that protochl is itself the absorbing pigment. However, the relative

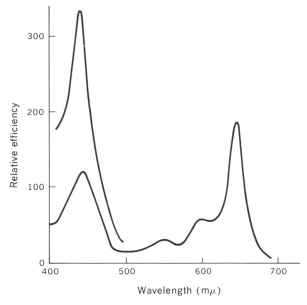

*Fig. 2-11*   Action spectra for the transformation of proto-chlorophyll to chlorophyll *a* (Koski et al., 1951). The unit of relative quantum efficiency is

$$\frac{10^{17}}{\text{quanta sec}^{-1} \text{ cm}^{-2} \text{ for 20 percent conversion}}$$

The lower curve was obtained with normal etiolated corn leaves; the upper curve was obtained with albino mutants.

heights of the blue and red action peaks are much different from the absorption peaks of protochl.

One of the bogeys of action spectra is screening pigments. One should not be surprised if carotenoids in the etiolated plants had screened protochl from blue light, but not from red light. In fact, when one employs an albino mutant (one deficient in carotenoids) of corn seedlings, the blue action peak rises to a quite respectable height (Fig. 2-11).

## Absorption Spectra

If the hypothesis is correct, we should expect that as protochl is transformed to chl *a,* the absorption spectra of the tissues would show the disappearance of the protochl spectrum and the simultaneous appearance of the chl *a* spectrum.

The spectra for protochl and chl *a* are shown in Fig. 2-12. The exact position and height of the peaks depend somewhat on the solvent, but in general it is clear that protochl has substantially stronger absorption in the blue (432 m$\mu$) than chl *a* has. The blue peak is called the Soret band and is typical of porphyrins. One would expect that protochl, with more opportunities for resonance than chl *a,* would absorb more strongly there. The red band

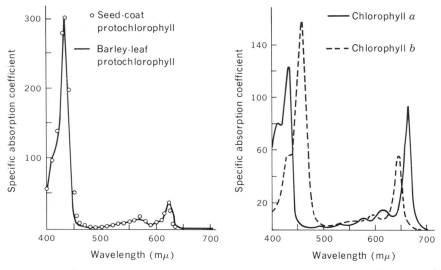

Fig. 2-12 Absorption spectra of protochlorophyll, chlorophyll *a,* and chlorophyll *b* (Smith and Young, 1956). (*Left*) Barley leaf and pumpkin seed-coat protochlorophyll in ether; (*right*) chlorophylls *a* and *b* in ether.

of protochl is at 623 m$\mu$ as compared with about 660 m$\mu$ in chl *a* and is substantially weaker than the chl *a* absorption.

Absorption spectra of intact etiolated tissues vary somewhat and it is difficult to correct for light scattering. There must also be some doubt about the environment in which protochl finds itself—whether it is bound to protein, in a lipid matrix, etc.—and the consequences of the environment on the absorption spectrum. Despite all of these qualms, the absorbance of etiolated bean leaves before and after illumination shows very clearly a decrease in absorption at 648 m$\mu$ and a strong increase at 684 m$\mu$, which shifts to 677 m$\mu$ over a period of subsequent darkness (Fig. 2-13).

Although the bands of protochl and chl are displaced from those observed in organic solvents, these changes are clearly consistent with the hypothesis.

### Temperature Dependence

A photochemical reaction which involves only an electronic transition would be temperature independent. The conversion of protochlorophyll to chlorophyll shows only a slight temperature dependence down to —77° but the reaction failed at —195° (Smith and Benitez, 1954; cf. Kupke and French, 1960).

Thus the reaction cannot involve only an electronic transition. It is unlikely from the energy of activation that the production of the hydrogen donor or the rate of its reaction with the photochemical reaction product accounts for the temperature dependence. It may be that the effective electronic transition

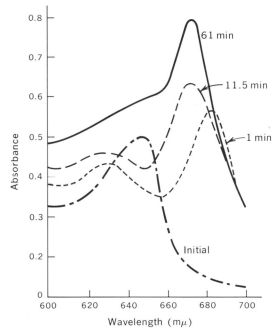

*Fig. 2-13*  Absorption changes in etiolated bean leaves (after Shibata, 1957).

can occur only from certain vibrational or rotational states. One can imagine specifically that a molecular vibration is required to permit protochlorophyllide to react with an adjacent hydrogen donor.

### Hydrogen Donor

About all that is known about the hydrogen donor is that the reaction can also occur in the dark (enzymatically) in certain plant tissue (avocado fruit, many cryptogams, and many but not all seeds of conifers). In extracts of a mutant of *Arabidopsis* the transformation is promoted by NADH (Röbbelen, 1956).

### Fluorescence

The fluorescence spectra of isolated chl *a* and protochl are shown in Fig. 2-14. The differences are sufficient that one could expect to detect the presence of one in reasonable concentrations of the other.

Virgin (1955) studied the time course of greening of etiolated barley leaves in this way. The results (Fig. 2-15) show even more clearly than the absorption changes that the formation of chl *a* coincides with the disappearance of protochl.

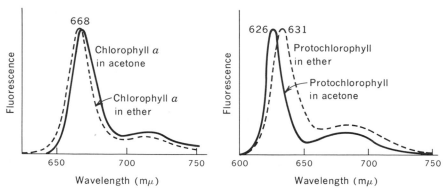

Fig. 2-14   Fluorescence spectra of chlorophyll *a* and protochlorophyll (after French, 1960).

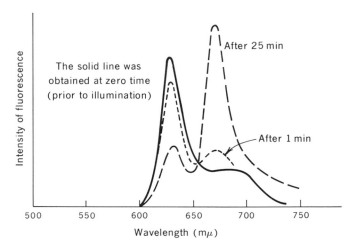

Fig. 2-15   Fluorescence spectra of etiolated barley leaves during the transformation of protochlorophyll to chlorophyll *a* (Virgin, 1955). Fluorescence spectra of etiolated barley leaves were obtained at various times after illumination. The fluorescence is measured in arbitrary units. The solid line was obtained at zero time (prior to illumination).

### Biochemical Reconstruction

Every kind of evidence described thus far in this case study has been essentially biophysical. The photochemical properties of plants have been compared with photochemical properties of isolated pigments. The fact remains however that protochl is *not* transformed to chl *a* by light outside the plant. The hypothesis that protochlorophyll undergoes a purely photochemical reaction must be incorrect or at least incomplete.

On the basis of small differences between the absorption spectra of protochl in organic solvents on the one hand and the action spectra and absorption changes *in vivo* on the other, Smith and his colleagues reasoned that protochl must exist in some special form in the cell. They ultimately succeeded in extracting from plants a pigment-protein complex,[1] which they designated "protochlorophyll holochrome" (Smith, 1958, 1961). The substance of molecular weight 600,000 has an absorption spectrum shifted toward the red (Fig. 2-16). Most important, the protochl-protein complex, unlike pure protochlorophyll, can be transformed to chl *a* photochemically (Fig. 2-17).

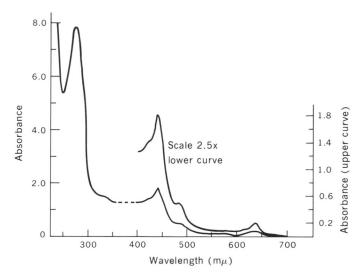

*Fig. 2-16* Absorption spectrum of protochlorophyll holochrome (Smith, 1961). Note that the pigment has an absorption maximum near 280 m$\mu$, characteristic of proteins, and maxima near 430 and 640, similar to the maxima in the action spectrum.

### Conclusion

The problem of the photochemical formation of chl *a* cannot be considered completely solved. For one thing, the "holochrome" is by no means completely pure and the identity of the hydrogen donor is not established. Nonetheless, the photochemistry of this process probably is on as firm ground as that of any other biological process.

### 2.5 CASE STUDY: THE LIGHT REACTIONS OF PHOTOSYNTHESIS

> *"Breathes there a man with soul so tough,*
> *To whom two photosystems are not enough?"*
> ——*R. F. Clayton*

From our first discussion of photosynthesis (Sec. 1.6) we concluded that for many plants the Calvin-Benson hypothesis is approximately correct. In order

---

[1] A similar separation was achieved by Krasnovsky and Kosobutskaya; cf. review of Boardman, 1966.

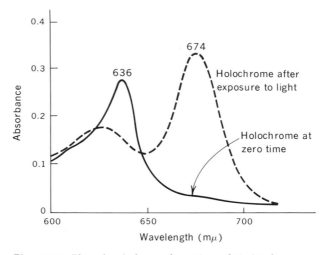

*Fig. 2-17*   Photochemical transformation of isolated proto-chlorophyll holochrome to chlorophyll *a* (Smith, 1958). Protochlorophyll holochrome was isolated, its absorption spectrum in the red region determined, then exposed to sufficient light to effect a transformation to chlorophyll *a,* and the spectrum determined again.

for carbon dioxide to be reduced to the level of sugar in this scheme, there must be a source of ATP and NADPH. We inferred further that we must look to the light reaction of photosynthesis for a means of generating these two coenzymes. Directly or indirectly there must be reactions

$$ADP + P + h\nu \rightarrow ATP \qquad\qquad \text{Eq. 2-9}$$

$$NADP^+ + 2H + h\nu \rightarrow NADPH + H^+ \qquad\qquad \text{Eq. 2-10}$$

in which photon energy $h\nu$ drives these two reaction. In most green plants the source of hydrogen in Eq. 2-10 is water, and we may rewrite Eq. 2-10 as follows:

$$NADP^+ + H_2O + h\nu \rightarrow NADPH + \tfrac{1}{2}O_2 + H^+ \qquad\qquad \text{Eq. 2-11}$$

In this way we see that oxygen, a familiar product of photosynthesis, arises as a waste product.

From the study of the comparative biochemistry of photosynthesis, van Niel (1949) proposed that bacteria, algae, and higher plants shared a more general process of photosynthesis, which we may write as

$$NADP^+ + H_2A + h\nu \rightarrow NADPH + A + H^+ \qquad\qquad \text{Eq. 2-12}$$

To include the reduction of $CO_2$ we must write

$$CO_2 + 2H_2A \rightarrow (CH_2O) + H_2O + 2A \qquad\qquad \text{Eq. 2-12}a$$

For bacteria and some algae, the hydrogen donor $H_2A$ need not be water, but can be any of several substances: thiosulfate, $H_2S$, hydrogen gas, or organic substances such as succinate.

The fundamental questions to be asked are the same as those for any photochemical reaction. What is the absorbing pigment? What is the fate of the pigment in the excited state? What are the immediate chemical products of the reaction? In the present case we must also pursue what may be the essentially biochemical question of the formation of ATP and NADPH from the immediate products of photochemistry.

It has long been recognized that photosynthesis is one of the most remarkable processes in nature. At least 30 percent of the energy of light can be transformed by photosynthesis to chemical energy, an efficiency which is some 10 times greater than any other known photochemical reaction. Because of the enormous theoretical importance of the light reaction for biology, physics, and chemistry, it has attracted some of the most vigorous minds of these three disciplines.

Let us now review the information that has been accumulated (reviews of Smith and French, 1963; Clayton, 1965; Bendall and Hill, 1968; Hind and Olson, 1968; Hoch and Knox, 1968; Fork and Amesz, 1969).

First, a word about experimental systems: the two broad categories of plant material that have been commonly used are suspensions of unicellular algae and chloroplasts isolated from higher plants. Students of photosynthesis have employed algae from every algal phylum to advantage, and also photosynthetic bacteria. Unfortunately, the algae have not yielded active chloroplasts so readily; consequently, *in vitro* studies have rested for the most part on chloroplasts from a few especially cooperative plants, such as spinach, swiss chard, and poke weed (*Phytolacca americana*). The chromatophores of bacteria, such as *Rhodospirillum rubrum*, have also been helpful, although these may be strongly modified during cell disruption.

Because gas exchanges can be measured sensitively and conveniently, the rates of the light reaction are usually measured as $CO_2$ absorption and oxygen consumption. They are not necessarily equal. In some cases ATP and NADPH formation have been measured directly.

## Hypotheses

The simplest photochemical hypothesis of photosynthesis was that chlorophyll is the absorbing pigment and that excited chlorophyll causes the dissociation of water to yield reducing equivalents and oxygen.

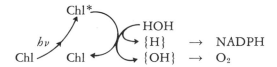

Braces were placed discreetly about the H and OH to indicate our innocence of their state. The scheme was elaborated further by the addition of X and Y to serve as intermediate acceptors for H and OH. For example,

$$HOH + XY \rightarrow XH + YOH$$

The supposition of X and Y has had some heuristic value, but has not of itself advanced our understanding of photosynthesis.

The fundamental notion here is that of charge separation. The first photochemical act can be thought of as the separation of a negative and positive charge, as could occur in a photoionization. If the photosynthetic machinery can be thought of as a crystalline array, we can substitute the idea of electrons and holes of a semiconductor. The proposed action of the excited pigment is analogous to that of a compressed spring (Fig. 2-18). Excitation energy

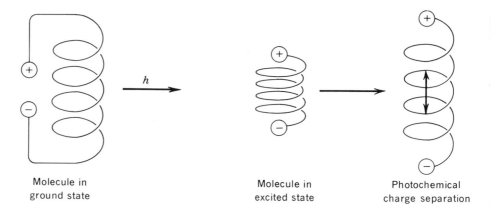

Molecule in ground state

Molecule in excited state

Photochemical charge separation

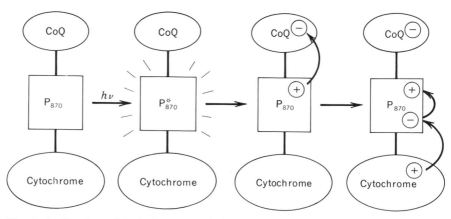

*Fig. 2-18* Simple model of photochemical charge separation.

compresses the spring; the spring then pulls the charges apart. The negative and positive charges or electrons and holes are then stabilized as reductant (NADPH) and oxidant ($O_2$), or alternatively they recombine along an electron transport pathway, which is coupled to the phosphorylation of ADP.

This simple picture serves as a fairly consistent model for bacterial photosynthesis. According to one hypothesis, a long wavelength form of bacteriochlorophyll termed $P_{870}$ (short for "pigment absorbing at 870 m$\mu$") is the photoreactive pigment. It forms a complex with an electron acceptor or oxidant, possibly coenzyme $Q$ (cf. Sec. 1.7), and a cytochrome (Fig. 2-18). Upon excitation, $P_{870}$ photoionizes and transfers an electron to CoQ. An electron is then donated to the now-oxidized $P_{870}^+$ by the accompanying cytochrome. The separated charges then recombine along the postulated electron transport chain with the coupled generation of ATP.

The idea of photochemical charge separation has been found as fruitful with green plants as with bacteria, but here the hypothesis of a single photoreactive pigment supplying energy to a single electron transport system is clearly inadequate. Let us consider a model that incorporates the idea of charge separation in *two* photochemical systems.

### Two-pigment Hypothesis

The two-pigment hypothesis is directly traceable to Robert Emerson's studies on action spectra (see below). He inferred from his investigations that two pigments had to be excited for photosynthesis to occur. Among the possible arrangements of a system of two pigments, Hill and Bendall (1960) proposed a "series formulation" or "z-scheme" very close to that currently accepted by most workers (Fig. 2-19). In this scheme one pigment, called $P_{700}$, absorbs at about 700 m$\mu$, and another, which we shall call $P_{670}$, absorbs at 670 m$\mu$. $P_{700}$ and an associated electron transport chain compose "system I," $P_{670}$ and its electron transport chain compose "system II," and the reducing end of system II acts as a bridge between the two.

The better part of system I is directly analogous to the whole of the bacterial charge separation system. There is an important difference, however: the electron acceptor, or oxidant, X is sufficiently electronegative to reduce NADP. The movement of electrons and "holes" or positive charges proposed for system I is

$$\text{cyto } f \xleftarrow{\oplus} P_{700}^* \xrightarrow{e^-} X \xrightarrow{e^-} \text{ferredoxin} \xrightarrow{e^-} \text{NADP}$$

Ferredoxin is an iron-containing protein with an unusual structure (see Fig. 2-32 below).

System II is thought to consist of components with more positive (that is, more oxidizing) redox potentials, whose function is to oxidize $OH^-$ to $O_2$. This strongly oxidizing system contains a manganese component, labeled here as $E_{Mn}$. The electron acceptor is called "Q" for "quencher." The electron transport system following Q probably contains plastoquinone (PQ), which then

Oxidation potential
$E_0'$ (mv)

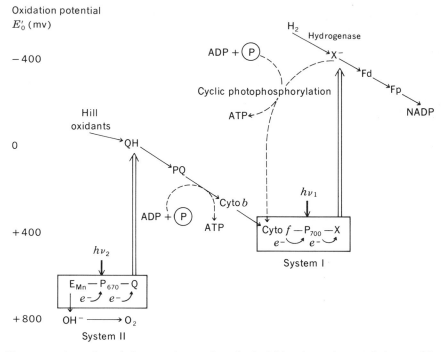

*Fig. 2-19* A version of the two-pigment hypothesis. This scheme is one of the possible representations of photosynthesis in green plants. System I is analogous to the scheme for bacterial photosynthesis (Fig. 2-18). P$_{700}$, a long wavelength form of chlorophyll, is complexed with cytochrome $f$ and electron acceptor X. When P$_{700}$ is excited, the electron path is P$_{700}$ → X → ferredoxin → NADP. P$_{700}$ subsequently receives an electron from cytochrome $f$. In system II excitation leads to charge separation with the electron ultimately given to cytochrome $f$ and the positive charge reacting with OH$^-$ to yield O$_2$.

reduces cytochrome. PQ is one of a number of CoQ-like quinones found in green plants. Note that the role proposed for quinones is substantially different from that in bacterial photosynthesis. The electron and hole transport system of II is

$$OH^- \overset{\oplus}{\leftarrow} E_{Mn} \overset{\oplus}{\leftarrow} P_{670}^* \overset{e^-}{\rightarrow} Q \overset{e^-}{\rightarrow} PQ \overset{e^-}{\rightarrow} \text{cyto } b \overset{e^-}{\rightarrow} \text{cyto } f$$

Phosphorylation may in principle be coupled to one or both of the electron transport systems, but most workers place it on the electron transport "bridge" connecting Q and cytochrome $f$.

The systems of the two-pigment hypothesis can be conceived of as photochemical specialization. The generation of reduced NADP and the utilization of water as a source of electrons presented two separate evolutionary problems. Once Nature had mastered the trick of photochemical charge separation in purple bacteria, she evolved distinct but interacting photochemical systems toward the solution of these separate problems. The result was green plants with a photosynthetic versatility far surpassing that of bacteria (cf. Olson, 1969).

Bearing in mind that these schemes, in addition to being incomplete, may also be wrong, let us turn to the evidence.

### Bunsen-Roscoe Law

Obedience to the Bunsen-Roscoe law is an essential characteristic of true (or at least uncomplicated) photochemical reactions. If we are to study the photochemistry of processes in which a photoreaction is accompanied by "dark" chemistry, it is essential to limit ourselves to experimental conditions under which the dark reactions are not limiting to the overall process.

Plants typically show a fairly wide range of light intensities over which the rate of photosynthesis (measured as $CO_2$ reduction) is linear (Figs. 2-20

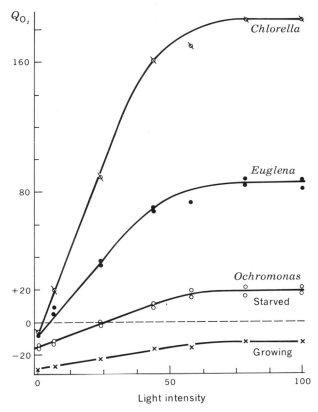

Fig. 2-20  Photosynthesis in several algae as functions of light intensity (Gaffron, 1960, p. 63). Photosynthesis in *Chlorella* and *Euglena* is saturated at less than 100 ft-candles and the maximum rate is many times greater than the rate of respiration (oxygen absorption at zero light intensity). *Ochromonas malhamensis* by comparison photosynthesizes at rates less than or only slightly greater than its respires. *Ochromonas* may employ its photosynthetic apparatus only to generate $O_2$ in stagnant environments.

and 2-21). Beyond some critical level of light which depends on the concentration of carbon dioxide, temperature, nutrition, etc., photosynthesis is no longer increased. In 1905 Blackman proposed that the fact of interaction between light intensity and $CO_2$ levels was evidence for dark or enzymatic reactions intervening in photosynthesis.

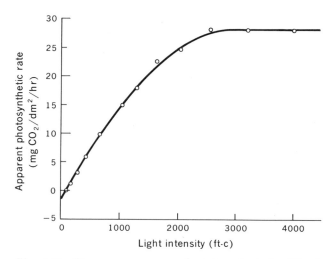

*Fig. 2-21*  Dose-response curve of photosynthesis in *Mimulus cardinalis* (from Milner and Hiesey, 1964). Photosynthesis is linear with light intensity to about 1,000 ft-candles and saturates at about 2,500 ft-candles.

More conclusive evidence for dark reactions was provided by Otto Warburg (cf. Gaffron, 1960, p. 60), who found that the yield from high intensities of light was considerably enhanced by giving the light in brief flashes interspersed with relatively long dark intervals.

Robert Emerson and William Arnold (1932) pursued this and found that the photochemical yield was constant when dark intervals of 20 msec or longer intervened between light flashes of 0.1 msec (Fig. 2-22). One might have imagined that a plant would not become saturated with light until all of its chlorophyll had become excited. But surprisingly the *number of quanta* in a saturating flash was found to correspond to only one three-hundredth of the number of chlorophyll molecules. Yet in studies with *low* light intensities, photosynthesis behaved as if all the chlorophyll were at least potentially active. Emerson and Arnold proposed that this paradox could be resolved if a single quantum of light were "collected" in units of about 300 chlorophyll molecules.[1]

---

[1] Not exactly. Emerson and Arnold proposed a larger unit for collecting 8 quanta of light, enough to liberate one molecule of $O_2$. The important point for this discussion is the *notion* of a collecting unit.

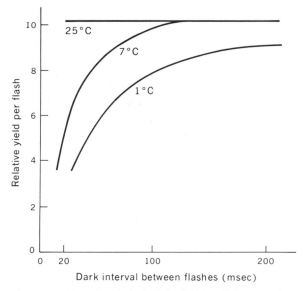

*Fig. 2-22* Photochemical yield in flashing light as a function of the dark time (from Emerson and Arnold, 1932,
as plotted by Gaffron, 1960). At 25°, the yield is constant
even when the dark interval is decreased to 20 msec. At 7°,
the time required for completion of the dark reaction is
about 100 msec. At 1°, the time for the dark reaction is
over 200 msec.

Such a "photosynthetic unit" would then process its collected light energy with
a common set of enzymes.

For our present purposes we may conclude that dark reactions do intervene
and that the Bunsen-Roscoe law is obeyed under conditions of brief light flashes
followed by relatively long dark periods.

### Action Spectra

An example of an action spectrum in a green alga is shown in Fig. 2-23.
With a *single* beam of light, the effectiveness of light (dotted line) shows
two principal peaks, one near 435 m$\mu$ and one near 675 m$\mu$; the action then
falls sharply to zero near 700 m$\mu$. The action spectra obtained with a single
beam correspond only in a general way to the absorption spectrum of chlorophyll. Action is higher in green light (500 to 600 m$\mu$) than we should have
expected from chlorophyll absorption, and light beyond 700 m$\mu$ is much less
effective than the corresponding absorption of chlorophyll *in vivo*. The data
are clearly at variance with the single chlorophyll pigment hypothesis first
considered.

The two-pigment hypothesis has its origin in Emerson's (Emerson et al.,
1957) studies of action spectra obtained with *two* wavelengths (cf. French,

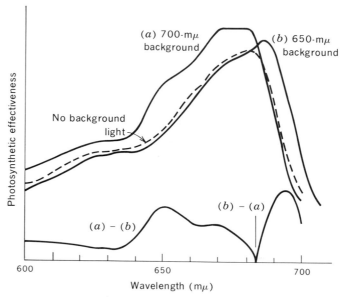

*Fig. 2-23* Action spectrum for photosynthesis and enhancement of photosynthesis in the green alga *Chlorella* (French et al., 1960). Photosynthetic $O_2$ evolution was measured with and without background illumination of 700 or 650 m$\mu$. Without any background illumination, that is, with a single beam of light, photosynthetic efficiency is very low near 700 m$\mu$. The addition of a background beam at 650 m$\mu$ increases the effectiveness of far-red light. This increased efficiency caused by a second beam of light is called "enhancement."

1961; Smith and French, 1963) (Fig. 2-24). Two features are observed: (*a*) light in the far red which cannot support photosynthesis alone *enhances* the photosynthesis when a shorter wavelength is added and (*b*) photosynthesis with two wavelengths together is typically more efficient than the sum of the two separately. A sharper contrast between action spectra obtained with one and two beams of light is found with many red, blue-green, and brown algae. In Fig. 2-24 the action spectrum with 546-m$\mu$ supplementary light looks very similar to the absorption spectrum of chl *a*, but that with supplementary light of 536 m$\mu$ shows a peak in the green and minimal activity in the blue and red where chlorophyll absorbs.

The inference from this phenomenon of "enhancement" or the "second Emerson effect" is that at least two photochemical reactions must occur in photosynthesis; that is, $P_{700}$ *and* $P_{670}$ must be excited in order for photosynthesis to be completed. An alternative view, that the second beam of light is required for a physical process such as excitation to a higher state, is vitiated by the finding of Blinks (1960) that the two beams of light may be separated in time. Let us look at this crucial piece of evidence more closely.

Blinks discovered that, if two beams of light at two wavelengths are given successively rather than simultaneously, transient changes in the rate of $O_2$ production, called *chromatic transients,* are observed. If darkness intervenes between the two beams, the chromatic transients still occur, but the magnitudes decrease with the length of the dark interval. The changes observed upon illumination at 680 m$\mu$ show a halflife of 5 sec in the dark (Figs. 2-25 to 2-27).

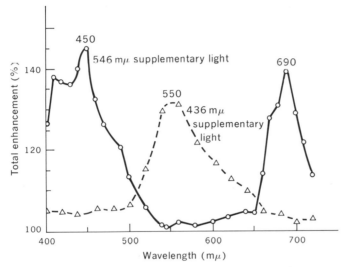

*Fig. 2-24* Action spectrum for enhancement in the red alga *Porphyra perforata* (Fork, 1963). Enhancement was measured as additional oxygen production obtained by shining a beam of light with a background light of 546 or 436 m$\mu$. Enhancement occurs with 546-m$\mu$ background in the red and blue with maxima characteristic of chlorophyll *a*. Enhancement spectrum with 436-m$\mu$ light shows maximum at 550 m$\mu$, which is near the absorption maximum of the phycobilins.

Note that the two enhancement spectra are more distinct from one another and the percentage increases are larger than with green plants, in which chlorophylls *a* and *b* have similar absorption spectra.

The lifetimes of these transients can be interpreted to mean that steady states of the pigment systems are reached within a relatively few seconds and that the intermediates which link the two systems have a lifetime of about the same duration.

The action spectra for enhancement of photosynthesis obtained with two simultaneous light beams agree nicely with the action spectra for exciting chromatic transients (Blinks, 1960; Myers and French, 1960) (Figs. 2-26 and 2-27). One can say, then, that enhancement occurs when the two beams of light are given either simultaneously or within a few seconds of each other.

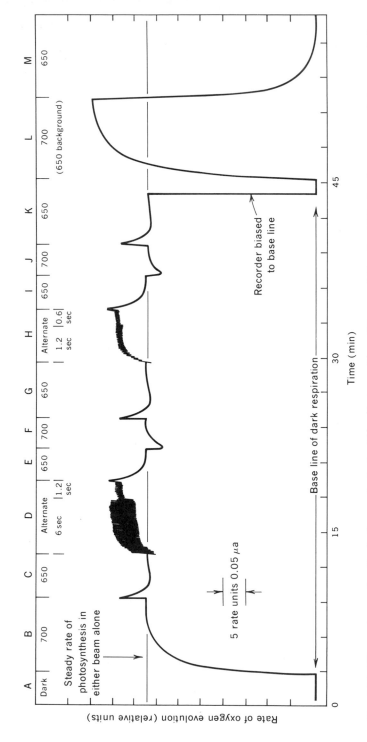

*Fig. 2-25* "Chromatic transients" of photosynthesis in *Chlorella pyrenoidosa* (Myers and French 1960). Steady-state rates of oxygen evolution are made equal in different colors of light by adjusting the intensities of the lights. Transient increases and decreases in the rates of photosynthesis occur when the lights are interchanged.

There are two additional bits of intersecting evidence. A number of algae show "photoreduction," in which $H_2$ serves as $[H_2A]$ in Eq. 2-12, and in which no $O_2$ is evolved. The action spectrum for photoreduction corresponds to the $P_{700}$ system and does *not* show enhancement. From this it was inferred that $P_{670}$ is concerned principally with driving $O_2$ evolution (Bishop and

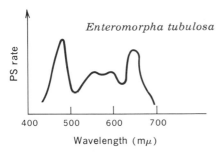
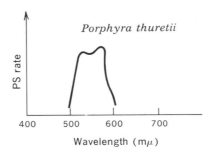

*Fig. 2-26*  Action spectra for chromatic transients (Blinks 1960). The action in these experiments was the transient increase in the rates of oxygen evolution that occurs upon shifting from 700-m$\mu$ light to light of a lower wavelength. In each case the intensity of the second beam was adjusted so that the *steady state* of photosynthesis was equal to that of the 700-m$\mu$ light alone.

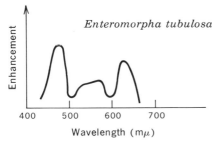
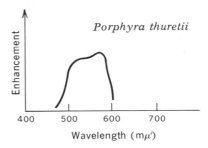

*Fig. 2-27*  Action spectra for enhancement (Blinks, 1960). Although these action spectra look the same as those in Fig. 2-26, the action in these experiments was the *additional* oxygen evolution that occurs when a second beam of light is imposed simultaneously with a 700-m$\mu$ beam.

Gaffron, 1962). Enhancement is also absent from bacterial photosynthesis (Clayton, 1963; 1965), where again $O_2$ evolution does not occur.

## Light Absorption

If chlorophyll undergoes a photochemical transformation in photosynthesis, we may expect to see changes in the absorption spectra of photosynthesizing organs, especially in the red region where there are few absorbing pigments other than the chlorophylls.

Although absorption spectra have been determined for the several isolated chlorophylls (Fig. 2-11; French, 1960), we know that these spectra may be substantially altered by the physical state of chlorophyll, the solvents in which they are dissolved, or the substrates on which they are adsorbed. Part of our problem therefore is to determine the precise location and height of the absorption peaks of chlorophyll *in vivo*.

Robert Emerson, Stacey French, and their associates have concerned themselves for a number of years with the problem of absorption as well as action spectra in relation to photosynthesis (French, 1960, 1961; Allen et al., 1960). Several absorption spectra of photosynthetic organisms are shown in Fig. 2-28.

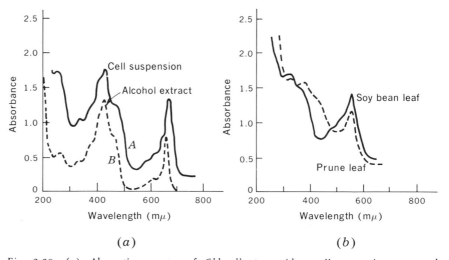

*Fig. 2-28*   (*a*) Absorption spectra of *Chlorella pyrenoidosa* cell suspension compared with an ethanol extract (Shibata et al., 1954). (*b*) Absorption spectra of soy bean and prune leaves (Shibata et al., 1954).

Typical green plants such as *Chlorella* and the leaves of higher plants show the unmistakable red absorption band of chlorophyll, even though the red peak may range from 670 to 690 m$\mu$ and even though relatively large amounts of other pigments may be present. Among members of the brown, golden brown, red, and blue-green algae, chlorophyll may be dwarfed by the presence of "accessory" pigments, the phycobilins (cf. Fig. 2-29).

Regardless of the kind of pigments present, we may look for differences in absorption (that is, difference spectra) between illuminated and darkened plants. Extremely small changes in absorption have in fact been detected.

In 1952 Duysens observed very small decreases in absorption by a long wavelength form of bacteriochlorophyll, P$_{890}$. Proportionally larger effects (Fig. 2-30) were subsequently obtained by Clayton (1965), with a green mutant which had very little of the bulk bacteriochlorophyll, presumably needed for light collection and protection of the photochemical reaction centers.

Similar difference spectra were obtained in higher plants by Kok and

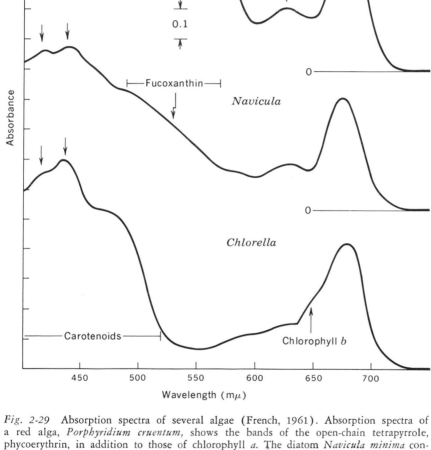

*Fig. 2-29* Absorption spectra of several algae (French, 1961). Absorption spectra of a red alga, *Porphyridium cruentum,* shows the bands of the open-chain tetrapyrrole, phycoerythrin, in addition to those of chlorophyll *a.* The diatom *Navicula minima* contains the carotenoid fucoxanthin, and chlorophyll *a.* The green alga *Chlorella vulgaris* contains chlorophylls *a* and *b* together with several carotenoids.

Hoch (1961). Bleaching (that is, decreases in absorption) occurs in both the blue and the far-red upon illumination. Looking at the 703-m$\mu$ changes of green plants with extremely fast detectors, Witt's group (Rumberg et al., 1963) found that upon illumination with a flash of light, bleaching occurred in

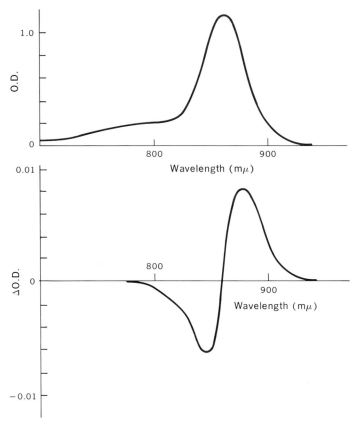

*Fig. 2-30*  Absorbancy changes in *Rhodopseudomonas spheroides* (R. F. Clayton, 1965). A green mutant of one of the purple bacteria was selected for having minimum amounts of masking pigments. In the upper curve the absolute absorbance of a cell suspension was recorded. In the lower curve, the difference spectrum was determined after illumination (O.D. darkened — O.D. illuminated). The decrease in absorbance at 890 m$\mu$ in the illuminated cells corresponds to an oxidation of a long wavelength form of bacteriochlorophyll, called P$_{890}$.

less than $10^{-5}$ sec and returned to normal in the ensuing few hundredths of a second (Fig. 2-31). Moreover, the bleaching at 703 m$\mu$ occurs with negligible temperature dependence. The reaction thus has the characteristics of a primary photochemical event. The fact that a 435-m$\mu$ band behaves identically lends

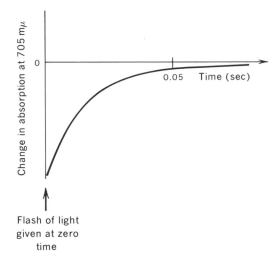

*Fig. 2-31* Fast changes in light absorption at 705 m$\mu$ after illumination with a flash of light. *Chlorella* cells were illuminated with a brief flash of light and the change in absorption at 705 m$\mu$ determined with extremely fast detectors. Bleaching occurs within $10^{-5}$ sec. Absorption returns to normal in the dark within 0.05 sec (Rumberg et al., 1963).

credence to the idea that this $P_{700}$, as the substance is called, is a long-wavelength form of chlorophyll *a*.

Other difference spectra under different conditions were obtained by Kok, Duysens, Witt, Chance, Clayton, and a number of others (cf. collections of papers in Kok and Jagendorf, 1963). These spectra implicate cytochromes *f* and *b* (forms of *c* and *b* peculiar to green plants) and plastoquinone as early participants along with $P_{700}$ in photochemistry.

In principle, one should be able to sort out the sequence of interacting substances from the sequence of absorbancy changes after illumination. In practice, the changes observed and even their directions have been influenced by the intensity and color of light, the choice of organism, and their physiological state. Some of the results, achieved with subcellular particles treated in various ways, have not been duplicated with intact cells. The result is that contradictory interpretations of the two-pigment hypothesis compete with one another, each supported by evidence which is inconclusive and sometimes of uncertain relevance to the situation *in vivo*.

The earliest observable change after illumination in the cytochrome region is a decrease in absorption, corresponding to oxidation of cytochrome *f* (Duysens, 1954; cf. Duysens, 1963). Witt's group observed that after a brief flash of light, the cytochrome oxidation coincided with the decay of 705-m$\mu$

bleaching (Fig. 2-31). Cytochrome oxidation also shows negligible temperature dependence (Chance and Bonner, 1963). Finally cytochrome $f$ remains oxidized until the cells are exposed to short wavelength light (e.g., 650 m$\mu$) and this reduction *is* temperature dependent.

These facts taken together are the basis for proposing that cytochrome $f$, $P_{700}$, and X are part of a single physical complex in which charge transfers can occur without molecular vibrations or collisions, and that the sequence of electron and hole migration is as proposed.

Similarly, a decrease in 255-m$\mu$ absorption is interpreted as a reduction of plastoquinone and its action spectrum is characteristic of system II.

The general sequence of electron transport is also supported by the occurrence and absence of light-induced absorbancy changes in mutants of the green algae *Scenedesmus* and *Chlamydomonas* (cf. Levine, 1969). There are, however, specific points of contradiction: for example, in a *Chlamydomonas* mutant lacking the copper protein, plastocyanin, there appears to be an interruption of electron transport *between* cytochrome$_{553}$ and $P_{700}$. This would place plastocyanin within the system I charge-transfer complex.

Typical of the conflicts in the data on absorption changes are those recorded in behavior of bacteriochlorophyll in the purple bacteria. At low light intensities, we see bleaching (*oxidation*) of $P_{890}$ and reduction of CoQ. At high light intensities, absorption changes are quite different and can be interpreted as the *reduction* of bacteriochlorophyll, and in this case it is coupled with the oxidation of cytochrome $c$. Thus, depending on the evidence selected, one can construct different and, in a sense, opposite charge separation sytems (R. F. Clayton, 1965). The model selected for Fig. 2-18 is an eclectic one. Obviously other schemes might have been selected with equal justification.

## Light Emission

If the photosynthetic pigments can be persuaded to emit some of their excitation energy, we can hope to identify them both from their fluorescence emission spectra and action (that is, fluorescence excitation) spectra. In fact plants can be seen to fluoresce, especially when photosynthesis is inhibited or prevented, just as if the excitation energy were switched from photochemistry to light emission.

The emission spectrum of green leaves and algae at room temperature normally shows a single maximum at 685 m$\mu$. At liquid nitrogen temperature the emission band splits into two or three bands at 685, 695, and 720 m$\mu$. The location and widths of the absorption bands at 685 and 695 m$\mu$ are characteristic of chlorophyll $a$ in organic solvents (cf. Fork and Amesz, 1969). In fact fluorescence of chl $b$, carotenoids, or phycobilins is rarely observed *in vivo!* There are, however, long-wavelength fluorescence bands which might be due to long-wavelength forms of chl $a$ in $P_{700}$ (Smith and French, 1963).

In sharp contrast to the unanimity of the emission spectra of plants, the excitation spectra are far more diverse. Chl *a* fluorescence can be induced by exciting chl *b*, carotenoids, phycobilins, and chl *a* itself. If these other pigments are excited and chl *a* emits the fluorescence, we can infer only that there must be energy transfer (exciton migration) among them and that chl *a*, or at least some form of it, is the energy sink.

Quantitatively, the excitation spectra show that most fluorescence comes from system II. Indeed, the oxidant of system II was named "quencher" (Q) because in its normal, oxidized state it competes with or quenches fluorescence for the energy of $P_{670}$ (Duysens and Sweers, 1963). Consistent with this view, one finds that fluorescence is decreased by absorption of light by system I, as this causes QH to revert rapidly to Q, and is promoted by DCMU[1] which blocks the oxidation of QH.

As suggested by Duysens and Amesz (1962) (cf. review of Smith and French, 1963), the model can be refined further: the two pigment systems are each arrays of pigments with overlapping dipoles. System I is composed of about 300 (chl *a* 683): 1 (cyt *f*): 1 (chl *a* 700) at the energy sink, or collection center. System II is composed of chl *b* or, in the case of blue-greens, red, and brown algae, phycobilins. The energy sink in this case is another form of chlorophyll, chl *a* 673, at the collection center. Hundreds of pigment molecules might thus be combined into two separate kinds of units with overlapping dipoles [as in the photosynthetic units of Emerson and Arnold (1932)], so that a photon absorbed by any one of the molecules in the unit is passed on to a few molecules of chlorophyll *a* at the collection center.

The average distance between chlorophyll molecules in chloroplasts is computed to be 15 to 20 Å. At this distance the model for energy transfer (page 91) predicts nearly 100 percent efficiency. Additional light-gathering or "antenna" pigments include $\beta$-carotene, which probably transfers only to system I. Xanthophylls appear not to transfer at all (cf. Fork and Amesz 1969).

What if one collection center becomes inoperative, due, for example, to a bleached reaction center? Can photons be passed from one unit to another of the same kind? The answer appears to be yes. On the basis of fluorescence measurements, units of system I or system II can pass photons to other units with an efficiency of about 0.5 (Fork and Amesz, 1969).

A more difficult problem, both conceptually and experimentally, is that called "spill-over," the transfer of energy between different kinds of units. If system II were to receive more light energy than it could efficiently use, could it relay the excess energy to system I? The seeming necessity for some regulated spill-over comes from the nearly constant quantum yield of photosynthesis over a range of wavelengths where the two systems would be expected

---

[1] DCMU is a member of a large class of substituted urea herbicides. All appear to act at this same point in photochemical electron transport.

to absorb very unevenly. The experimental evidence points both ways (Fork and Amesz, 1969).

Experiments with chromatophores of *Chromatium,* one of the photosynthetic bacteria, support this general scheme (cf. review of Smith and French, 1963). These bacteria have at least two kinds of bacteriochlorophyll, $B_{890}$ and $B_{850}$. When the chromatophores have been extracted with detergents, fluorescence of $B_{850}$ is enhanced whereas that of $B_{890}$ is decreased, as if the detergent had interrupted exciton migration to the long-wavelength energy sink.

Blue fluorescence characteristic of NADPH or NADH emission also occurs (Duysens, 1963). The action spectra for NADPH fluorescence corresponds to that of $P_{700}$.

### Electron Spin Resonance

As described in Sec. 2.3, electron spin resonance can be used to detect unpaired electrons, such as those that might exist in free radicals or triplet states. Signals

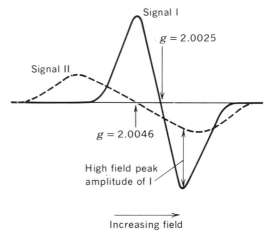

Signal I

$g = 2.0025$

Signal II

$g = 2.0046$

High field peak amplitude of I

Increasing field

*Fig. 2-32* EPR signals from *Chlamydomonas reinhardi.* The two kinds of electron paramagnetic resonance signals detected in photosynthetic organisms are distinguished by their decay curves. Signal I decays within seconds whereas signal II decays slowly, over the course of hours. Signal II is associated with oxygen evolution and signal I is associated with the $P_{700}$ form of chlorophyll *a* (Weaver, 1968).

from photosynthesizing cells were first detected by Commoner's group (cf. review of Weaver, 1968).

Two kinds of ESR signals (originally denoted I and II) are seen in green plants (Fig. 2-32). Signal I has a *g*-value of 2.0025 and a $\Delta H$ of 7 to 9 gauss. A closely similar signal occurs in photosynthetic bacteria. The occurrence of signal I correlates exactly with $P_{700}$ (or the corresponding bacterial reaction centers) with respect to the kinetics of their response to light, their response to redox potential (Beinert and Kok, 1963), and their deletion from *Scenedesmus* mutants (Weaver and Bishop, 1963) and at least approximately with the estimated numbers of reaction centers. Although chlorophyll alone will give a similar signal *in vitro,* there is no correlation *in vivo* with bulk chlorophyll.

Signal II just as fortuitously correlates with system II. It has a *g*-value of 2.0046, and its $\Delta H$ value is about 20 gauss. It is induced by light, as is signal I, but decays over several hours in the dark. Signal II is absent from bacteria. From the absence of signal II from *Screnedesmus* mutants lacking plastoquinone (Weaver and Bishop, 1963; Weaver. 1968), and the similarity of the ESR signal of plastoquinone *in vitro,* it is not unreasonable to think that a semiquinone of plastoquinone may be the source of signal II.

### Conclusions on Photochemistry

The accumulated evidence is consistent with the overall view of the two-pigment hypothesis. The participation of a variety of chloroplast pigments as collectors of photon energy is clearly indicated, but special forms of chl *a,* probably bound to specific proteins, appear to be the photoreactive components. Thus the identities of the absorbing pigments are tentatively and approximately established.

Photoionization of the excited chl *a* centers, followed by stabilization of the charge separation, is established with about the same level of uncertainty. It seems clear from the lack of temperature dependence of the absorption changes corresponding to cytochrome oxidation that electron transport is set in motion as an early event after excitation. However, we must reserve judgment on the precise nature of the photoreaction.

The behavior of system I, including $P_{700}$, is reasonably clear for the simple reason that it can be set in motion by light which leaves system II cold, so to speak.

The most promising approaches to the isolation of system II are through excitation of the phycobilins of red algae where the overlap between systems I and II is minimal, and through the use of algal mutants, in which the interaction has been rendered inoperative.

Until complete stories are filed on several organisms, we must view our formulations with cheerful skepticism.

### Biochemical Evidence

We should now assemble the biochemical evidence for the pathways of ATP and NADPH formation (Eqs. 2-9 to 2-11; review of San Pietro and Black, 1965). Such evidence has been obtained largely with isolated chloroplasts and fragments of chloroplast lamellae, loosely called grana, from higher plants. Additional evidence has come from the method of mutations (cf. Bishop, 1964; Levine, 1969).

Robert Hill (1939) first demonstrated that oxygen could be liberated from water by illuminated chloroplast provided with a variety of hydrogen acceptors.

$$B + H_2 \xrightarrow[\text{chloroplasts}]{h\nu} BH_2 + \tfrac{1}{2}O_2$$

where B could be ferricyanide, benzoquinone, or any of a variety of dyes. Judging by the action spectra of the Hill reaction, ferricyanide must normally accept electrons from system II. It thus provides a lever for separating the $O_2$-liberating end of photosynthesis from the photochemistry of system I. But the Hill reaction is at once more complicated and more general than system II.

It now seems that Hill oxidants, depending on their redox potentials, may act as electron acceptors either to $X^-$ or to the $P_{670}$ electron acceptor or to other intermediates in electron transport. Hill's general strategy of feeding oxidants to photoreducing systems helped others to identify the electron transport path from $X^-$ to NADP through the isolation and characterization of ferredoxin (Fd) and its flavoprotein reductase (fp). Ferredoxin has a most unusual structure for a protein (Fig. 2-33). These steps have been worked out by San Pietro, Jagendorf, Arnon, and others (San Pietro and Black, 1965).

Formation of the other required photoreaction product, ATP (cf. Eq. 2-9), can also be demonstrated in isolated chloroplasts and bacterial chromatophores (cf. Avron and Neumann, 1968). The process, called *photophosphorylation,* was first described by Arnon et al. (1954) with chloroplasts and by Frenkel (1954) with fragments ("chromatophores") of photosynthetic bacteria. Arnon identified both "cyclic" and "noncyclic" photophosphorylation. In the cyclic process, phosphorylation is somewhat coupled to an electron transport chain that starts with one of the photoacts (probably $P_{700}$); the electrons proceed around a closed path, returning to chlorophyll. In the noncyclic process, the electron chain terminates in a pool of oxidant, e.g., NADP.

Which of these two processes fuels the Calvin-Benson cycle of $CO_2$ fixation? Tanner et al. (1969) have approached this question by showing that noncyclic (but not cyclic) photophosphorylation proceeds *pari passu* with $CO_2$ fixation with respect to light intensity and the response to certain kinds of inhibitors. While cyclic phosphorylation surely exists *in vivo* and indeed seems to be the sole contributor of ATP to several transport processes, its magnitude appears to be too small to contribute significantly to normal rates of photosynthesis.

The difficulty, as in the case of absorption changes, is that different workers obtain substantially different results depending on the conditions and the choice of experimental material. Action spectra for photophosphorylation obtained in different laboratories show peaks at the following wavelengths: 650, 670,

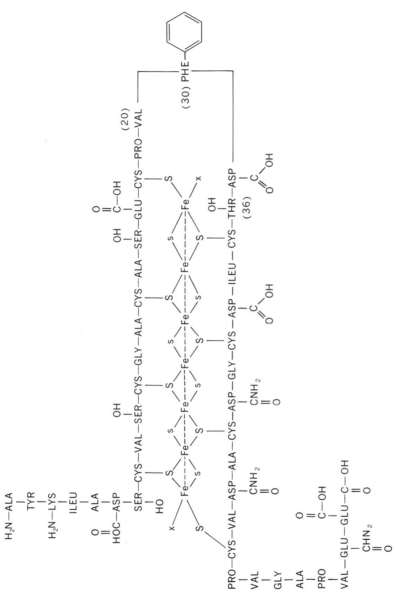

*Fig. 2-33* **Proposed structure of *Clostridium ferredoxin*.** Tanaka et al. (1965) proposed a structure for the ferredoxin of *Clostridium pasteurianum*, a carrier which is involved in nitrogen fixation and is similar to ferredoxin from higher plants. Plant ferredoxin differs (cf. San Pietro and Black, 1965) in having much more tyrosine, less iron, and less sulfhydryl.

677, 680, and 710 m$\mu$ (Smith and French, 1963). In other words, the action spectra correspond to system I, system II, and both!

The simplest conclusion for the present is that photophosphorylation may be coupled to any of the electron transport chains in the chloroplast.

## 2.6   CASE STUDY: PHYTOCHROME AND RED–FAR-RED RESPONSES

Certain varieties of lettuce seeds, conditioned in a specific way, will germinate only when exposed to red light in the region of 660 m$\mu$. The promotion by red light is completely reversed by a subsequent exposure to far-red light in the region of 730 m$\mu$ (Fig. 2-34). That *both* responses to light are completely

*Fig. 2-34* Control of lettuce seed germination by red and far-red light (photograph kindly supplied by Sterling B. Hendricks and Harry Borthwick). Lettuce seeds were exposed to successive brief exposures to red (R) and far-red (I) light and then allowed to germinate. The photograph shows that red light promotes germination and far-red inhibits it and that the last exposure received is the deciding one.

reversible is shown by the fact that, after many alternate exposures to red and far-red light, germination either occurs or does not occur depending solely on the color of light in the *last* treatment. It is exactly as if some physiological switch were turned off and on.

The germination of lettuce seeds is only one of a rapidly lengthening list of biological responses to red and far-red light. A partial accounting is presented in Table 2-2. The flowering of certain plants in short days, as first shown by Garner and Allard in 1920, is one of the most widely encountered of these phenomena. In these cases of *photoperiodism,* plants can be induced to flower when the length of the night exceeds some critical length (cf. Leopold, 1965). For example, the cocklebur *Xanthium* spp. can be induced to flower

Table 2-2    Photoreversible Physiological Processes Controlled by Red and Far-red Light*

| | |
|---|---|
| Seed germination | Flower induction |
| Fern spore germination | Flower development |
| Moss spore germination | Formation of cleistogamous (self-pollinating) flowers |
| Etiolation | Formation of rhizomes |
| Fern sporeling elongation | Formation of bulbs |
| Seed respiration | Formation of gemmae (budlike reproductive structures) |
| Chlorophyll formation | Sex expression |
| Chloroplast orientation | Leaf abcission |
| Stem stiffening | Phylloidy of bracts |
| Hair formation | Epinasty |
| Anthocyanin synthesis | Crassulacean metabolism (cf. Sec. 1.7) |
| Cuticle coloration | Bud dormancy |

* Compiled by Hendricks and Borthwick, 1965.

when given a single night longer than 12 hr (Fig. 2-35). A species of *Perilla*, on the other hand, requires nine long nights.

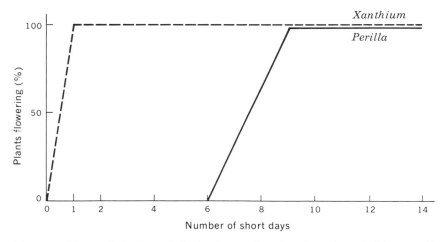

*Fig. 2-35*    Photoperiodic flower induction in two short-day plants (compiled by Leopold, 1965). *Xanthium* and *Perilla* plants were first grown on long days so that they would not normally flower, placed on short days for different lengths of time, and then returned to long days. The number of plants that subsequently developed floral primordia was then determined.

That these and other photoperiodic phenomena are in fact manifestations of the red–far-red system can be shown through interruption of the long night. Light given in the middle of the night will completely prevent flower induction (Hamner and Bonner, 1938). Hendricks and Borthwick (cf. their 1965 review) later found that the most effective wavelengths are in the region of 600 to 650 m$\mu$ and that inhibition by this red light can be completely reversed by far-red light. Moreover, only extremely small amounts of light are required for night interruption or for reversal of night interruption, on the order of $10^{-3}$ erg cm$^{-2}$ or $3 \times 10^{12}$ photons cm$^{-2}$.

Although most of the phenomena controlled by red and far-red light have been observed in plants, completely analogous phenomena involving, for example, reproduction have been described in animals. The details of the animal responses are less well established.

If organisms respond to light, then there must be an absorbing pigment or pigments. We shall consider the hypothesis that the responsible pigment, called *phytochrome,* exists in two photoreversible forms. We shall also see how the pursuit of evidence on this hypothesis led to the separation and partial purification of a protein pigment with the predicted properties of phytochrome.

Hendricks and Borthwick reasoned that, since the physiological effects of red and far-red light are mutually reversible, the simplest model must assume that the two forms of the hypothetical pigment phytochrome are photochemically interconvertible

$$P_r \underset{h\nu \text{ (far red)}}{\overset{h\nu \text{ (red)}}{\rightleftharpoons}} P_{fr}$$

### Action Spectra

Two action spectra are found in the typical red–far-red system. The physiological response in one direction is promoted by light with a peak in the region of 660 m$\mu$, and the response is canceled or pushed in the opposite direction by light with a peak near 730 m$\mu$ (Fig. 2-36). There are often effects in the blue as well.

Some of the physiological responses to red and far-red light are atypical, or at least different. The formation of anthocyanin in red cabbage, for example, shows a peak at 690 m$\mu$. There are obviously several possible interpretations of the atypical spectra, but let us pursue for now the typical ones.

We stated in Sec. 2.3 that an action spectrum should have the same shape as the absorption spectrum of the absorbing pigment. In this way action spectra provide a means of predicting the absorption spectrum and hence the identity of the absorbing pigment. We qualified this axiom for the presence of masking pigments (cf. Sec. 2.4). Now we must qualify it again. In a photoreversible system the shape of the action spectrum depends also on whether or not the absorption bands of the two pigments overlap. If they do not, we have the classical situation of Eq. 2-7. If they do overlap, the shape depends on how much of the second pigment is present.[1]

Let us take the example of driving the system $A \leftrightarrow B$ in the direction of

---

[1] Hendricks et al. (1956), in developing a quantitative treatment of their action spectra and photoconversions, stipulated that one must be able to "drive" a system wholly into the red and far-red forms.

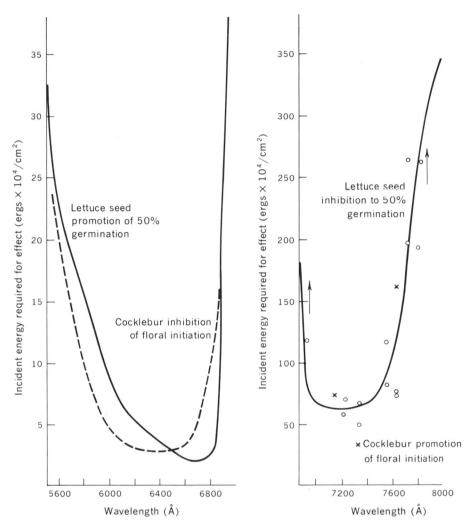

*Fig. 2-36*  Action spectra for the promotion and the inhibition of germination of lettuce seeds (Butler and Norris 1962). Lettuce seeds are driven by red or far-red light into states in which they would normally germinate or not germinate. The seeds were then exposed to a subsequent treatment in which carefully measured quantities of monochromatic light were shown on the seeds and the resulting effects on germination were measured. The ordinate is in reciprocal units of light energy per $cm^2$. Note that the scale for inhibition is one-tenth that for promotion.

B (Fig. 2-37). If, say, equal amounts of B are present at the start, light in the region of overlap will drive A to B, but will also drive B to A. In that case the action spectrum would have the shape of the difference spectrum of the two pigments.

One can see furthermore that, if absorption overlap exists throughout

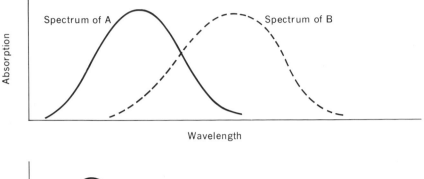

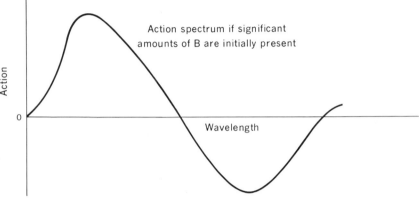

*Fig. 2-37* Relations between absorption and action spectra for photoreversible systems with overlapping absorption bands. Imagine a photoreversible system $A \rightleftharpoons B$ in which the absorption spectra for A and B have a region for overlapping absorption bands. The action spectrum for the conversion of A to B could vary depending on how much of B was initially present. If the concentration of B was initially negligible, the action spectrum would correspond to the classic situation in which action and absorption spectra have the same shape. If, however, large amounts of B were present, the action spectrum could shift to the point where it resembled the difference spectrum.

the absorption spectra of the two pigments, it would be impossible ever to drive the system wholly into one form by purely photochemical means.

Thus the action spectra of Fig. 2-38 might reasonably correspond to the absorption spectra of the two pigments, to the difference spectrum, or to something intermediate.

## Absorption Spectra

We can irradiate a plant with far-red light in sufficient intensity to drive the phytochrome system all the way into the red form, determine its absorption spectrum, then irradiate it with red light, and determine the absorption spectrum again. The difference between the first and second spectra would be the difference spectrum $P_r - P_{fr}$.

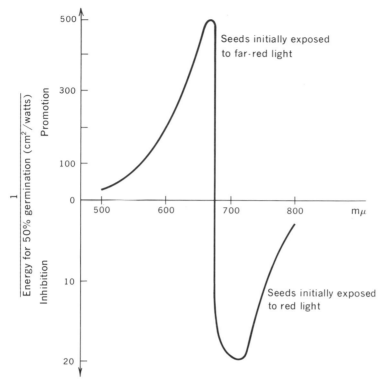

*Fig. 2-38* By substituting a continuous ordinate of quantum efficiency of germination, we can construct a continuous action spectrum for germination. This curve may be related to the difference spectrum of the two pigments (see text).

A technical consideration: the light required to determine the absorption spectra can in principle also transform the pigments. In practice it happens that the *detecting* light can be two orders of magnitude less than the *actinic* light required for the phototransformation. Although the detecting light probes the system, it does not significantly alter it.

The observed difference spectra of red–far-red transformations in several plants are shown in Fig. 2-39. The spectra show the same features as the difference spectra calculated from the action spectra: a peak near 660 m$\mu$, a cross-over point near 690 m$\mu$, and a second peak near 730 m$\mu$.

Within the limitations imposed by multiple reflections and scattering, the *change* in the difference in absorbancy between 600 and 730 m$\mu$ should be a measure of the amount of phytochrome transformed by the actinic light. This quantity is known as the $\Delta\Delta$O.D., and has served as a monitor for phytochrome changes in several physiological processes (Butler and Lane, 1965). We should bear in mind that the measurements are inherently inaccurate.

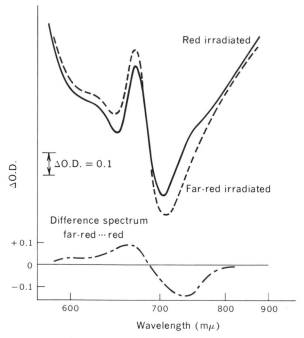

Fig. *2-39*   Absorption spectra of corn coleoptiles after exposure to red and far-red light (Butler and Norris, 1962).

### Extinction Coefficient from Physiological Measurements

Hendricks reasoned (Hendricks et al., 1956), that, if the model of a photoreversible pigment was correct, the situation would be analogous to that of reversal of carbon monoxide inhibition of cytochrome oxidase by light. Warburg and Negelein (1929) had earlier derived for this case a means of estimating the extinction coefficient of the absorbing pigment. Hendricks et al. (1956) set out to make the corresponding estimate for the red–far-red pigments. The reasoning is as follows:

Let us set the sum of the two pigments arbitrarily to equal unity. Let the fraction in the 660 form be designated $F$; the fraction in the far-red form is then $1 - F$.

If we illuminate the system with far-red light, the rate of conversion of the $P_{730}$ to $P_{660}$ will be proportional to the amount of $P_{730}$ remaining, or

$$\frac{dF}{dE} = k(1 - F) \qquad\qquad \text{Eq. 2-13}$$

where $E$ is the number of photons absorbed by the system and $k$ is a constant. In other words, the photoconversion is a first-order process.

Integrating both sides we obtain

$$kE = \log \frac{1}{1 - F}$$   Eq. 2-14

In order to eliminate $k$, let us treat the system, initially all in the far-red form, with two quantities of photons, $E_1$ and $E_2$. (We must assume that the pigments do not overlap or that it is possible to drive the system wholly into one form or the other.) We shall label the fractional conversions corresponding to $E_1$ and $E_2$ as $F_1$ and $F_2$; so that

$$E_1 \log \frac{1}{1 - F_1} = E_2 \log \frac{1}{1 - F_2}$$

If we let $\alpha = E_1/E_2$, then

$$\alpha \log \frac{1}{1 - F_1} = \log \frac{1}{1 - F_2}$$   Eq. 2-15

The same reasoning that led to Eq. 2-15 can be applied to the formation of the far-red form from the red form. In the reverse direction then

$$\beta \log \frac{1}{F_2} = \log \frac{1}{F_1}$$   Eq. 2-16

For small conversions we can reasonably expect that the fractional response (in the case of lettuce seed, the fractional germination) is equal to the fractional pigment conversion. We can thus evaluate $F$. We can also set $F$ for small pigment conversions as equal to the product of the extinction coefficient, the quantum efficiency and the number of quanta reaching the pigment. It is convenient to measure the number of photons as einsteins per centimeter squared. The quantum efficiency $\phi$ is a dimensionless fraction less than or equal to 1. To express the extinction coefficient in equivalent units, we shall make it $\omega$ [cm$^2$ (gram molecular weight)$^{-1}$]

$$F = \omega \phi E$$   Eq. 2-17

Since masking pigments may be present, Eq. 2-17 will yield a minimum value. To convert $\omega$ to the more familiar $\epsilon$, we note that

$$\epsilon \text{ liter mole}^{-1} \text{ cm}^{-1} = \text{cm}^2 \text{ mole}^{-1} \frac{\text{liter}}{10^3 \text{ cm}} \ln 10$$

$$\epsilon = 2.303 \times 10^{-3} \omega$$

Values of $\epsilon\phi X$ of several red–far-red responses are shown in Table 2-3. Since $\phi X$ can only be less than 1, the molar-difference extinction coefficient $\epsilon$ for both the red and far-red forms was inferred to be in the region of $10^5$ moles cm$^{-1}$ liter$^{-1}$. The best experimental values, obtained subsequently from the photoconversion of the purified pigment, are lower, but within an order of magnitude of the physiologically determined values. Convinced that the

Table 2.3  Computed Values of $\epsilon\phi X$ for Several Phytochrome-mediated Processes*

|  | Pinto Bean Stem Elongation | *Lepidium virginicum* Germination | *Lepidium sativum* Germination | Photoconversion of Purified Pigment† |
|---|---|---|---|---|
| $P_r$ at 650 m$\mu$ | $0.021 \times 10^5$ | $0.015 \times 10^5$ | $1.15 \times 10^5$ | $0.16 \times 10^5$ |
| $P_{fr}$ at 730 m$\mu$ | $0.025 \times 10^5$ | $0.44 \times 10^5$ | $0.025 \times 10^5$ | $0.07 \times 10^5$ |

* Hendricks, 1959.
† Butler et al., 1965 (cf, Fig. 2-41).

hypothetical phytochrome must have large extinction coefficients in the red and far red, Borthwick Hendrick's group devised means of searching *in vivo* and *in vitro* for pigments with the predicted optical properties. Within a few years they succeeded in separating phytochrome from plants (cf. Butler et al., 1965).

### Phytochrome: Purification and Properties

Proteins with the properties of the hypothetical phytochrome have been isolated from a number of plants. Substantial purification has been achieved by standard methods of protein chemistry and the pigments are seen to have molecular weights in the order of 100,000 and to be almost identical to allophycocyanin (Fig. 2-40). This open-chain tetrapyrrole is also the chromo-

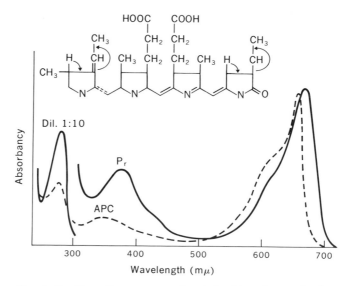

*Fig. 2-40* Absorption spectra of red-absorbing forms of phytochrome ($P_r$) and allophycocyanin (APC) (Siegelman and Hendricks, 1965). Absorption spectrum of phytochrome is similar but not identical with that of the open-chain tetrapyrrole pigment from blue-green algae. Also shown is the proposed structure of the chromophore of the $P_r$ form; arrows indicate increases in conjugation predicted for $P_{rr}$ (Siegelman et al., 1969).

phore of one of the biliproteins found among red and blue-green algae (Siegelman et al., 1969).

The isolated protein can be converted photoreversibly into a red and far-red form, exactly as would be expected of phytochrome. The action spectra for photoconversion (Fig. 2-41) should be compared with the action spectra for the physiological processes (Fig. 2-37) and to the absorption spectra of the $P_r$ and $P_{fr}$ forms shown in Fig. 2-41.

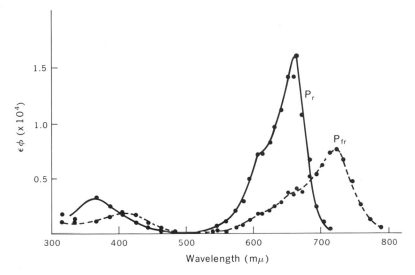

*Fig. 2-41* Action spectra for the photoconversion of phytochrome *in vitro* (Butler et al. 1965). Purified phytochrome was driven into the red ($P_r$) or far-red ($P_{fr}$) form and the action spectra for interconversion determined. The ordinate in this case is $\epsilon\phi$; where $\epsilon$ is the extinction coefficient and $\phi$ is the quantum yield.

It seems inescapable that these proteins are in fact the absorbing pigments for the red–far-red transformations.

There is little direct information on the photochemistry of phytochrome transformations, nor is there any information on the means by which phytochrome controls physiological processes. Further explorations of some of the physiological processes themselves will be considered in Chap. 5.

## SUGGESTED CASE STUDIES

In addition to chlorophyll formation, photosynthesis, and photoperiodism, plants also respond to light through a variety of positive and negative phototropisms and phototaxes. The general concepts of photophysiology apply to these processes and also to a reverse process, bioluminescence.

### Phototropisms in the *Avena* Coleoptile

The *Avena* coleoptile, perhaps the most harried and pampered object in plant physiology, can also be made to elicit phototropic responses. At very low light intensity the coleoptile bends toward light (positive phototropism); at progressively higher intensities it bends first away from the light and then again toward the light. Detailed information is available on the response of the organ to intensity, time, and wavelength of the light. While it is widely agreed that the direct agent of the bending is differential cell elongation under the control of auxin (cf. Sec. 5.6), the identity and function of the absorbing pigment has been hotly debated. Scientific study of the problem started with Charles Darwin and continues strong.

*References:* Bara and Galston, 1968; Shen-Miller et al., 1969; Thimann, 1967.

### Phototaxes

Phototaxis is the movement of a free-living organism toward or away from light. Physically, this must be a more complicated process than phototropism, since the organism must somehow orient itself with respect to the light in three dimensions. Phototaxes appear to be of primary importance to marine and fresh-water algae in seeking optimum conditions for photosynthesis; diurnal vertical movements of phytoplankton are a regular feature of marine ecology. As with other photophysiological processes, we may ask what pigment is responsible and what is its fate.

*References:* Halldal, 1962; Haupt, 1965.

### Bioluminescence

Transformation of chemical energy to light occurs in a number of unrelated organisms, from bacteria to higher animals, and including many algae. The concepts and methodology appropriate to photochemistry and photophysiology apply equally to the problem of how the excited state can be generated by purely chemical (as opposed to thermal or photochemical) processes. Progress on the biophysics of bioluminescence has been matched by great progress on the biochemistry of the process: the identification of enzymes and substrates leading to and fueling luminescence, characterization of light-emitting organelles, and some educated thoughts on the functional significance of luminescence.

*References:* Hastings, 1968; Seliger and Morton, 1969.

## REFERENCES

Allen, M. B., 1964. Absorption spectra, spectrophotometry, and action spectra, in A. C. Giese (ed.), "Photophysiology," vol. I, pp. 83–110, Academic Press, New York.

———, C. S. French, and J. S. Brown, 1960. Native and extractable forms of chlorophyll in various algal groups, in M. B. Allen (ed.), "Comparative Biochemistry of Photoreactive Systems," pp. 33–52, Academic Press, New York.

Arnon, D. I., M. B. Allen, and F. R. Whatley, 1954. Photosynthesis by isolated chloroplasts, *Nature,* **174:**394–396.

Avron, M., and J. Neumann, 1968. Photophosphorylation in chloroplasts, *Ann. Rev. Plant Physiol.,* **19:**137–166.

Bara, M., and A. W. Galston, 1968. Experimental modification of pigment content and phototropic sensitivity in excised *Avena* coleoptiles, *Physiol. Plantarum,* **21:**109–118.

Beinert, H., and B. Kok, 1963. Relationship between light induced EPR signal and pigment $P_{700}$, in B. Kok and A. Jagendorf (eds.), "Photosynthetic Mechanisms in Green Plants," pp. 131–137, NAS-NRC Publication #1145.

Bendall, D. S., and R. Hill, 1968. Haem-proteins in photosynthesis, *Ann. Rev. Plant Physiol.,* **19:**167–186.

Bishop, N. I., 1964. Mutations of unicellular green algae and their application to studies on the mechanism of photosynthesis, *Record Chem. Progr.,* **25:**181–195.

——— and H. Gaffron, 1962. In R. F. Clayton (ed.), "Molecular Physics in Photosynthesis," p. 53, Blaisdell Publishing Co., New York.

Blackman, F. F., 1905. Optima and limiting factors, *Ann. Botany, London,* **19:**281–295.

Blinks, L. R., 1960. Action spectra of chromatic transients and the Emerson effect in marine algae, *Proc. Nat. Acad. Sci.,* **46:**327–333.

Blois, M. S., Jr., and E. C. Weaver, 1964. Electron spin resonance and its application to photophysiology, in A. C. Giese (ed.), "Photophysiology," vol. I, pp. 35–63, Academic Press, New York.

Boardman, N. K., 1966. Protochlorophyll, in L. P. Vernon and G. R. Seely (eds.), "The Chlorophylls," pp. 437–479, Academic Press, New York.

Butler, W. L., S. B. Hendricks, and H. W. Siegelman, 1965. Purification and properties of phytochrome, in T. W. Goodwin (ed.), "Chemistry and Biochemistry of Plant Pigments," pp. 197–210, Academic Press, London.

——— and H. C. Lane, 1965. Dark transformation of phytochrome *in vivo,* II, *Plant Physiol.,* **40:**13–17.

——— and K. H. Norris, 1962. Plant spectra: Absorption and action, in H. F. Linskens and M. V. Tracey (eds.), "Modern Methods of Plant Analysis," vol. V, pp. 51–72, Springer-Verlag, Berlin.

Chance, B., and W. D. Bonner, Jr., 1963. The temperature insensitive oxidation of cytochrome *f* in green leaves—a primary biochemical event of photosynthesis, in B. Kok and A. Jagendorf (eds.), "Photosynthetic Mechanisms in Green Plants," pp. 66–81, NAS-NRC Publication #1145.

Claesson, S., 1964. Principles of photochemistry and photochemical methods, in A. C. Giese (ed.), "Photophysiology," vol. I, pp. 19–33, Academic Press, New York.

Clayton, R. F., 1963. Photosynthesis: Primary physical and chemical processes, *Ann. Rev. Plant Physiol.,* **14:**159–180.

———, 1965. "Molecular Physics in Photosynthesis," Blaisdell Publishing Co., New York.

Duysens, L. N. M., 1963. Studies on primary reactions and hydrogen or electron transport in photosynthesis by means of absorption and fluorescence difference spectrophotometry of intact cells, in B. Kok and A. Jagendorf (eds.), "Photosynthetic Mechanisms in Green Plants," pp. 1–17, NAS-NRC Publication #1145.

——— and J. Amesz, 1962. Function and identification of two photochemical systems in photosynthesis, *Biochim. Biophys. Acta,* **64:**243–260.

Egle, Karl, 1960. Biogenese des Chlorophylls, Vorstufen, Beziehung zum Hämin, Protochlorophyll, *Handb. Pflanzenphysiol,* **5:**323–353.

Emerson, R., and W. Arnold, 1932. A separation of the reactions in photosynthesis by means of intermittent light, *J. Gen. Physiol.,* **12:**609–622.

———, R. V. Chalmers, and C. Cederstrand, 1957. Some factors influencing the long-wave limit of photosynthesis, *Proc. Nat. Acad. Sci.,* **43:**133–143.

Fork, D. C., 1963. Observations on the function of chlorophyll *a* and accessory pigments in photosynthesis, in B. Kok and A. Jagendorf (eds.), "Photosynthetic Mechanisms in Green Plants," pp. 352–361, NAS-NRC Publication #1145.

——— and J. Amesz, 1969. Action spectra and energy transfer in photosynthesis. *Ann. Rev. Plant Physiol.,* **20:**305–328.

French, C. S., 1960. The chlorophylls *in vivo* and *in vitro,* in W. Ruhland (ed.), "Encyclopedia of Plant Physiology," vol. V, pp. 252–297, Springer-Verlag, Berlin.

———, 1961. Light, pigments, and photosynthesis, in W. D. McElroy and B. Glass (eds.), "Light and Life," pp. 447–472, Johns Hopkins Press, Baltimore.

———, J. Myers, and G. C. McLeod, 1960. Automatic recording of photosynthesis action spectra used to measure the Emerson enhancement effect, in M. B. Allen (ed.), "Comparative Biochemistry of Photoreactive Systems," pp. 361–365, Academic Press, New York.

Frenkel, A. W., 1954. Light-induced phosphorylation by cell-free preparations of photosynthetic bacteria, *J. Am. Chem. Soc.,* **76:**5568–5569.

Gaffron, H., 1960. Energy storage: Photosynthesis, in F. C. Steward (ed.), "Plant Physiology," **1b,** pp. 3–277, Academic Press, New York.

Garner, W. W., and H. A. Allard, 1920. Effect of length of day on plant growth, *J. of Agr. Res.*, **18**:553–606.

Giese, A. C., 1964–1968. "Photophysiology," I–IV. Academic Press, New York.

Halldal, P., 1962. In R. A. Lewin (ed.), "Physiology and Biochemistry of Algae," pp. 583–593, Academic Press, New York.

Hamner, K. C., and J. Bonner, 1938. Photoperiodism in relation to hormones as factors in floral initiation, *Botan. Gaz.,* **100**:388–431.

Hastings, J. W., 1968. Bioluminescence, *Ann. Rev. Biochem.,* **37**:597–630.

Haupt, W., 1965. Perception of environmental stimuli orienting growth and movement in lower plants, *Ann. Rev. Plant Physiol.,* **16**:267–290.

Hendricks, S. B., 1959. The photoreaction and associated changes of plant morphogenesis, in R. B. Withrow (ed.), "Photoperiodism," pp. 423–438, AAAS Publication 55.

———, H. A. Borthwick, and R. J. Downs, 1956. Pigment conversions in the formative responses of plants to radiation, *Proc. Natl. Acad. Sci.,* **42**:19–26.

——— and ———, 1965. The physiological functions of phytochrome, in T. W. Goodwin (ed.), "Chemistry and Biochemistry of Plant Pigments," pp. 405–436, Academic Press, London.

Hill, R., 1939. Oxygen produced by isolated chloroplasts, *Proc. Roy. Soc (London)*, **B127**:192–210.

——— and F. Bendall, 1960. Function of the two cytochrome components in chloroplasts: A working hypothesis, *Nature,* **186**:136–137.

Hind, G., and J. M. Olson, 1968. Electron transport pathways in photosynthesis, *Ann. Rev. Plant Physiol.,* **19**:249–282.

Hoch, G., and R. S. Knox, 1968. Primary processes in photosynthesis, in A. C. Giese (ed.), "Photophysiology," vol. III, pp. 225–251, Academic Press, New York.

Jones, D. F., J. MacMillan, and M. Radley, 1963. Plant hormones: III. Identification of gibberellic acid in immature barley and immature oats, *Phytochemistry,* **2**:307–314.

Kamen, Martin (ed.), 1963. "Primary Processes in Photosynthesis," Academic Press, New York.

Kok, B., and G. Hoch, 1961. Spectral changes in photosynthesis, in "Light and Life," pp. 397–416.

——— and A. Jagendorf, 1963. "Photosynthetic Mechanisms of Green Plants," Natl. Acad. of Science, Washington, D.C.

Koski, V. M., 1950. Chlorophyll formation in seedlings of *Zea mays* L., *Arch. Biochem.,* **29**:339–343.

———, C. S. French, and J. H. C. Smith, 1951. The action spectrum for the transformation of protochlorophyll to chlorophyll *a* in normal and albino corn seedlings, *Arch. Biochem.,* **31**:1–17.

Kupke, D. W., and C. S. French, 1960. Relationship of chlorophyll to protein and lipoids; molecular and colloidal solutions: Chlorophyll units, in W. Ruhland (ed.), "Encyclopedia of Plant Physiology," vol. V, pp. 298–322, Springer-Verlag, Berlin.

Leopold, A. C., 1965. "'Plant Growth and Development," 466 pp., McGraw-Hill Book Company, New York.

Levine, R. P., 1969. The analysis of photosynthesis using mutant strains of algae and higher plants, *Ann. Rev. Plant Physiol.,* **20**:523–540.

Milner, H. W., and W. M. Hiesey, 1964. Photosynthesis in climatic races of *Mimulus:* I. Effect of light intensity and temperature on rate, *Plant Physiol.,* **139**:208–213.

Myers, J., and C. S. French, 1960. Relationships between time course, chromatic transient, and enhancement phenomena of photosynthesis, *Plant Physiol.,* **35**:963–969.

Pauling, L., 1960. "The Nature of the Chemical Bond," 644 pp., Cornell University Press, Ithaca, New York.

Röbbelen, G., 1956. Über die Protochlorophyllreduktion in einer Mutante von *Arabidopsis thaliana* (L.) Heynh, *Planta,* **47**:532–546.

Rumberg, B., P. Schmidt-Mende, J. Weikard, and H. T. Witt, 1963. Correlation between absorption changes and electron transport in photosynthesis, in B. Kok and A. Jagendorf (eds.), "Photosynthetic Mechanisms in Green Plants," NAS-NRC Publication #1145.

San Pietro, A., and C. Black, 1965. Enzymology of energy conversion in photosynthesis, *Ann. Rev. Plant Physiol.,* **16**:155–174.

Seliger, H. H., and R. A. Morton, 1969. A physical approach to bioluminescence, in A. C. Giese (ed.), "Photophysiology," vol. IV, pp. 253–314, Academic Press, New York.

Shen-Miller, J., P. Cooper, and S. A. Gordon, 1969. Phototropism and photoinhibition of basipolar transport of auxin in oat coleoptiles, *Plant Physiol.,* **44**:491–496.

Shibata, K., 1957. Spectroscopic studies on chlorophyll formation in intact leaves, *J. Biochem. Tokyo,* **44**:147–173.

———, A. A. Benson, and M. Calvin, 1954. The absorption spectra of suspensions of living organisms, *Biochim. Biophys. Acta,* **15**:461–470.

Siegelman, H. W., and S. B. Hendricks, 1965. Purification and properties of phytochrome: A chromoprotein regulating plant growth, *Fed. Proc.,* **24**(4):863–867.

———, D. J. Chapman, and W. J. Cole, 1969. The bile pigments of plants, *Biochem. Soc. Symp.,* **28**:107–120.

Smith, J. H .C., 1958. Quantum yield of the protochlorophyll-chlorophyll transformation, in "The Photochemical Apparatus, its Structure and Function," *Brookhaven Symposia in Biology,* **11**:296–302.

———, 1960. Protochlorophyll transformations, in M. B. Allen (ed.), "Comparative Biochemistry of Photoreactive Systems," pp. 257–277, Academic Press, New York.

———, 1961. Some physical and chemical properties of the proto-chlorophyll holochrome, in T. W. Goodwin and O. Lindberg (eds.), "Biological Structure and Function," Academic Press, London.

———, 1963. Chlorophyll formation and photosynthesis, *Proc. Fifth Intern. Cong. Biochem.*, **6**:151–162.

——— et al., 1957. The natural state of protochlorophyll, "Research in Photosynthesis," pp. 464–474, Interscience Publishers, New York.

——— and A. Benitez, 1954. The effect of temperature on the conversion of protochlorophyll to chlorophyll *a* in etiolated barley leaves, *Plant Physiol.*, **29**:135–143.

——— and Violet M. K. Young, 1956. Chlorophyll formation and accumulation in plants, in A. Hollaender (ed.), "Radiation Biology," vol. III, "Visible and Near-visible Light," pp. 393–442, McGraw-Hill Book Company, New York.

——— and C. S. French, 1963. The major and accessory pigments in photosynthesis, *Ann. Rev. Plant Physiol.*, **14**:181–224.

Tanaka, M., A. M. Benson, and K. T. Yasunobu, 1965. A proposed structure of *C. pasteurianum* ferredoxin, in A. San Pietro (ed.), "Non-heme Iron Proteins," pp. 221–224, Antioch Press, Yellow Springs, Ohio.

Tanner, W., M. Loffler, and O. Kandler, 1969. Cyclic photophosphorylation *in vivo* and its relation to photosynthetic-$CO_2$-fixation, *Plant Physiol.*, **44**:422–428.

Thimann, K. V., 1967. Phototropism, in M. Florkin and E. H. Stotz (eds.), "Comprehensive Biochemistry," pp. 1–29, Elsevier, Amsterdam.

van Niel, C. B., 1949. The comparative biochemistry of photosynthesis, in J. Franck and W. E. Loomis (eds.), "Photosynthesis in Plants," pp. 437–496, Iowa State College Press, Ames, Iowa.

Virgin, H. I., 1955. The conversion of protochlorophyll to chlorophyll *a* in continuous and intermittent illumination, *Physiol. Plantarum,* **8**:389–403.

Warburg, O., and E. Negelein, 1929. Über das Absorptionsspektrum des Atmungsferments, *Biochem. Z.,* **214**:64–100.

Weaver, E. C., 1968. EPR studies of free radicals in photosynthetic systems, *Ann. Rev. Plant Physiol.,* **19**:283–294.

——— and N. I. Bishop, 1963. EPR and optical studies on *Scenedesmus* mutants, in B. Kok and A. Jagendorf (eds.), "Photosynthetic Mechanisms in Green Plants," NAS-NRC Publication #1145.

# 3
# The Movements of Water and Other Fluids

*"Growing plants are extremely responsive to their environment. Sunlight, soil, climate, and moisture dictate the nature and extent of plant communities, whether they be kelp beds, sagebrush, cactus, prairie grass, tundra, swamps, or deep pine forests. Nature's inventions, expressed by living green plants from one-celled algae to thousand-year-old redwoods, have successfully adapted their life forms and cycles to an inconceivable variety of environments. Each square foot of the earth's surface has had a pattern of plant life that may have evolved as a community over countless ages, only to be revolutionized by men within the last few centuries."*
——The Plant Sciences, *NAS Publication* #1405.

## 3.1  INTRODUCTION

Plants require water; deprived of it, they will wilt and die. In fact, the typical land plant consumes prodigious quantities of water, vastly more than any of the other substances that enter it, and some 100 times more than its content of water at maturity. Stated differently, the rate of water loss is 100 times greater than the rate of entry of carbon dioxide. Most of this water therefore does not remain within the plant, but passes through it to the atmosphere. The process is called *transpiration*. The transpiration stream carries to the aerial parts of the plant nutrient ions absorbed by the roots, although the rate of transpiration is not necessarily a limiting factor in the absorption of nutrients.

The seemingly wasteful consumption of water through transpiration is nonetheless essential to the growth of plants. In nature and even in agriculture, it is a rare and fortunate plant that enjoys an optimum supply of water through-out its life. Without irrigation, most crops are inhibited, often seriously, from achieving full growth and development.

If we are concerned with the processes controlling the growth of plants in nature or if we are concerned with the world-wide problems of agriculture, then we must recognize that there is perhaps no single factor that is more crucial and none more amenable to our intervention (at least in principle) than that of control over transpiration in plants.

At one stage, the movement of water in plants involves its diffusion as a gas. We shall therefore want to consider some of the laws of gaseous diffusion, including the behavior of $CO_2$ and oxygen as well as water.

We shall see that long-distance water movement occurs primarily through a system of capillaries called *xylem*. In contrast, the movement of organic solutes occurs in an independent set of extraordinary channels called *phloem*. We shall examine this process under the term *translocation*.

## 3.2  THE PATH OF TRANSPIRATION

Let us first survey the phenomena of transpiration, looking especially at the anatomy of water movements. We shall here take as facts certain notions which we may wish to examine more critically in the sections that follow.

Water is absorbed from soil through young roots, especially by root hairs (Figs. 3-1 and 3-2). The water appears to move through the cortex of roots and through their cell walls. Cortical cells are closely interconnected by strands of protoplasm called *plasmodesmata*. The collection of interconnecting proto-plasm is often called the *symplast*.

The peculiar anatomy of the endodermis, composed of cells with radially thickened and suberized cell walls, has led many physiologists to infer that it is a barrier to water movement, that water at this point must pass through

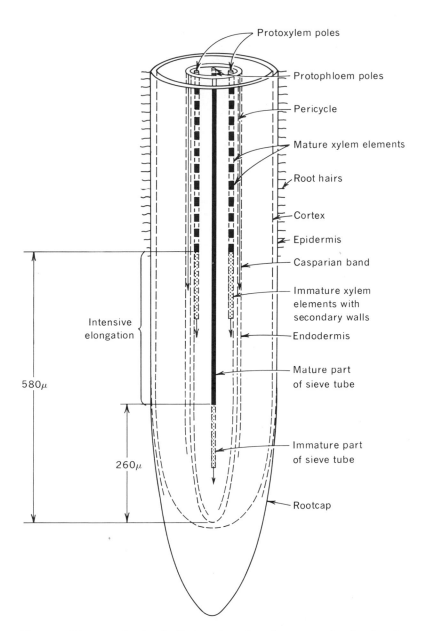

Fig. 3-1 Diagram of a longitudinal section of a tobacco root tip, showing the regions important for water absorption (Esau, 1965).

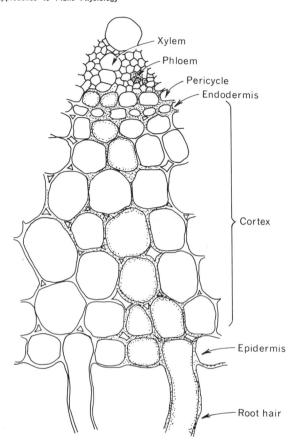

*Fig. 3-2*  The path of water across a root tip (Kramer, 1959). Water is thought to enter root hairs, follow cell walls, and traverse the cortex to the xylem.

the cells rather than the cell walls. Although this notion is of only marginal relevance to the problems of water movement, it becomes of major importance for the leading hypothesis on ion transport (cf. Sec. 4.4).

Whatever are the more subtle aspects of radial movements of water across the root, once water enters the stele its distribution through the plant appears to be a matter of plumbing, rather elegant plumbing, but nonetheless plumbing.

Pteridophytes, gymnosperms, and angiosperms evolved conductive cells called *tracheids* (Fig. 3-3). These cells are elongated, ranging typically from a few tenths of a millimeter to a few millimeters, and exceptionally to 10 or 20 mm in length, with strongly lignified walls. Since water must move across many more cross walls in tracheids than in vessels (see below), the efficiency of water conduction is substantially less. This inefficiency is somewhat compensated by frequent lateral connections, known as bordered pits (Fig. 3-4).

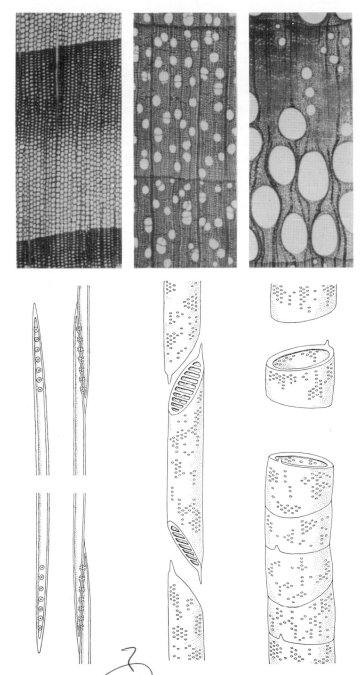

*Fig. 3-3* Different kinds of plumbing in plants (Zimmerman, 1963). (*a*) The conductive units of pine are shown on the left. As a gymnosperm, it has only tracheids, shown in the photomicrograph above and in longitudinal diagram below. Tracheids connect with one another through bordered pits. (*b*) The center units are vessels of birch; the end walls are fully perforated and conduct freely. (*c*) The more efficient conduits on the right are vessels of oak. Here the end walls are completely eliminated with the result that water moves through long rigid tubes.

These pits appear to be cleverly designed check valves, which may be able to seal off a tracheid which has suffered an embolism.

Angiosperm wood, in addition to tracheids, also contains the hydrodynamically more efficient xylem *vessels*. These structures may be up to several meters long and a few tenths of a millimeter in diameter and can conduct impressive volumes of water. The vessels are the mortal remains of vertical chains of cells whose lateral walls are sealed with lignin and other amorphous polysaccharides and strengthened with helical thickenings of cellulose. The end plates

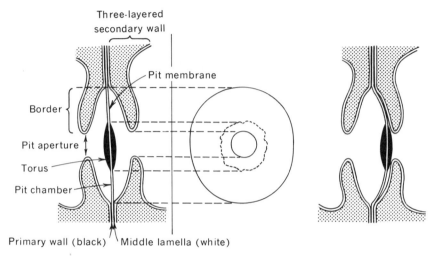

*Fig. 3-4* The function of bordered pits (after Esau, 1965). In their normal state (on the left) the hydrostatic pressure in cells connected by bordered pits is nearly equal. Water diffuses across the thin portion of the membrane. As shown on the right, if the pressure in one cell is lost, as through a cut, the membrane is forced to the side and the central portion of the membrane or torus, is pushed against the opposite opening, thus sealing off the embolism.

have dissolved away to form long capillary channels (Fig. 3-3). The protoplasm has disappeared.

Water in vessels is typically under negative pressure. Since the introduction of bubbles into these capillaries will destroy the conductive system, the capillary column will often be broken in winter when the water freezes. In order to succeed in so-called temperate climates, dicots learned simply to abandon air-filled vessels and cut off new vessels from the cambium each spring before new leaves expand.

All along the conductive system a modest amount of lateral transport supplies the living cells of the stem with inorganic nutrients and replenishes the relatively small amounts of water lost by evaporation. Most of the transpiration stream continues upward to the leaves. There, small amounts of water and the remainder of the nutrients enter the leaf parenchyma (Fig. 3-5). The bulk of the water passes along the parenchyma cell walls and evaporates into the intercellular spaces. Finally the water diffuses out into the atmosphere through the *stomata*.

The water system of a tall tree or liana (woody vine) is an astonishing device, especially when viewed in developmental sequence. The water columns are first filled by capillarity and osmosis as the living protoxylem cells expand. The water remains as the end walls dissolve and the protoplasm disappears. The apical meristems and leaves are carried tens of meters above their source of water as the plant grows. At their upper ends the capillary columns "hang"

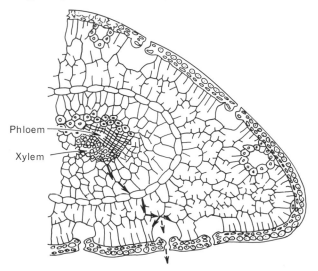

Phloem

Xylem

*Fig. 3-5*  Path of water across a pine needle (Zimmerman, 1963). Water is carried into the leaves through branches of xylem, moves through the cell walls of the leaf parenchyma, evaporates into the intercellular spaces, and thence diffuses through the stomata to the outside.

on the evaporative surfaces of the leaves. Consequently the columns are under such enormous negative pressures (tensions) that if the tiniest bubble forms, the columns snap and cease to conduct. Nonetheless, the water system remains in a kind of secular equilibrium; old vessels lose their function only to be replaced by new ones periodically formed from the cambium.

We have said that the transpiration stream brings inorganic nutrients to the aerial parts of the plant, that the volume of water transpired is larger than required for this purpose, and also that most of the water evaporates through the stomata. It is obvious that, if the rate of water loss is not subject to some kind of control, the plant's tissues will suffer from desiccation, as would those of any organism. Even though the leaves and stems are enveloped in a waxy cuticle which is almost impermeable to water, the stomata are so ingeniously designed that, when they are open, water can evaporate through them almost as rapidly as if the entire leaf surface were open to the atmosphere! This seeming death wish on the part of plants becomes understandable when we recognize that the stomata are programmed to admit carbon dioxide for photosynthesis. Paradoxically in order to carry on maximal rates of photosynthesis, plants must accept a potentially dangerous loss of water. Land plants are thus forever poised between the Scylla of carbon starvation and the Charybdis of desiccation!

Considering the seemingly limitless capacity of plants for chemical synthesis, we can only conclude that the failure of plants to develop an epidermis that would admit carbon dioxide but retain water vapor is some cosmic oversight.

In the sections below we shall inquire more rigorously into the movement of water in plants. We shall then consider as case studies some of the limiting factors governing water movements in the several stages of transpiration.

## 3.3  HYDROSTATICS AND EQUILIBRIUM RELATIONS OF WATER IN PLANTS

Exactly as in the case of chemical transformations (cf. Sec. 1.4) net movements of water can occur only in the direction of a decrease in the free energy of the system. We must therefore inquire into the factors within the soil-plant-atmosphere system which may affect the free energy of water.

Although investigators concerned with soil physics have reported states of water in terms of its free energy, plant physiologists have usually used terms such as "diffusion pressure deficit," which are difficult to relate rigorously to thermodynamic quantities. Recently there has been general agreement (cf. Kramer et al., 1966) to use the term *water potential,* $\psi$, defined as the chemical potential per unit molal volume of water (Eq. 3-1).

$$\psi \equiv \frac{\Delta\mu_w}{\bar{V}} \qquad\qquad \text{Eq. 3-1}$$

where $\bar{V}$ is the partial molal volume of water (18 cc mole$^{-1}$) and $\Delta\mu_w$ is the difference between the chemical potential of water in the solution and the chemical potential of pure water at $0°$ and 1 atm pressure. Chemical potential has its usual thermodynamic meaning of the change in free energy per mole at constant temperature and pressure (Eq. 3-2).

$$\mu_w \equiv \text{chemical potential of water} \equiv \left(\frac{\partial G}{\partial n}\right)_{T,P} \qquad\qquad \text{Eq. 3-2}$$

The term water potential at once keeps physical chemists content and has intuitive appeal to biologists; as with other potentials, water will tend to flow from higher to lower water potentials. Similarly, the water of solutions with identical $\psi$ values will be in equilibrium; that is, the net flux will be zero.

Water potentials can be thought of as composed of three terms, attributable to osmotic pressure, hydrostatic pressure, and the so-called matric potential or matric effect (Eq. 3-3).

$$\psi = \psi_o + \psi_P + \psi_M \qquad \text{Eq. 3-3}$$

$\psi$ and the terms that compose it have the units of pressure. Under standard conditions (pure water at $0°$ and 1 atm), $\psi = 1$ atm.

We shall give here a brief account of how each term is generated.

### $\psi_o$ and Osmotic Pressure

If a solution contains solutes, the water potential will be less than that of pure water. This happens quite simply because the presence of solutes decreases the concentration of water itself. More precisely, it is the *activity* of water which is decreased. (Activity is another of the useful fictions invented by physical chemists: it is the concentration of a solute or solvent in an ideal solution.)

Quantitatively, the decrease in water potential due to solutes, $\psi_o$, is related to the activity of water by Eq. 3-4.

$$\psi_o = -\frac{RT}{\bar{V}} \ln \frac{a°}{a} \qquad \text{Eq. 3-4}$$

where $R$ is the universal gas constant, $T$ the absolute temperature, $\bar{V}$ the partial molal volume of water, and $a°$ and $a$ the activities of pure water and the water in the solution.

$\psi_o$ is also the negative of the osmotic pressure $\pi$.

$$\psi_o = -\pi \qquad \text{Eq. 3-5}$$

If the solution is "ideal,"

$$\psi_o = -\frac{RT}{\bar{V}} \ln \frac{1}{N_w} \qquad \text{Eq. 3-6}$$

where $N_w$ is the mole fraction of water.

For solutions which are both dilute and ideal,

$$\ln \frac{1}{N_w} \simeq \Sigma n_s$$

where $n_s$ is the number of moles of solute. This leads to the van't Hoff relation

$$\pi = \frac{\Sigma n_s RT}{\bar{V}} \qquad \text{Eq. 3-7}$$

Note that since the osmotic pressure $\pi$ must always be positive, the $\psi_o$ term is always negative.

### $\psi_P$ and Hydrostatic Pressure

The pressure term, $\psi_P$, is numerically equal to the hydrostatic pressure.

$$\psi_P = P \qquad \text{Eq. 3-8}$$

Hydrostatic pressure may be either positive or negative. The inward pressure on a cell exerted by the cell wall is called the *turgor pressure* and is positive. Negative pressure or tension can exist in xylem elements.

### $\psi_M$ or Matric Potential[1]

The peculiar properties of water are largely accountable for the tendency of water molecules to form hydrogen bonds with other water molecules, with solutes, and with hydrophilic colloids. Water has a large electric dipole moment, so that the electron-rich (oxygen) region of water molecules tends to align toward electron-poor centers (cations), while the hydrogen ends are oriented around anions. Ions typically attract a shell of water up to 2 Å deep, depending on the electric field strength. The shell is called water of hydration (Fig. 3-6).

The hydrogen bonding of water with oxygen-, nitrogen-, or sulfur-containing radicals of large molecules, such as proteins and polysaccharides, may be extremely extensive. The association of water with large molecules results in the phenomenon of *imbibition,* whereby seeds absorb large amounts of water even from relatively dry soils.

Finally, water may be bound in tissue capillaries by van der Waals' forces.

All of these forces tend to lower the potential of water and are lumped as $\psi_M$. The determination of $\psi_M$ is largely empirical and is measured as the water potential after the hydrostatic pressures and solute effects have been eliminated or subtracted out (cf. Wiebe, 1966).

### Calculations Involving Water Potentials

In order to predict the directions or rates of water movements in the soil-plant-atmosphere system, we should gain some experience in calculating water potentials of various plant compartments.

EXAMPLE 3-1.  What is the water potential of a 1 molal solution of sucrose at 25° and 1 atm pressure?

Let us assume that a 1 molal solution is "ideal." Starting with Eq. 3-3

$$\psi = \psi_o + \psi_P + \psi_M$$

we may take as given that $\psi_P$ and $\psi_M = 0$.

From Eq. 3-6

$$\psi_o = -\frac{RT}{\bar{V}} \ln \frac{1}{N_w}$$

$$R = \frac{82.06 \text{ cc atm}}{\text{deg mole}} \qquad T = 298°\text{K}$$

$$\bar{V} = 18.07 \text{ cc}$$

$$N_w = \frac{55.5}{56.5}$$

[1] Reference: Ogston, 1966.

Radii of ions
in crystal lattice

Hydration numbers
determined from diffusion

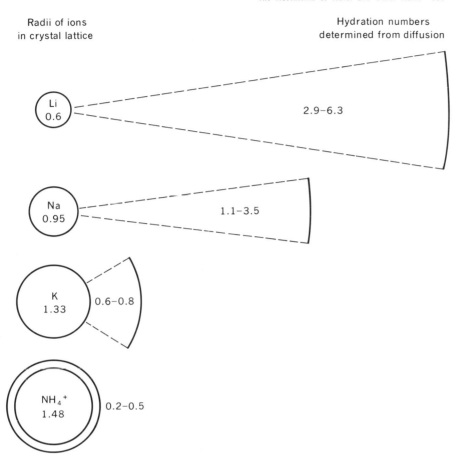

2.9–6.3

1.1–3.5

0.6–0.8

0.2–0.5

*Fig. 3-6* Relative size of some alkali ions. The radii of ions can be determined accurately from measurements of crystal lattices. But the effective size of an ion in aqueous solution is determined by its charge which attracts water molecules around it. The smaller ions have a more intense charge and are able to organize one or more layers of less oriented water molecules. In this way a sodium ion during diffusion or ionophoresis behaves as a larger particle than does a potassium ion. The actual size of these hydration shells cannot be determined unambiguously or in absolute units. The numbers and the radii shown are only relative.

(that is, the mole fraction of water in the solution is equal to the number of moles of water divided by the total number of moles present).

$$\psi_o = -\frac{82.06 \cdot 298}{18.07} \ln \frac{55.5}{56.5} \text{ atm}$$

$$\psi = \psi_o = -24.5 \text{ atm}$$

EXAMPLE 3-2.   What hydrostatic pressure is required to cause the water potential of water in a 1 molal solution of sucrose to equal that of pure water at 25° and 1 atm?

We want $\psi_P$. We determined in Eq. 3–1 that $\psi_0 = -24.5$ atm. Since $\psi_M = 0$, we have

$$0 = -24.5 \text{ atm} + \psi_P + 0$$
$$\psi_P = 24.5 \text{ atm}$$

Hence, the pressure required would be 24.5 atm.

*Note:* The *hydrostatic pressure* required to bring a solution into equilibrium with pure water *is defined as the osmotic pressure of* that solution.

### Elasticity of Cell Walls

Since the $\psi$ of root cells is generally less than that of soil water, water tends to enter the root cells. The root cells swell and a hydrostatic pressure develops as the cellulose walls are stretched. This is called turgor, and the hydrostatic pressure is referred to as turgor pressure. The cells walls are neither perfectly nor infinitely elastic so that the change in cell volume is a complex function of the external water potential (Fig. 3-7).

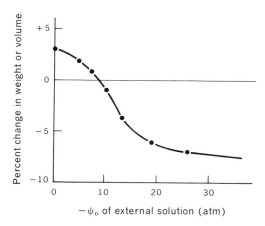

*Fig. 3-7* Cell volume as a function of external osmotic pressure (Bennet-Clark, in Steward, 1959). As the osmotic pressure increases, water moves out of plant cells. Since the cell walls are somewhat elastic, the cell walls contract and the volume of the cell decreases.

EXAMPLE 3-3.   The volume of a cell is measured, and then the cell is placed in water at 25°, in which it is observed to swell. It is then placed in successively increasing concentrations of mannitol; each time it shrinks some. When it is put into 0.2 *M* mannitol it is found to be the same size as it was initially. It continues to shrink in stronger solutions, until in 0.5 *M* mannitol very slight plasmolysis is seen, while in 0.6 $M$ mannitol it is strongly plasmolyzed. In 0.5 $M$ mannitol the volume of the cell is 5 percent less than it was initially. Calculate, in atmospheres:

(a) The cell's $\psi_o$ at zero turgor
(b) The cell's $\psi_o$ in its initial state
(c) The cell's $\psi_P$ in its initial state
(d) The cell's $\psi$ in its initial state

As the cell is placed in progressively more concentrated mannitol, water will begin to leave the cell, causing a decrease in volume until the $\psi_o$ of the external solution reaches the $\psi_o$ of the cell. At this point the elastic cell walls begin to relax; the hydrostatic pressure $\psi_P$ falls. At zero turgor where $\psi_P = 0$ and (ignoring $\psi_M$) $\psi_o$ (outside) = $\psi_o$ (inside), the cell contents will just begin to pull away from the cell walls. This is called *incipient plasmolysis*.

(a) In the example, incipient plasmolysis occurs in 0.5 $M$ mannitol; hence, from Eq. 3–7,

$$\psi_o = -0.5 \times 0.08206 \times 298 = -12.2 \text{ atm}$$

(b) Since the cell volume in 0.5 $M$ mannitol is 95 percent of the initial volume, it follows that the initial $\psi_o$ must have been

$$\psi_o = -0.95 \times 0.5 \times 0.08206 \times 298 = -11.6 \text{ atm}$$

(c) The effect of 0.2M mannitol is equivalent to the original hydrostatic pressure; hence,

$$\psi_P = 0.2 \times 0.08206 \times 298 = 4.9 \text{ atm}$$

(d) The initial $\psi = \psi_o + \psi_P = -11.6 + 4.9 = -6.7$ atm.

## Air

The calculation of water potentials of water vapor in air is similar to the situation in solutions. We substitute the notion of partial pressures of the gas for the concentrations. (More precisely, we should substitute *fugacity*, the idealized partial pressure, for activity in Eq. 3-4.)

$$\psi_o = -\frac{RT}{\overline{V}} \ln \frac{p^\circ}{p}$$

Eq. 3-9

where $p^\circ$ is the partial pressure in equilibrium with pure water and $p$ is that for the solution in question.

EXAMPLE 3-4.   Calculate the water potential of air at 90 percent relative humidity at 25°.

$$\psi = \psi_o + \psi_P + \psi_M$$
$$\psi_P = \psi_M = 0$$
$$\psi_o = \frac{RT}{\bar{V}} \ln \frac{p^o}{p}$$
$$\psi_o = -\frac{82.06 \cdot 298}{18.07} \ln \frac{100}{90} \; \text{atm}$$

Hence,

$$\psi = \psi_o = -141 \; \text{atm}$$

The astonishingly low water potential of even moist air means that equilibrium during the day between water in the leaf and in the air is virtually impossible.

## 3.4 HYDRODYNAMICS: LAWS GOVERNING THE RATES OF FLUID MOVEMENT

In Sec. 3.3 we considered the factors governing water potentials in and around plants. Differences in water potential will determine the *direction* taken by water movements, but exactly as in the case of chemical transformations (cf. Sec. 1.4), the *rate* of water movement is determined by additional factors.

The movement of liquids in capillaries is well understood. From Poiseuille's law (Eq. 3-10), we can predict rates of volume flow $J_v$

$$J_v = \frac{\pi r^4 \, \Delta P}{8 \eta l}$$

Eq. 3-10

given the radius of the capillary, $r$; the viscosity $\eta$; and $\Delta P/l$, the mean pressure gradient along the capillary. From the $r^4$ factor we may infer that in comparison to a tracheid of $10\mu$ radius, a vessel with a radius of $100\mu$ would have $10^4$ greater specific conductivity!

We can similarly predict the flow of gases through stomata from Fick's law (Eq. 3-11).

$$J = -DA \frac{\partial s}{\partial x} \quad \text{or} \quad \frac{J}{A} = -D \frac{\partial s}{\partial x}$$

Eq. 3-11

where $D$ is a diffusion coefficient, $\partial s/\partial x$ is the concentration gradient of a solute $s$ with respect to distance $x$, and $A$ is the cross-sectional area.

Serious difficulties arise when we come to a membrane. The property of membranes to pass fluids is called *permeability*.

Nearly all substances move in or out of cells less rapidly than would be expected on the basis of free diffusion. Some substances (for example, proteins) appear to move across cell boundaries at essentially zero rates. We conclude that cells and tissues possess restricted permeability. We say that membranes are differentially permeable or, ambiguously, that they are "semipermeable."

Let us imagine an idealized membrane (Fig. 3-8) of thickness $\Delta x$ separating an outer solution I from an inner solution II. The solutions contain

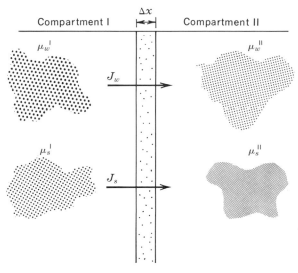

*Fig. 3-8* Model for membrane transport. Two homogeneous (that is, stirred) compartments I and II are separated by a membrane of thickness $\Delta X$.

$\mu_w^{\text{I}}$ = chemical potential of water in solution I
$\mu_s^{\text{I}}$ = chemical potential of solute in solution I
$\mu_w^{\text{II}}$ = chemical potential of water in solution II
$\mu_s^{\text{II}}$ = chemical potential of solute in solution II
$J_w$ = flow of water
$J_s$ = flow of solute

In the diagram the $\mu_w^{\text{I}}$ is shown as being greater than $\mu_w^{\text{II}}$ so water would tend to flow from left to right. On the other hand, $\mu_s^{\text{II}} > \mu_s^{\text{I}}$ and the solute would tend to flow from right to left, so that the value of $J_s$ would be negative.

water $w$ and solute $s$. The chemical potentials $\mu_i$ of the water and solute are shown as $\mu_w^{\text{I}}$, $\mu_w^{\text{II}}$, $\mu_s^{\text{I}}$, $\mu_s^{\text{II}}$, etc.

Intuitively (that is, viewing the system as simple plumbing), we can imagine characterizing the membranes in part by the rate of flow of fluid as a function of the pressure drop across the membrane.

Let $J_v \equiv$ total flow across the membrane. If the osmotic pressures of I and II are equal, we could expect that flow would be simply proportional to the pressure drop $\Delta P$:

$$J_v = L_p \, \Delta P \hspace{5cm} \text{Eq. 3-12}$$

where we define $L_p$ as the hydraulic or filtration coefficient for unit pressure difference. We can also imagine characterizing the membrane by the rate of diffusion of solute across the membrane. With zero electrical potential difference, we expect according to Fick's law (Eq. 3-11)

$$J_s = -\omega \frac{ds}{dx} \qquad \text{Eq. 3-13}$$

where $\omega$ is the diffusion coefficient for the solute and $ds/dx$ is the concentration gradient across the membrane. Assuming a homogeneous membrane, $ds/dx$ should be (intuitively) proportional to the concentration difference, $s_I - s_{II}$. Hence Eq. 3–13 could be rewritten

$$J_s = -k(\pi_I - \pi_{II}) \qquad \text{Eq. 3-13}a$$

where $k$ is an arbitrary constant.

Both of these kinds of measurements are of limited application since flows arise in the intact plant from differences of both osmotic pressure and hydrostatic pressure acting simultaneously. This problem was attacked intuitively by Ursprung and Blum, who hypothesized that the rates of flow across membranes are proportional to differences in water potential (Eq. 3-14)

$$J_v = k(\psi_I - \psi_{II}) \qquad \text{Eq. 3-14}$$

or, ignoring matric potentials

$$J_v = k[(P_I - \pi_I) - (P_{II} - \pi_{II})]$$

This same relation is known in animal physiology as Starling's hypothesis.

Unqualified acceptance of Eq. 3-14 resulted in certain distressing contradictions, which will be discussed below.

Since Eq. 3-14 was not derived rigorously, biophysicists have searched for a theoretical approach to permeability based on thermodynamics. Katchalsky (1961) assayed the problem as follows:

"Modern research on biological permeability is mainly concerned with the molecular organization of the membranes. The ultimate goal of the investigation is the development of structural models which can be used for prediction of the physico-chemical properties of the membranes.

"During the last few years, another approach is emerging whose main concern is not the construction of adequate models, but the establishment of equations capable of organizing the experimental results in a consistent formal framework. This approach is based on the thermodynamics of irreversible processes. The thermodynamic approach admittedly cannot yield quantitative numerical predictions as derived from the treatment of a tangible model. However, it has the advantage of being general, free from contradiction and able to predict new correlations. There is, moreover, a possibility of bridging the two approaches by expressing the thermodynamic flow parameters in terms of physical coefficients, such as frictional coefficients, amendable to kinetic interpretation."

The following is an account of the basic transport equation greatly abbreviated from Katchalsky (1961; cf. also Dainty, 1963).

We shall use the same model as that in Fig. 3-8.

According to the thermodynamics of irreversible processes, the rate of flow of water is

$$J_w = L_{ww}X_w + L_{ws}X_s \qquad \text{Eq. 3-15}$$

The symbols deserve some explanation: $X_w$ and $X_s$ are the gradients of chemical potential (cf. Eq. 3-2) across the membrane.

$$X_w \equiv \frac{-d\mu_w}{dx}$$

$$X_s \equiv \frac{-d\mu_s}{dx}$$

$L_{ww}$ and $L_{ws}$ are called Onsager coefficients. It seems intuitively reasonable in Eq. 3-15 that the flow of water should be proportional to the gradient of chemical potential of water. What we are unprepared for in the dogma of irreversible thermodynamics is that the flow of water *may* also be affected by the gradient of solute potential. I say *may* because the $L_{ws}$ factor may be positive but it may also be zero.

By the same token, the flow of solute across the membrane, $J_s$, is also the resultant of two terms:

$$J_s = L_{ss}X_s + L_{sw}X_w \qquad \text{Eq. 3-16}$$

From Eq. 3-16 we see that the movement of solutes is a function of the solute potential gradient $X_s$ but may also be affected by the gradient of water potential $X_w$.

According to the disciples of irreversible thermodynamics, the necessary and sufficient characterization of any membrane would consist in a specification of the coefficients $L_{ww}$, $L_{ss}$, and $L_{ws}$ ($= L_{sw}$) for each solute.

In dealing with water movements we should like to use the same terms $P$ and $\pi$, which proved useful for water equilibria.

We can write a pair of equations analogous to Eqs. 3-15 and 3-16, but with $P$ and $\pi$ substituting for $X_s$ and $X_w$ as the driving forces. $J_w$ and $J_s$ must also be replaced with $J_v$ and $J_D$.

$$J_v \equiv \text{total flow across the membrane} = L_p \Delta P + L_{pD} \Delta\pi \qquad \text{Eq. 3-17}$$

$$J_D \equiv \text{relative flow of solute vs. water} = L_{pD} \Delta P + L_D \Delta\pi \qquad \text{Eq. 3-18}$$

where $\Delta P$ is the pressure difference and $\Delta\pi$ is the difference in osmotic potential across the membrane. $L_p$, $L_{pD}$, and $L_D$ are a corresponding set of Onsager coefficients.

We can appreciate the physical significance of the new Onsager coefficients by letting the two forces become zero separately. When $\Delta\pi$ is zero, $Jv$ is the flow resulting from only a pressure difference, and $L_p$ is seen to be the same coefficient of filtration that we encountered earlier.

$$J_v = L_p\,\Delta P \hspace{6cm} \text{Eq. 3-12}$$

When $\Delta P = 0$, $J_D$ is the exchange flow, so that $L_D$ is the coefficient of diffusion across the membrane,

$$J_D = L_D\,\Delta\pi \hspace{6cm} \text{Eq. 3-19}$$

The significance of the cross coefficient $L_{pD}$ can be seen in Eq. 3–17 when $\Delta P = 0$. Then $J_v = L_{pD}\,\Delta\pi$ which means that $L_{pD}$ is the coefficient for volume flow with unit osmotic gradient.

Another way of seeing the significance of $L_{pD}$ is simply to adjust the pressure so that there is no volume flow; that is, $J_v = 0$. Then

$$\frac{\Delta P}{\pi} = -\frac{L_pD}{L_p} = \sigma \hspace{5cm} \text{Eq. 3-20}$$

But for an ideal membrane $P = \pi$ under equilibrium conditions; therefore, $\sigma$ is the ratio of the *apparent* osmotic pressure to the *theoretical* osmotic pressure. The ideality of a membrane is thus expressed by $\sigma$, the reflection coefficient.

$$\sigma \equiv -\frac{L_pD}{L_p}$$

The traditional (that is, ideal) differentially permeable membrane will have unit volume flow for unit osmotic gradient, so that $\sigma = 1$. But for a completely nonselective membrane under these conditions, osmotic flow is zero, and hence $L_{pD} = 0$. For real membranes, therefore, $\sigma$ varies between 0 and 1. Equation 3-17 may now be rewritten

$$J_v = L_p(\Delta P - \sigma\,\Delta\pi) \hspace{4cm} \text{Eq. 3-21}$$

This leaves $L_D$ as the only coefficient which cannot be conveniently measured. It turns out that the solute permeability coefficient $\omega$ is related to the Onsager coefficient by Eq. 3–22

$$\omega = \frac{L_pL_D - L_{pD}{}^2}{L_p}\,\bar{C}_s \hspace{4cm} \text{Eq. 3-22}$$

where $\bar{C}_s$ is the mean concentration of solute in the membrane. Thus a membrane can (in principle) be completely characterized by two conventional coefficients and a new one arising from the Onsager equations:

$L_p$ = hydraulic or filtration coefficient
$\omega$ = solute permeability coefficient
$\sigma$ = reflection coefficient

### Evaluation of Membrane Properties and the Ideal Semipermeable Membrane

The hydraulic coefficient can be evaluated fairly directly for the relatively spacious membranes of giant algae, such as the *Characea* (Dainty and Hope, 1959), but with much greater difficulty and uncertainty in cellular organisms (Kohn and Dainty, 1966). Estimates of $L_p$ for several plant membranes are shown in Table 3-1 on page 164.

Measurements of the solute permeability coefficient $\omega$ have also been made, again most directly with giant algal cells. The permeability of membranes toward nonelectrolytes follows a very curious rule, first identified by Overton, an English physiologist working in Switzerland in the last century. According to Overton's rule, natural membranes behave as if they contained a layer of lipid (Fig. 3-9).

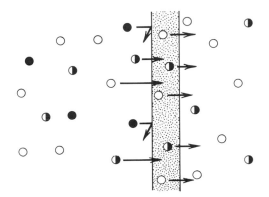

*Fig. 3-9* Diagram of natural membranes according to the lipid hypothesis of Overton. According to Overton's hypothesis, solutes traverse biological membranes only by passing through a lipid phase in the membrane. Solutes that are insoluble in lipids (●) cannot pass. Solutes that are soluble (○) or partially soluble (◐) in lipids permeate the membrane in proportion to their distribution coefficients between the lipids of the membrane and water.

Since a substance could traverse the membrane only by traversing the lipid layer, the permeability is related to the lipid solubility of the substance, more specifically the distribution coefficient between lipid and water.

Lipid solubility turns out not to be the only factor in permeability. Molecular weight also influences permeability to about the 1.5 power. Collander has determined the permeability of the cell membrane of *Nitella*, one of the *Characea*, toward a large number of substances (Fig. 3-10) (cf. Coe and Coe, 1965).

One of the striking exceptions to the Overton rule is the permeability of all membranes toward water (see Fig. 3-10). The observed values are as much as 1,000 times greater than would be predicted from the Overton rule.

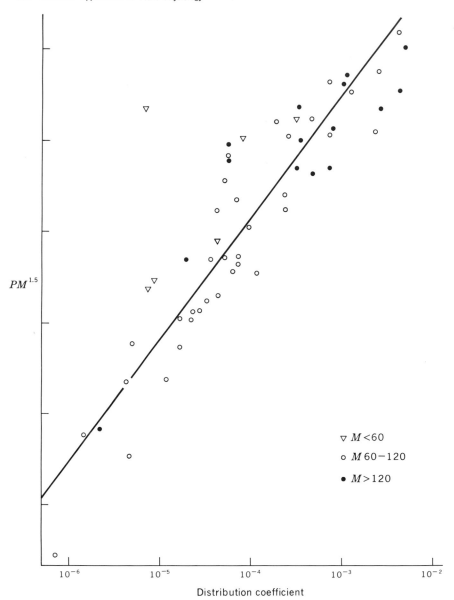

$PM^{1.5}$

▽ $M < 60$

o $M\,60-120$

● $M > 120$

Distribution coefficient

*Fig. 3-10* Permeability of *Nitella mucronata* cells as a function of oil-water distribution coefficients and molecular weights of nonelectrolytes (from Collander, 1957). The ordinate is $PM^{1.5}$, where $P$ is the permeability coefficient in cm sec$^{-1}$ and $M$ is the molecular weight. The selection of the exponent is empirical. The abscissa is the distribution coefficient of the substances between olive oil and water. The empirical relation is $P \propto K^{1.3}M^{-1.5}$. Small molecules, notably water, deviate toward higher permeability coefficients.

A determination of the reflection coefficient $\sigma$ will indicate the ideality of a membrane. In an ideal semipermeable membrane,

$$\sigma = 1$$
$$L_{pD} = -L_p$$

In such a case the volume flow $J_v$ in Eq. 3-17 becomes

$$J_v = L_p(\Delta P - \Delta \pi) \qquad\qquad \text{Eq. 3-23}$$

or from Eqs. 3-1, 3-5, and 3-8

$$J_v = L_p \, \Delta \psi$$

This states that volume flow is simply proportional to the difference between the pressure and osmotic gradients across the membrane. This of course is our old friend, Eq. 3-14, from Ursprung, Blum, and Starling.

It was only reasonable that a gradient of osmotic pressure should cause differences in the activities of water on the two sides of a membrane and hence a *net diffusion* of water from higher to lower activity. But it was *not* obvious how osmotic pressure would generate a flow *equivalent* to that produced by hydrostatic pressure.

One can define an osmotic permeability coefficient $P_o$ for water as

$$P_o = \frac{L_p RT}{\overline{V}_w}$$

with the units of cm sec$^{-1}$.

If osmotic flow occurs only by diffusion, the coefficient for the movement of water due to diffusion can be related to hydraulic flow by the equation

$$\frac{L_p RT}{\overline{V}_w} = P_d = P_o \qquad\qquad \text{Eq. 3-24}$$

where $P_d$ is the diffusion permeability coefficient of the membrane for water. Attempts were made, therefore, to test Eq. 3-23 through the prediction of Eq. 3-24. $P_o$ was evaluated by measuring the bulk flow of water induced by differences in osmotic pressure; $P_d$ was evaluated by measuring the movement of labeled water, DHO or THO, into or out of cells that were at hydrostatic equilibrium. The results (Table 3-1) showed that $P_o$ exceeded $P_d$ by a factor of 10.

Solomon, Ussing, and others (cf. Ray, 1960; Dainty, 1963; Stein, 1969) argued that these results could be rationalized if the transport of water (generated by $\Delta P$ or $\Delta \pi$) occurred not by diffusion alone but also by bulk flow through microcapillary pores in the membrane.

It is possible that pores may be detected by electron microscopists armed with greater resolution, but there is a simpler solution: it may be that membranes tested are not ideal, that $\sigma < 1$. Doubts of the existence of these pores are

Table 3-1   Ratio of Osmotic and Diffusion Permeability Coefficients of Water of Some Natural Membranes*

If osmotic flow of water occurs by diffusion across membranes $P_o$, the osmotic permeability coefficient, should equal $P_d$, the diffusion permeability coefficient. Most of the $P_d$ values reported are in fact much lower than $P_o$.

| Tissue | $P_o$, cm sec$^{-1}$ $10^{-4}$ | $P_d$, cm sec$^{-1}$ $10^{-4}$ | $P_o/P_d$ | References |
|---|---|---|---|---|
| *Avena* coleoptiles | | | 9.3 | Ordin and Bonner, 1956 |
| *Tolypellopsis stelligera* single cells | | | 5.3 | Wartiovaara, 1944 |
| *Nitella mucronata* | 370 | 6.33 | 59 | Collander, 1954 |
| Potato tuber discs | | | 6.7 | Thimann and Samuel, 1955 |
| *Vicia faba* root segments | | | 35 | Ordin and Kramer, 1956; Brewig, 1937; Brouwer, 1954 |
| *Valonia ventricosa* single cells | 2.36 ± 0.17 | 2.4 ± 0.29 | 0.98 | Gutknecht, 1967 |

* Modified from Ray, 1960.

based on two lines of reasoning: (1) Eq. 3-21 stands on thermodynamic, not empirical or intuitive, foundations, so that no special mechanism is needed to "explain it"; (2) the existence of unstirred layers of solution on the order of 10 $\mu$ thick (Dainty, 1963) is sufficient to make calculations of $P_d$ extremely doubtful. Such unstirred layers are far less likely to exist under conditions of bulk flow.

In fact, the one known case where $P_o$ equals $P_d$ is that of *Valonia* (Table 3-1), a giant unicellular alga. The dimensions of the alga provide two principal advantages: (1) the effects of the unstirred layer will be small compared with the dimensions of the cell, and (2) the cells can be perfused internally, which facilitates the measurement of the permeability of rapidly permeating substances (Gutknecht, 1967). Thus, where Eqs. 3-23 and 3-24 can be evaluated with confidence, the results are consistent with a simple, nonperforate membrane model.

The same problem of unstirred layers prevents the accurate estimation of $\sigma$-values. Thus until improved techniques are at hand, e.g., until we can subject isolated membranes to direct physical chemical analysis, the full characterization of natural membranes remains incomplete.

## Membrane Structure

The molecular structure of membranes is also disputed. There are two extreme models—the lipid sandwich or lipid bilayer of Davson, Danielli, and Robertson (cf. Robertson, 1960, 1967) and the lipoprotein subunit hypothesis of Benson (1964). These alternate models and some variants are represented in Fig. 3-11.

The lipid bilayer model finds support in electron microscopy where membranes can sometimes be visualized as a triple layer of two 20-Å electron-dense layers, thought to be protein, enclosing a 35-Å electron-transparent layer, thought

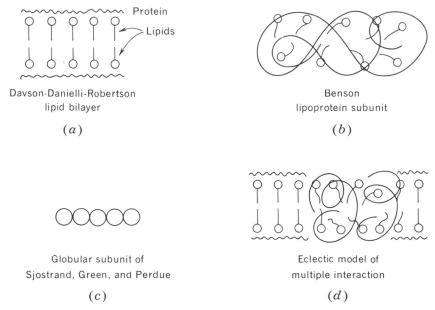

Davson-Danielli-Robertson
lipid bilayer

(a)

Benson
lipoprotein subunit

(b)

Globular subunit of
Sjostrand, Green, and Perdue

(c)

Eclectic model of
multiple interaction

(d)

*Fig. 3-11*  Molecular models of membranes (after Rothfield and Finkelstein, 1968). (*a*) A sheet of protein forms a monolayer over a double layer of lipids. The polar ends of the lipids face the protein, and the nonpolar ends bond hydrophobically to one another. (*b*) A lipoprotein in a random coil bonds lipids in various orientations. (*c*) Globular lipoprotein subunits form a two-dimensional sheet—an extension of the lipoprotein model. (*d*) Some regions of the membrane are oriented as lipid bilayers and others as randomly coiled lipoproteins.

to be lipid. But the optical rotatory dispersion spectra of a number of natural membranes show substantial helical structures among the proteins, which is inconsistent with the spread form or monolayer predicted from the bilayer model.

Benson finds support for his alternate view of lipoprotein subunits in the pock-marked appearance of freeze-etched chloroplast membranes. Sjostrand, Green, and Perdue (cf. Rothfield and Finkelstein, 1968), in an extension of the lipoprotein subunit model, visualize a two-dimensional sheet of lipoproteins in which different proteins are imbedded in the same sets of lipids, forming a functional mosaic.

It is safe to conclude that the several models of membrane structure remain unproven.

## 3.5 KINDS OF EVIDENCE REQUIRED FOR THE CHARACTERIZATION OF FLUID MOVEMENTS IN PLANTS

Since we cannot at this time expect to achieve the kind of characterization of membrane properties required by thermodynamics, the best we can do is a semiquantitative treatment as follows:

1. Information on the variables affecting water potential in a given plant compartment
2. Data on the anatomical pathways of fluid movement, including the ultrastructure of the membranes, plasmodesmata, and cell walls involved
3. Measurement of parameters, e.g., vessel radii, affecting fluid movements through nonmembranous paths, and factors affecting these parameters

## 3.6 CASE STUDY: WATER MOVEMENT IN SOIL AND ROOTS[1]

The water potential of soils depends in a characteristic way on the water contents of the soil. The relation is approximately hyperbolic (Fig. 3-12) (Veihmeyer,

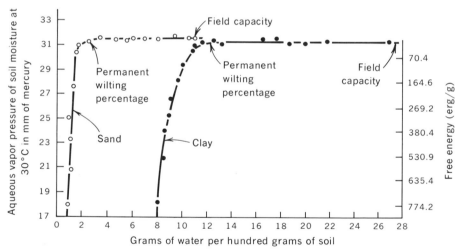

*Fig. 3-12* Osmotic pressure and related values of soil water, as functions of water contents (Veihmeyer, 1956). The water potential of a soil, normally equal to the osmotic pressure, is a complex function of the water content and is dependent on hydrogen bonding and stronger forces between water and the soil constituents. In the figure, it is clear that clay particles bind water relatively more tightly than sand does.

1956; Kramer, 1959). We may note, for example, that the water content of sand must be decreased to one-tenth of *field capacity*[2] before the potential

[1] General references: Briggs, 1967; Kramer, 1959; Veihmeyer, 1956.
[2] Field capacity is the amount of water remaining after liquid water has been allowed to drain through a column of soil.

falls appreciably, whereas the water potential in the clay soil begins to fall at about one-half of field capacity. Clearly, substantial amounts of water are bound by clay. The "permanent wilting percentages" of the soils are also noted in Fig. 3-12. At first blush we might imagine that at these values the water potentials of water in the soil and in the roots are equal. Roots of mesophytic plants in fact have water potentials down to about —7 atm, which is much *less* than the $\psi$ of soils at the "permanent wilting percentage." Clearly some factor or factors other than water potential determine the point of irreversible wilting of plants.

We presume that the rate of water movement into roots follows Eq. 3-23 (or Eq. 3-21 if the membranes are not truly differentially permeable). From this equation we can deduce some of the physiological variables that should govern the movement of water into roots and test if roots in fact behave in this way.

The movement of water into roots should vary proportionally with the area of the absorbing surface, $A$, and inversely with the path length $l$.

Furthermore, if the membranes were ideal, water movement should be proportional to the difference in water potential:

$$\Delta\psi = \psi_{\text{roots}} - \psi_{\text{soil}} = P_{\text{roots}} - \Delta\pi$$

The system at first blush seems enormously heterogeneous: different parts of the root may have different membrane properties, different $\pi$, and different $P$. Moreover the external surface is not the only, and perhaps not even the, limiting barrier to water movement. From any one point on the root surface, water on its journey to the xylem may have to pass a dozen or more membranes with different parameters (Fig. 3-1).

However, certain generalizations are possible: root hairs comprise a major fraction of the absorption surface. These structures seem to be most abundant under moist conditions, when they are needed least. But in plants that bear root hairs, they account even at the minimum for over 60 percent of the total root surface (Kramer, 1959). It seems agreed that maximum water absorption takes place on the parts of the root where root hairs are most abundant. This region also coincides with that of xylem differentiation and lack of suberization.

Rates of water movement for several plant roots are shown in Table 3-2. The prodigious extent of roots is almost legendary. Annual crop plants in good soil typically penetrate to 2 m depth; trees and some xerophytes may penetrate to 10 m more.

The general applicability of Eq. 3-17 can be tested with the systems shown in Fig. 3-13. Brouwer (1953) tested *Vicia faba* roots with one such system in which the water deficits of the medium and the tissue could be varied independently. He found that the conductivity was constant with varying water

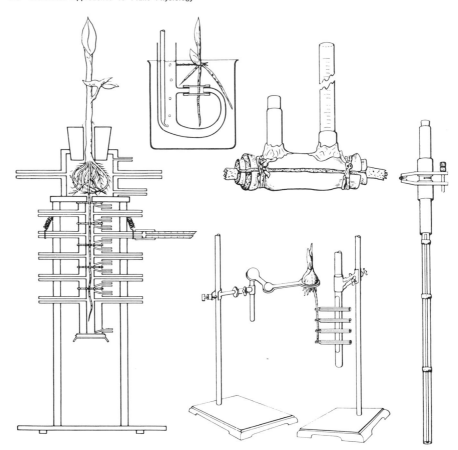

*Fig. 3-13*  Apparatus for measuring water conductivities of roots (collected by Kramer, 1956).

potential of the medium but was not constant with changing internal water potential (Fig. 3-14).

This is a remarkably general finding (cf. reviews of Kramer, 1959; Bennet-Clark, 1959). Increased water deficits, normally resulting from increased transpiration, depress somewhat the conductivity of tissue near the tip but greatly increase the conductivity of older tissue further from the tip. The result (Fig. 3-14) is that with increased transpiration, there is a shift in the region of maximal water absorption toward older tissues.

Water conductivity is also strongly depressed by anaerobiosis and cyanide poisoning. Hence conductivity contains metabolically sensitive components (cf. Glinka and Reinhold, 1964).

If we reflect on the nature of water conduction, we may inquire what might constitute the metabolically sensitive elements in this expression. There

Table 3-2   Rates of Water Movement into Roots of Various Plants*

| Material | Rate, mm³ cm² hr⁻¹ | Investigator |
|---|---|---|
| Corn, young roots in water | 20.0 | Hayward et al., 1942 |
| Sour orange, suberized roots in water | 5.0 | |
| Onion, young roots in water | 50.4 | Rosene, 1941 |
| Radish, root hairs | 33.3 | Rosene and Walthall, 1949 |
| Corn, root hairs | 28.3 | |
| Coffee tree, entire root system in soil | 0.25 | Nutman, 1934 |
| Short leaf pine (*Pinus echinata*), attached, suberized roots in soil | 2.63 | Kramer, 1946 |
| Short leaf pine, attached, roots in water | 3.37 | |
| Short leaf pine, excised, suberized roots, 0.4 atm tension | 9.0 | |
| Dogwood (*Cornus florida*), same as above | 15.6 | |
| Yellow poplar (*Liriodendron tulipfera*), same as above | 101.4 | |

* Collected by Kramer, 1956.

are several possibilities: metabolism may control the path length (thickness of the membrane) or the membrane characteristics.

Of the terms that compose the water potentials of root tissues, the hydrostatic pressure $P$ is typically more important than the osmotic pressure $\Delta\pi$. Because of evaporation from the leaf surfaces and the development of large water tensions in the xylem, $P$ is normally negative. In contrast, $\Delta\pi$ (and hence $-\psi_o$) is rarely larger than 6 atm. The importance of $P$ can be seen by com-

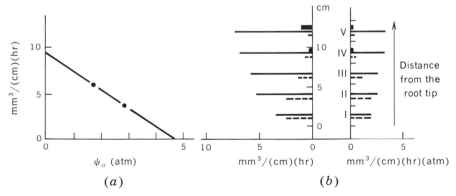

*Fig. 3-14* Water conductivities of *Vicia faba* roots under conditions of varying water potential (Brouwer, 1954a). The data were obtained using the apparatus shown in Fig. 3-13. (*a*) Rates of water movement as a function of applied osmotic pressure. (*b*) Rates of water flow (left) and conductivities (right) in different root zones. Water potentials were −1.3 atm (dotted lines) and −2.5 atm (light solid lines). The heavy solid lines are for roots poisoned with $10^{-3}$ *M* KCN and at $\psi = -2.5$ atm.

Table 3-3   Rates of Water Absorption by Intact and De-topped Plants*

| Species | Transpiration† | | Exudation‡ | | Exudation as Percentage of Transpiration |
| | First Hour | Second Hour | First Hour | Second Hour | |
|---|---|---|---|---|---|
| *Coleus blumei* | 8.6 | 8.7 | 0.30 | 0.28 | 3.2 |
| *Helianthus annuus* | 4.3 | 5.0 | 0.02 | 0.02 | 0.4 |
| *Hibiscus moscheutos* | 5.8 | 6.7 | −0.01 | 0.05 | 0.7 |
| *Impatiens sultanii* | 2.1 | 1.9 | −0.22 | −0.06 | |
| Tomato (*Lycopersicon esculentum*): | | | | | |
| Ser. 1 | 10.0 | 11.0 | −0.62 | 0.07 | 0.6 |
| Ser. 2 | 7.5 | 8.7 | 0.14 | 0.27 | 3.1 |

\* Kramer, 1959.
† Rates are in milliliters of water per half hour and are averages of 5–8 plants.
‡ Tops removed.

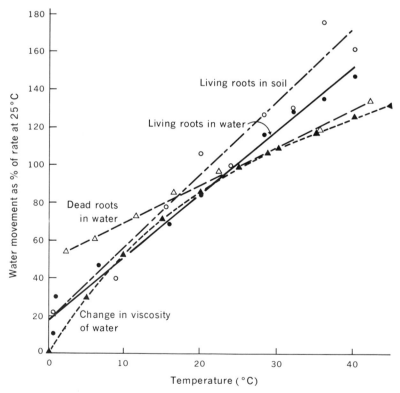

*Fig. 3-15*   Water movements in living and dead *Helianthus* roots as a function of temperature (Kramer, 1940). The corresponding change in the viscosity of water is shown for comparison.

paring the rates of water absorption in intact with the 100-fold lower rates of de-topped plants (Table 3-3).

The conductivity of water in roots is very sensitive to temperature. In fact the water movement through killed *Helianthus* roots at different temperatures almost exactly parallels the viscosity of water (Fig. 3-15). Living roots show a steeper response to temperature (Fig. 3-16), especially in the case of

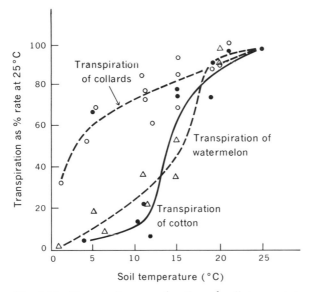

*Fig. 3-16*  Transpiration as a function of soil temperature in different species (Kramer, 1942). Note that the transpiration of collards (*Brassica oleracea* var. *acephela*), which is a cool-weather crop, is much less affected by temperature than transpiration is in the two plants accustomed to warm soils.

warm-weather plants. That there should be species differences in the response to temperature strongly supports the notion of a metabolically controlled component in conductivity. The correlation with viscosity reminds us of the microcapillary model for bulk flow through membranes.

Paradoxically, we may think of the roots as organs of water absorption but equally well as barriers to absorption. Indeed, below some critical ratio of root to leaf area, transpiration decreases linearly with root area (Fig. 3-17), but if the roots are cut off and the stems placed in water, transpiration is more rapid than in the intact plant.

Thus, in sum, it seems that roots are more than merely pieces of plumbing, but along with the stomata (Sec. 3.8), roots are carefully controlled devices designed to respond to adverse conditions of water supply.

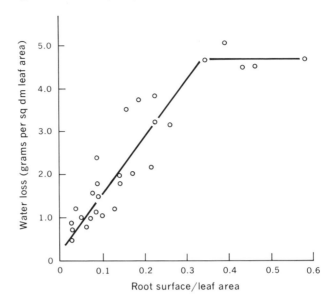

*Fig. 3-17*   Transpiration as a function of the ratio of root surface to leaf area in rooted lemon cuttings (Bialoglowski, 1936).

## 3.7   CASE STUDY: WATER MOVEMENT IN STEMS

The water potentials of stem tissues are lower than roots, but of the same order of magnitude. The striking characteristic of water in stems, particularly of tall vines and trees, is the existence of *negative pressure* (tension) in the xylem. Since xylem sap is fairly dilute (that is, has a small $\psi_o$) and since it must be in near equilibrium with surrounding tissues, the hydrostatic pressure must become negative as the water potentials of these tissues fall.

The conductive tissue of xylem consists of tracheids and vessels. These structures resemble capillaries, and we may consider the physics of water movement through them accordingly.

The factors governing the rate of movement of water through capillaries are expressed in Poiseuille's law (the same as Eq. 3-10):

$$J_v = \frac{\pi r^4 \, \Delta P}{8 \eta l}$$

Eq. 3-25

where $\eta$ is the viscosity, $\Delta P/l$ is the mean pressure gradient along the capillary, $r$ is the radius of the capillary, and $l$ is the length.

As we noted above (Sec. 3.4), a vessel with a radius of $100\mu$ would have a $10^4$ greater specific conductivity than a tracheid of $10\mu$ radius would have.

But does the water in xylem obey Poiseuille's law? We restate Eq. 3-25 as the sum of all conducting elements in a stem:

$$J_v = \frac{\pi}{8\eta} \frac{\Delta P}{l} \sum_{i=1}^{i=n} r_i^4$$

Eq. 3-26

A comparison of actual vs. expected conductivities is shown for several species in Table 3-4. The water conduction of the wood of lianas (first three species)

**Table 3-4   Efficiency of Water Conductivity in Various Plants***

Comparison of the observed rates of water movement ($v_o$) compared with rates predicted ($v_t$) from Poiseuille's law.

| Plant | Conductive Efficiency<br>$v_o/v_t \times 100$ |
|---|---|
| *Vitis vinifera* | 100 |
| *Aristolochia sipho* | 100 |
| *Atragene alpina* | 100 |
| Root wood of oak | 84 |
| Root wood of beech | 37.5 |
| *Helianthus annuus* | 32 |
| *Rhododendron ferrugineum* | 20 |
| *Rhododendron hirsutum* | 13 |

* Collected by Huber, 1965.

is precisely that expected. The conductivities of oak wood are nearly equal to theoretical, whereas those of the other species range down to a small fraction of the expected values. Generally close agreement between predicted and observed rates of flow through the xylem of tomato was also observed by Dimond (1966).

Deviations from Poiseuille's law should be anticipated: tracheids and vessels have end walls. Short vessels and especially tracheids must show conductivities less than that calculated for an open-ended tube.

The lengths of tracheids are fairly easy to determine in longitudinal sections (Table 3-5). Vessels present more of a problem. Huber's method (1956) is to macerate a given length of woody tissue and count the percentage of end walls. This procedure is directly analogous to end-group analysis in carbohydrates and proteins. Another is to measure the rise of mercury into the capillaries of the wood. Scholander et al. (1957) employed still another method with lianas: when segments of the vine were cut at progressively lower levels, water was sometimes released from the lower end. The release of water occurs when a vessel is cut for the first time. From the volume released and the diameter of the emptied vessels, the lengths of the vessels can be calculated. A summary of some typical vessel lengths is shown in Table 3-6. It has even been proposed (see Huber, 1956) that in some plants, single vessels may extend from the roots to the leaves.

Huber (1956) calculated an empirical quantity called the "relative conduction surface." It is the ratio of the area of the conducting elements to the

Table 3-5 Typical Lengths of Tracheids*

| Plant Type | Length, mm |
|---|---|
| Medullosae | 24 |
| *Araucaria* | 5(1.5–9) |
| Conifers | 2–3 |
| *Populus* | 1 |
| Fraxinus | 0.12–0.29 |

* Collected by Huber, 1956.

fresh weight of the plant. The distribution of values for this ratio is shown in Fig. 3-18. A very large fraction of the values fall between 0.2 and 0.8. Since the conductivity is related not to the area, but to the square of the area of the xylem elements, the degree of constancy of the relation is somewhat surprising. We should also remember that in general only the xylem elements

Table 3-6 Typical Lengths of Vessels*

| Type of Wood | Length, mm |
|---|---|
| Diffuse-porous wood | 80–150 |
| *Fagus* | 200 |
| Ring-porous wood | 500–1800 |

* Greenidge, 1952.

formed by recent growth are fully filled with water (Fig. 3-19); the remainder are stopped by air "embolisms." Therefore the measurement of total capillary cross sections overestimates the actual conduction area available to the plant.

An ingenious method for the measurement of the velocity of the transpiration streams (see review in Huber, 1956) employs a localized heating of the plant sap together with thermocouples for measuring the displacement by the transpiration stream of increased temperature along the stem (Fig. 3-20). Zimmermann (1964a) employed a variant of this scheme to prove that there was no interruption in water conduction in wood at temperatures in the vicinity of 0°, provided that no actual freezing occurred.

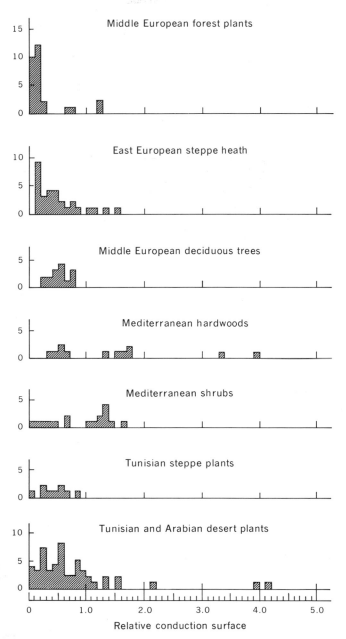

*Fig. 3-18* Distribution of relative conduction-surface values for a variety of plants (Huber, 1956).

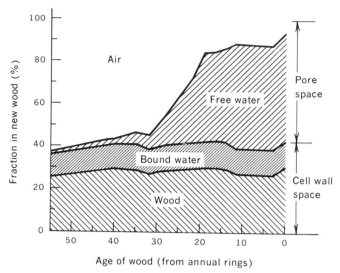

Fig. 3-19   Air embolism of xylem elements (Huber, 1956).

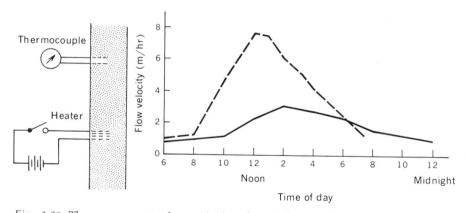

Fig. 3-20 The measurement of transpiration through heat conductivity (Huber, 1956; Zimmermann, 1963). (Left) A heating element is inserted into the xylem and a thermocouple some distance above it measures the steady-state difference in temperature. Faster transpiration is recorded as increased temperature. (Right) Transpiration rates in twigs (dotted line) and trunk (solid line) of a tree over the course of a day.

## 3.8   CASE STUDY: MOVEMENT OF GASES—WATER VAPOR AND $CO_2$—ACROSS LEAF SURFACES[1]

During the day, when stomata are open, leaf water potentials may become very low (Table 3-7) (Kramer, 1956, 1959). Values of —15 to —20 atm and lower are not unusual.

[1] General references: Heath, 1959; Zelitch, 1969.

Table 3-7   Osmotic Pressure and Water Potentials of Leaves*

| Time of Day | *Andropogon trifida* Osmotic Pressure, atm | *Ambrosia scoparius* Water Potentials, atm |
|---|---|---|
| Early morning | 20–22 | −7–10 |
| Noon | 22–24 | −12–15 |
| Mid-afternoon | 25–27 | −15–17 |

* Water potentials of *Ambrosia* leaves (Herrick, 1933; quoted by Kramer, 1956); osmotic pressure of *Andropogon* leaves (Stoddard, 1935; quoted by Meyer, 1956).

Possible values of water potentials from soil to air are shown in Fig. 3-21. Clearly the largest free-energy changes occur between leaf and atmosphere.

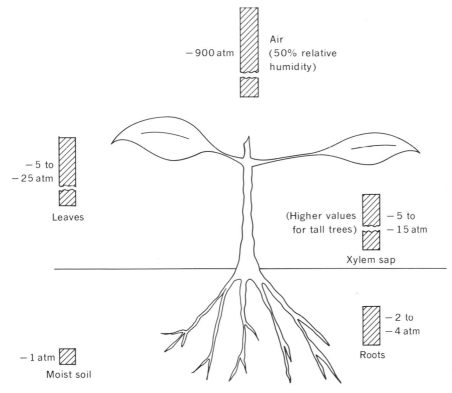

*Fig. 3-21*   Some possible values of water potentials in soil-plant-air systems. *Note:* The actual values of water potentials will obviously vary with the availability of water, transpiration rate, etc.

EXAMPLE 3-5. In a saturated atmosphere, is the decrease in water potential across the surface of an illuminated leaf sufficient to drive transpiration (cf. Veihmeyer, 1956)?

Let us grant that the sun raised the temperature of the leaf by only 1°, say from 25 to 26°. This would increase the vapor pressure of water in the leaf from 23.69 to 25.13 mm Hg. The $\Delta\psi$ for the movement of water from the heated interior of the leaf

$$\Delta\psi = -\frac{RT}{\overline{V}} \ln \frac{25.13}{23.69} = -80 \text{ atm}$$

is clearly greater than the $\Delta\psi$ for the movement of water into soil at the permanent wilting percentage (approx. −15 atm).

We calculated earlier that the principal free-energy decrease in the transpiration stream occurs across the leaf surface. The opening and closing of the stomata are therefore likely to be major factors in the control of water movement. The part the stomata play is determined, of course, not by free-energy changes alone (that is, hydrostatics) but by the factors governing the rates of water movement (hydrodynamics). We can construct a hydrodynamic model of the leaf-stomata-atmosphere system from a consideration of the path of water movement (Fig. 3-22). The intercellular air spaces of leaves are in effec-

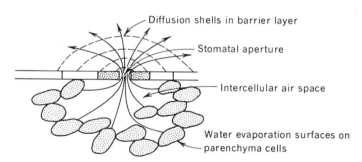

*Fig. 3-22* Path of water from parenchyma cells to the atmosphere. Water evaporates from the cell walls of leaf parenchyma, diffuses through the intercellular air spaces and through the stomatal aperture to the atmosphere. In still air, a "barrier layer" or "diffusion shells" lie between the leaf surface and the atmosphere.

tive contact with most of the leaf parenchyma, including the palisade as well as the spongy parenchyma. The leaf parenchyma cell walls therefore present an evaporation surface to these numerous intercellular air spaces of the leaf. Water vapor can also diffuse through the epidermis and cuticle.

In this segment of the transpiration stream, water movement is by diffusion, at least within and through the leaf. Fick's law (Eq. 3-11) in a time-varying system is terribly messy. Let us make the physiologically reasonable assumption of a steady state, with one concentration (activity or fugacity) of water in the atmosphere outside the leaf ($p_a$) and another at the evaporating surface of the parenchyma cells ($p_p$). Now, instead of trying to calculate the tortuous path of diffusion of water vapor through the intercellular spaces, let us lump all the stomata and all the intercellular air spaces into two "apparent path lengths" of a diffusion system of unit area (Fig. 3-23).

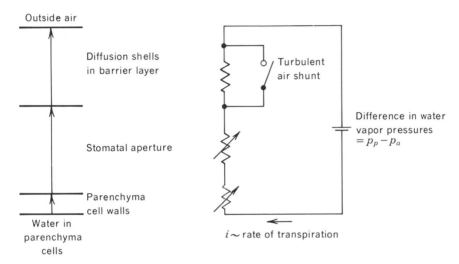

Fig. 3-23 Model for transpiration through stomata. The diagram on the left is a hydrodynamic model abstracted from Fig. 3-22. The effective path length through the intercellular spaces is negligible, whereas those associated with stomata or barrier layer are the controlling ones.

The corresponding electrical circuit is shown on the right. The stomatal resistance is variable and the resistance through the barrier layer may be effectively shunted in turbulent air.

Fick's equation (3-11) at steady state then reduces to Eq. 3-27

$$T = \frac{DA(p_a - p_p)}{L + S}$$ 
Eq. 3-27

where $T$ is the rate of transpiration, $p_a$ and $p_p$ are the partial pressures (or fugacities) of water in the atmosphere and at the evaporating surface, respectively, $D$ is the diffusion constant for water in air, $A$ is the cross-sectional area, and $L$ and $S$ are the apparent path lengths for diffusion in intercellular air spaces and through the stomata, respectively. Equation 3-27 gives transpiration as a function of stomatal opening, but we must carry the analysis further before we can evaluate it. The apparent path length through the stomata, $S$, can be evaluated by a consideration of stomatal geometry (Fig. 3-24).

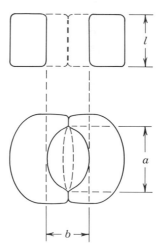

*Fig. 3-24* Geometry of stomatal diffusion path. The outline of an open stomate appears in the solid line and the closed stomate in dotted line. To calculate the effective diffusion path, the distance from inside to outside is designated $l$, and the cross section is taken to be an ellipse with major and minor axes of $a$ and $b$.

The fluid path length $S$ can be approximated by the relation

$$S = \frac{l/\pi ab + l/2\sqrt{ab}}{n}$$

Eq. 3-28

where $l$, $a$, and $b$ are the depth, width, and length of the stomatal aperture, and $n$ is the number of stomata per square centimeter.

Equation 3-27 can be inverted

$$\frac{1}{T} = \frac{L}{DA(p_a - p_p)} + \frac{S}{DA(p_a - p_p)}$$

Eq. 3-29

which, for steady-state conditions, is a linear equation. $L$ may now be evaluated by extrapolating $1/T$ for $S = 0$. Zelitch and Waggoner have shown that a linear relation obtains for tobacco leaves (Fig. 3-25).

We see in Fig. 3-25 that the reciprocal of transpiration $1/T$, can be a linear function of $S$. We took $L$, the apparent path length for diffusion in air, to be constant. Just outside the stomatal pore there will be a layer of quiet air, the barrier layer which represents part of the apparent path length $L$. The diffusion of water vapor from the pore through the barrier layer is commonly identified by the term "diffusion shells" in reference to the surfaces representing equal concentration of water vapor (Fig. 3-22). Brown and Escomb (1900) long ago demonstrated that the diffusion of gases through a system of pores through a membrane is remarkably efficient, up to 40 percent as great as diffusion from an open vessel of the same area.

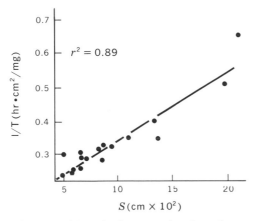

*Fig. 3-25* Transpiration as a function of stomatal opening (Zelitch and Waggonner, 1962a). According to Eq. 3-29, 1/T should be linear with stomatal aperture $S$. This is shown to be so for tobacco leaves. The plants were kept in turbulent air, so that the value of $L$ ($y$ intercept) is very small.

The thickness of the barrier layer is determined by the rate of movement of air outside the leaf. We can expect that at a sufficiently rapid rate of air movement, the component of $L$ due to the barrier layer becomes negligible and $L$ becomes equal to the diffusion path within the leaf. In fact with high air turbulence, $L$ virtually disappears (Fig. 3-25). The apparent diffusion path of water vapor in air is therefore due almost exclusively to diffusion through the barrier layer.

Up to now we have treated the stomata kinetically, but what controls the opening and closing of these pores? There is a general agreement that the stomata are controlled by the turgor of the guard cells. This can be demonstrated by placing a section of leaf in a hypertonic sucrose solution medium. But in nature these turgor changes are triggered by a bewildering variety of factors. If a generalization is possible, we may say that the guard cells are programmed to close in response to conditions that either are unfavorable to photosynthesis or might be indicative of impaired photosynthesis: low light intensity, high internal $CO_2$ concentration, low water supply, and poor nutrition. The detailed effects of a number of these factors are reviewed by Heath (1959).

The mechanism of the control of turgor changes is a subject of intense dispute. Attention has focused on the chloroplasts of the guard cells: functional guard cells always contain chloroplasts; the action spectrum for stomatal opening corresponds to typical chloroplast absorption spectra (Fig. 3-26); glycollate,

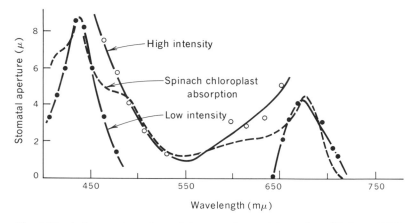

*Fig. 3-26*   Action spectrum for maintenance of stomatal openings (Kuiper, 1964). Strips were peeled from the lower epidermis of *Senecio odoris* (Defl.) leaves in the light and floated on water. The stomata remained open in white light and closed in the dark. The action spectrum at two levels of intensity ($\bigcirc$, $1.9 \times 10^{-9}$ and $\bullet$, $1.14 \times 10^{-9}$ einstein $cm^{-2}$ $sec^{-1}$) is compared with the absorption spectrum of spinach chloroplasts (dotted line).

a typical product of photosynthesis, is unusually effective in promoting opening, and inhibitors of glycollate catabolism are effective in inhibiting stomatal opening (Zelitch and Walker, 1964). If, as seems likely, the closing of stomata is caused by a passive loss of water from cells, and opening is active (that is, metabolically dependent), we should expect that low temperature and inhibitors should inhibit stomatal opening but not closing. Zelitch has shown that this is indeed so (Zelitch, 1963).

The immediate mechanism of opening is almost certainly osmotic, and the most likely immediate cause is a large increase in the potassium concentration in the guard cells (discussed by Zelitch, 1969). For the present one can only guess at the identity of the putative potassium pump and the means by which the various environmental factors control it.

Substances that combine with sulfhydryl groups are strikingly effective in inhibiting opening. Metal-complexing agents also inhibit opening, but, at certain concentrations, they inhibit closing as well (Fig. 3-27).

The fact that the phenylmercuri compounds when sprayed onto leaves penetrate only the epidermal layer provides an opportunity for a closer examination of stomatal physiology. After treatment with these substances, the stomata are "fixed" in various positions of closing for long periods of time with no direct effect on the parenchyma. Since different values of $S$ can be obtained in this way, it becomes possible to test several deductions from the hydrodynamic model of gas movements through the stomata.

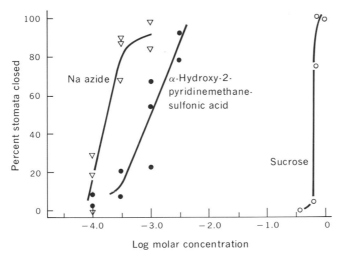

*Fig. 3-27*   Stomatal closure in different concentrations of metabolic inhibitors and sucrose (Zelitch, 1961). Stomata of tobacco leaf discs normally remain open when floating on water in the light. Additions of azide and inhibitors of glycollate oxidase, such as α-hydroxy-2-pyridinemethane sulfonic acid, cause rapid closure. That the effect of the inhibitors is not osmotic is shown by the much higher concentrations of sucrose required to cause closure.

One inference is that, since the principal $\Delta G$ change is across the leaf surface, resistance to water movement is through the stomata. If this were so, the rate of transpiration would be dependent only on $S$ (and $L$) and would be independent of soil water potentials. (In the absence of the inhibitor, water stress will influence stomatal opening.) Shimshi (1963) has shown that, contrary to expectation (Fig. 3-28), soil moisture alters the $1/T$ intercept (but not the slope) of $1/T$ vs. $S$ plots. Thus there is a significant resistance to water movement within the plant itself. This additional resistance, which we can calculate from the $X$ intercepts at different water potentials, may be in the parenchyma cell walls (cf. Gaff et al., 1964; Whiteman and Koller, 1964).

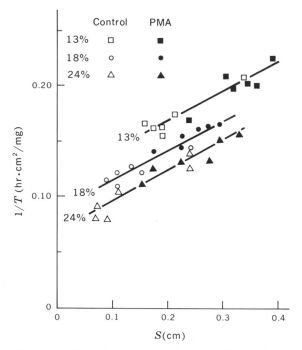

*Fig. 3-28* Transpiration as a function of $S$ and soil moisture (Shimshi, 1963). Maize plants were grown in soil of different water contents. The value $1/T$ was in each case linear with $S$, but the $1/T$ intercepts were different for the different $\psi$-values. Shimshi infers that with increasing moisture stress, there is a corresponding increase in the resistance to evaporation from the parenchyma cell walls.

Phenylmercuric acetate (PMA) was employed to obtain large values of $S$.

### CO₂ Movements through the Stomata

We have seen how the movements of water vapor through the leaf can be dealt with as a simple kinetic system. The movements of another gas, $CO_2$, also crucial to the life of the plant, should follow virtually the same scientific model.

We can modify Eq. 3-29 in two ways and obtain a mathematical model which should predict the movement of $CO_2$. The diffusion constant for $CO_2$ in air is different from that of water vapor; so that $D$ becomes $D'$. Also $CO_2$ must pass through the water layer surrounding the cell and diffuse to the chloroplasts, the presumed centers of $CO_2$ absorption. This requires another apparent diffusion path length, $M$. The partial pressures $p_a$ and $p_p$ must be changed to correspond to the partial pressures of $CO_2$ in the air and at the chloroplasts, respectively. Even though this distance $M$ may be very small, the diffusion constant through water is enormously less $(1/2,500)$ than that through air; consequently the $M$ term may well be significant.

The resulting equation (3-30) now expresses photosynthesis $p$ as a function of $M$, $L$, and $S$. It is of course essential in using this equation that we employ conditions under which the rate of photosynthesis is limited by the $CO_2$ concentration and moreover that photosynthesis is a linear function of the $CO_2$ concentration.

$$\frac{1}{p} = \frac{M}{D'_w A(p_a - p_c)} + \frac{L + S}{D'_a A(p_a - p_c)} \qquad \text{Eq. 3-30}$$

If we eliminate $L$ by the use of turbulent air, we can evaluate the $M$ term. It is in fact a significant component of the resistance to $CO_2$ movement (Fig. 3-29).

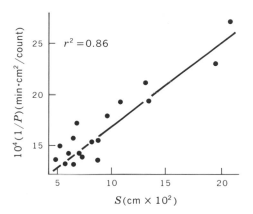

Fig. 3-29 Photosynthesis as a function of stomatal aperture (Zelitch and Waggoner, 1962). Photosynthesis in tobacco measured under the same conditions as Fig. 3-25.

One of the interesting conclusions from a comparison of the effects of stomatal aperture on transpiration and photosynthesis is that with moderate decreases in stomatal apertures, there is a disproportionately large decrease in transpiration (Fig. 3-30).

Much effort is now being spent in searching out the nature of stomatal control. Attempts are also being made to develop artificial cuticles (cf. Gale and Poljakoff-Mayber, 1967) with an eye toward rational measures of water conservation in arid land agriculture.

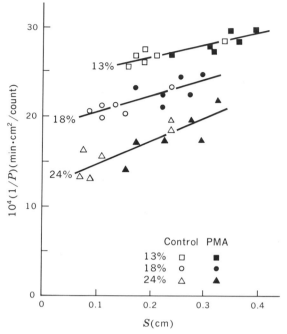

*Fig. 3-30*  Rates of photosynthesis as functions of $S$ and water contents (Shimshi, 1963). Photosynthesis was measured under the same conditions as those in Fig. 3-28. Again the $1/P$ values are linear with $S$, but they vary with water potential. The rates of photosynthesis are less sensitive to stomatal opening (curves are flatter) than the rates of transpiration are. Thus treatment with phenylmercuric compounds inhibits photosynthesis less than transpiration.

## 3.9  CASE STUDY: TRANSLOCATION

> " . . . but the only tune that he could play
> was 'Over the hills and far away' . . . . "

Up to now we have considered transpiration and the absorption of carbon dioxide as examples of fluid movements in plants. Another, the absorption of oxygen in respiration, we leave to the student (cf. Suggested Case Studies at the end of the chapter).

Another and quite different kind of fluid movement takes place in plants. The distribution to other parts of the plant of sugars elaborated in the leaves is thought to take place exclusively in the sieve tubes of the phloem. These structures form a continuous system of connecting channels parallel to the xylem between roots, leaves, and developing fruits. Sieve tubes possess a curious and unique sieve plate structure at the connecting ends of the tubes (Esau,

1965). The mature sieve tube has lost its nucleus, but unlike mature xylem elements which are devoid of protoplasm, the sieve tubes are filled with cytoplasm. There are analogous lateral structures in certain locations. Strands of protoplasm, termed *plasmodesmata,* lead through the plates from one cell to another.

In sharp contrast to the movement of water, which as we saw could be reduced to hydrodynamic models with reasonable predictive capacity, movement through the phloem appears to be "active," that is, dependent on metabolism. This fact should not discourage us, but up to the present nearly all of our information about phloem physiology is phenomenological.

We know, for example, that phloem translocation (again in contrast to movement in the xylem) is greatly depressed by local cooling of the stem and is abolished by steaming a narrow section of stem ("steam girdle"). We know also that the sugar translocated is an invariant characteristic of the species. In most plants the translocated sugar is sucrose, but in some it is another member of the raffinose family (Table 3-8). If a $C^{14}$-labeled substrate, such

Table 3-8 The Carbohydrates Transported in the Phloem Are Usually Members of the Raffinose Family

Each higher member of the raffinose family of sugars is formed by the addition of one 1,6-galactosyl residue.

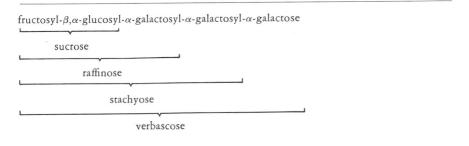

as $CO_2$, fructose, glucose, or sucrose, is fed to a leaf, the translocated sugar is nonetheless sucrose, labeled in both hexoses (Table 3-9) (Putman and Hassid, 1954). Sugar cane is an excellent test organism for studying translocation, since the sucrose is not converted to starch before or after translocation (cf. Hatch and Glasziou, 1964).

Although the sieve tubes *contain* protein as a normal constituent of cytoplasm, only small molecules are translocated. The sieve plates apparently retain molecules as large as proteins. Another intriguing bit of phenomena is that sugars are *secreted* from parenchyma cells into the phloem. We say this because the concentration of sucrose in the phloem is typically 0.5 $M$, which is an order of magnitude greater than the concentration in the leaf parenchyma as a whole. Unlike the possibly analogous case of sugar transport across the intestine where the molecule is phosphorylated in the membranes (cf. Wilson, 1962), we know next to nothing about the mechanism of this secretion.

Although aphids may know no more than we concerning the mechanism of sucrose secretion into phloem, they have long been aware of the existence of this treasure trove. These insects have learned to drive their stylet into a succulent stem, twist it past all manner of uninteresting cells, until it encounters a sieve tube; the stylet then penetrates the sieve tube and the aphid waits patiently for the phloem sap to come surging into his body. The aphid does not even have to suck up the fluid; turgor pressure takes care of that.

Table 3-9  Transformation of Various Sugars to Sucrose by Canna Leaf Discs*

Leaf discs of *Canna indica* were fed $C^{14}$-labeled sugars and the carbohydrates analyzed at various times for radioactivity in sucrose, fructose, and glucose. The data are presented for times after the transformations had gone to completion (about 100 min).

| Sugar Fed to Leaves | $C^{14}$ Found in Sugar Fractions as % of Total | | | | |
|---|---|---|---|---|---|
| | Free Fructose | Free Glucose | Sucrose | Fructose Moiety of Sucrose | Glucose Moiety of Sucrose |
| $C^{14}$-Glucose | | 15 | 50 | 18 | 32 |
| $C^{14}$-Fructose | 5 | | 55 | 35 | 20 |
| $C^{14}$-Sucrose | 15 | | 55 | 37 | 23 |

* From Putman and Hassid, 1954.

To the occasional disadvantage of aphids, man has learned to profit from the aphid's sagacity. The aphid may be anesthetized with a stream of $CO_2$ and carefully excised from his stylet. Pure phloem sap continues to flow from the stylet stump at a rate of about 5 $\mu$l hr$^{-1}$. By this means, one can obtain accurate measurements of the composition of phloem sap and the identity of translocated material (Zimmermann, 1962) (Table 3-10).

A reasonable generalization is that nothing moves out of the leaf except through the phloem. The physiology of phloem is consequently crucial in determining what nutrients can be recovered from senescent leaves in the fall and in determining the pattern of nutrient deficiencies. For example, nitrogen in the form of glutamate, glutamine, aspartic acid, and asparagine appear in the phloem in the fall, as do low concentrations of phosphate and potassium. Deficiency signs of nitrogen, potassium, and phosphate correspondingly appear first in the older leaves. In contrast calcium, which does not move in the phloem and therefore cannot be moved about, shows its first deficiency sign in new leaves and roots.

Table 3-10 Typical Composition of Phloem
Sap of White Ash (*Fraxinus americana* L.)*

Sample collected by the aphid stylet method
at a height of 10 m. Phloem sap of *Fraxinus*
is more complex than that of most plants.

| Component | Concentration, $M$ |
|---|---|
| Sucrose | 0.132 |
| Raffinose | 0.088 |
| Stachyose | 0.205 |
| Mannitol | 0.097 |
| Total osmotic pressure | 0.7–0.75† |

* Zimmermann, 1962.
† A noncarbohydrate component of as much
as 0.2 $M$ was present but unidentified.

The physiology of phloem for the same reasons is also crucial to the
movement of herbicides and systemic insecticides (Crafts, 1961).

There is one credible hypothesis concerned with phloem transport which
predicts the direction of movement. The *Münch hypothesis* (Münch, 1930)
states that the movement of fluids in phloem is an instance of bulk flow induced
by hydrostatic pressures. We are to picture the leaves as a *source* in which
a hydrostatic pressure is generated by secretion of sucrose into the phloem.
Another part of the plant, say the roots, can then serve as a *sink* in which
sucrose is consumed. The removal of phloem sap at the sink is then the direct
cause of a decrease in hydrostatic pressure in that region. These differences
in pressure at the two ends of the phloem result in a bulk flow of fluid
(Fig. 3-31).

The lines of evidence in favor of the Münch hypothesis are as follows:

1. Correlation with changes in sources and sinks. In the spring the phloem
   sap of hardwood trees moves upward and out from the phloem parenchyma,
   where sugars have been stored, to the expanding leaves (Fig. 3-32). As
   the leaves mature in late spring the sap movements cease, and then, as
   the leaves begin to produce more sugar than they consume, the direction
   of movement is reversed, and phloem flows back down the stem to newly
   formed and recent phloem parenchyma and to the roots. Also concentration
   gradients within the phloem are always in the direction of flow.
2. The rate of phloem transport may be on the order of 100 cm hr$^{-1}$. This
   rate is too fast for diffusion by a factor of at least $10^4$ and too fast for
   cyclosis (cytoplasmic streaming) by a factor of 10. Besides, cyclosis seems
   not to occur in mature phloem. This leaves mass flow as the only obvious
   alternative (cf. Zimmermann, 1964; Geiger, 1966).

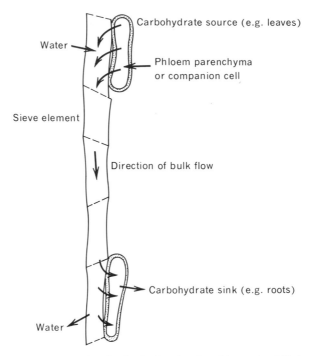

*Fig. 3-31* Model for bulk flow in the phloem (modified Münch hypothesis). At a source, carbohydrates are pumped into a sieve element from a phloem parenchyma or companion cell. The high solute concentration results in a high hydrostatic pressure. In another part of the plant (a sink) carbohydrates are consumed. The withdrawal of carbohydrates from the phloem causes a decreased hydrostatic pressure, resulting in a gradient of hydrostatic pressure in and consequent mass flow of liquid in the sieve elements.

3. The volumes of flow observed from fluid stylets are on the order of 5 $\mu$l hr$^{-1}$ (Zimmermann, 1964b) which means that the sieve elements must be refilled 3 to 10 times per second! This rate, too, calls for mass flow.

4. Münch predicted that, when sugars were removed from the phloem sap, water would have to be removed as well to the extent of about 5 percent of the volume of the transpiration stream. This prediction has been confirmed by Brown (1964).

There have been reports of translocation of different substances in two directions at once (cf. Biddulph and Cary, 1960). If we could establish that this occurred in the same element, it would certainly contradict the Münch hypothesis. However, these reports were based on relatively long term experiments in which the possibility of reverse movements in the xylem was not rigorously excluded.

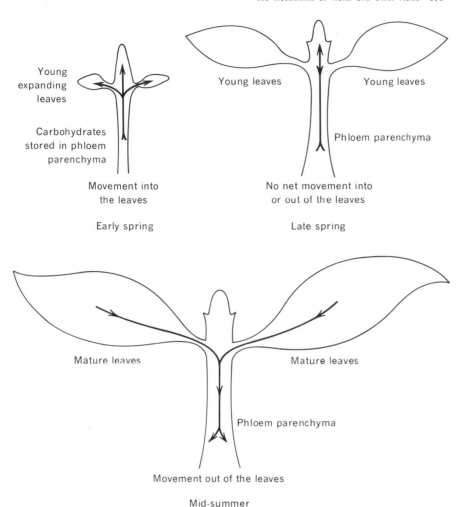

*Fig. 3-32* The direction of phloem movements during the growth of deciduous trees (cf. Zimmermann, 1964b). The direction of movement of phloem sap is always in the direction of carbohydrate consumption and away from sites of carbohydrate formation.

There is a remarkable specificity of movement in the phloem correlated with plant anatomy. The anatomy of phloem and xylem bundles can be traced from one node to another in strict species-characteristic patterns. This pattern is called the *orthostichy*. When $P^{32}O_4$ is fed to a single sunflower leaf, the radioactivity appears in a single arc of seeds in the head. Similarly the partial defoliation of a branch results in the almost immediate cessation of phloem movement in the phloem on that side (Zimmermann, 1958, 1961) (Fig. 3-33). We may infer therefore that, in striking contrast to the situation in the xylem, only a negligible amount of lateral movement occurs in the phloem.

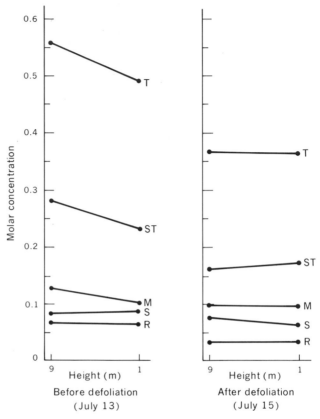

*Fig. 3-33* Cessation of concentration gradients in phloem sap of white ash (*Fraxinus americana* L.) following defoliation (Zimmermann, 1958). The concentrations of several carbohydrates in phloem sap were determined at 9 m and 1 m up the trunk of the tree.

M = mannitol
ST = stachyose
 S = sucrose
 R = raffinose
 T = total concentration

In conclusion, the model of phloem movements, as one of bulk flow governed by hydrostatic pressures, is an adequate one as far as it goes, but we are severely deficient in our understanding of the transport processes across the phloem membranes, the secretion of organic solutes from leaves, and their removal from the phloem at the sites of utilization.

### Polar Transport of Indol-3-Acetic Acid (IAA)

IAA, one of the naturally occurring growth substances (Sec. 5.5) is rapidly transported from the apex to the base of plant organs (e.g., coleoptiles), but not in the opposite direction. IAA transport is a highly specific process, dependent on metabolism, and important in the growth process itself.

What are the physical parameters of IAA uptake and transport? How does one distinguish between diffusion and transport, per se? What hypotheses would account for the high velocity, specificity, and polarity of the process? What evidence would you require to test these hypotheses?

*References:* Goldsmith, 1967ab, 1968; McCready, 1966.

### Diffusion-limited Respiration in Bulky Tissues

One can ask if the diffusion of oxygen into storage organs, such as carrots or potatoes, limits the rate of respiration. Data are available on the diffusion constants for gases through such tissues. Given these data and Fick's law, suitably modified to provide for a given rate of $O_2$ consumption in each layer from the exterior to the interior, one can ask how respiration will be controlled by the geometry of the tissue.

Approaches of this sort have shown excellent agreement between predictions from diffusion theory and observation. In fact, calculations of maximum permissible thickness of tissue slices were found to be essential to the rational testing of hypotheses concerning the identity of terminal oxidases.

*References:* Stiles, 1960. See also D. R. Goddard and W. D. Bonner, Jr., 1960: Cellular respiration, in F. C. Steward (ed.), "Plant Physiology," vol. IA, pp. 219–224, Academic Press, New York.

## REFERENCES

### General References

Bennet-Clark, T. A., 1959. Water relations of cells, in F. C. Steward (ed.), "Plant Physiology," vol. II, pp. 105–191, Academic Press, New York.

Briggs, G. E., 1967. "Movement of Water in Plants," Blackwell, Oxford.

Collander, R., 1959. Cell membranes: Their resistance to penetration and their capacity for transport, in F. C. Steward (ed.), "Plant Physiology," vol. II, pp. 3–102, Academic Press, New York.

Crafts, A. S., 1961. "Translocation in Plants," 182 pp., Holt, Rinehart & Winston, New York.

Dainty, J., 1963. Water relations of plant cells, in R. D. Preston (ed.), *Adv. Botan. Res.,* **1**:279–326.

Davson, H., and R. Danielli, 1952. "The Permeability of Natural Membranes," 365 pp., Cambridge Univ. Press, London.

Katchalsky, A., and P. E. Curran, 1965. "Nonequilibrium Thermodynamics in Biophysics," 248 pp., Harvard Univ. Press, Cambridge, Mass.

Zelitch, I., 1965a. Environmental and biochemical control of stomatal movement in leaves, *Biol. Rev.,* **40**:463–482.

———, 1969. Stomatal control, *Ann. Rev. Plant Physiol.,* **20**:329–350.

Zimmermann, M. H., 1961. Movement of organic substances in trees, *Science,* **133**:73–79.

———, 1963. How sap moves in trees, *Sci. Am.,* **208**:133–142.

### Specific References

Benson, A. A., 1964. Plant membrane lipids, *Ann. Rev. Plant Physiol.,* **14**:1–16.

Biologlowski, J., 1936. Effect of extent and temperature of roots on transpiration of rooted lemon cuttings, *Proc. Am. Soc. Hort. Sci.,* **34**:96–102.

Boyer, J. S., 1967. Matric potentials of leaves, *Plant Physiol.,* **42**:213–217.

——— and E. B. Knipling, 1965. Isopiestic technique for measuring leaf water potentials with a thermocouple psychrometer, *Proc. Nat. Acad. Sci.,* **54**:1044–1051.

Brouwer, R., 1953. *Koninkl. Ned. Akad. Wetenschap. Proc.* **56C**:106–115.

———, 1954a. Water absorption by roots of *Vicia faba* at various transpiration strengths: III. Changes in water conductivity artificially obtained, *Koninkl. Ned. Akad. Wetenschap. Proc.,* **57C**:68–80.

———, 1954b. The regulating influence of transpiration and suction tension on water and salt uptake of roots of intact *Vicia faba* plants, *Acta Botan. Neerl.,* **3**:264–312.

Brown, C. L., 1964. The influence of external pressure on the differentiation of cells and tissues cultured *in vitro,* in M. H. Zimmermann (ed.), "The Formation of Wood in Forest Trees," pp. 389–404, Academic Press, New York.

Brown, H. T., and F. Escombe, 1900. *Phil. Trans. Roy. Soc. London,* **B193**:223.

Buswell, A. M., and N. H. Rodebush, 1956. Water, *Sci. Am.,* **197**:77.

Clark, W. E., 1962. Prediction of ultrafiltration membrane performance, *Science,* **138**:148–149.

Coe, E. L., and M. H. Coe, 1965. A hypothesis relating oil:water partition coefficients and vapor pressures of nonelectrolytes to their penetration rates through biological membranes, *J. Theoret. Biol.,* **8**:327–343.

Collander, R., 1957. Permeability of plant cells, *Ann. Rev. Plant Physiol.,* **8**:335–348.

Dainty, J., and A. B. Hope, 1959. The water permeability of cells of *Chara australis* R. Br., *Aust. J. Biol. Sci.,* **12**:136–145.

Dick, D. A. T., 1964. The permeability coefficient of water in the cell membrane and the diffusion coefficient in the cell interior, *J. Theoret. Biol.,* **7**:504–531.

Dimond, A. E., 1966. Pressure and flow relations in vascular bundles of the tomato plant, *Plant Physiol.,* **41**:119–131.

Ekert, P. C., 1965. Evapotranspiration of pineapple in Hawaii, *Plant Physiol.,* **40**:736–739.

Esau, K., 1964a. Structure and development of the bark in dicotyledons, in M. H. Zimmermann (ed.), "The Formation of Wood in Forest Trees," pp. 37–50, Academic Press, New York.

———, 1964b. Aspects of ultrastructure of phloem, *ibid.,* pp. 51–63.

———, 1965. "Plant Anatomy," 2d ed., 767 pp., John Wiley & Sons, New York.

———, H. B. Currier, and V. I. Cheadle, 1957. Physiology of phloem, *Ann. Rev. Plant Physiol.,* **8**:349–374.

Firbas, F., 1931. Über die Ausbildung des Leitungssystems und das Verhalten der Spaltöffnungen im Frühjahr bei Pflanzen des Mediterrangebietes und der tunesischen Steppen und Wüsten, *Beit. bot. Cbl.,* **48**(1):451–465.

Gaastra, P., 1959. Application of diffusion equation, *Meded. Landb. Wag.,* **59**:1.

Gaff, D. F., T. C. Chambers, and K. Markus, 1964. Studies of extrafasicular movement of water in the leaf, *Aust. J. Biol. Sci.,* **17**:581–586.

Gale, J., and A. Poljakoff-Mayber, 1967. Plastic films on plants as antitranspirants, *Science,* **156**:650–652.

Geiger, D. R., 1966. Effect of sink region cooling on translocation of photosynthate, *Plant Physiol.,* **41**:1667–1672.

Glinka, Z., and L. Rinehold, 1964. Reversible changes in the hydraulic permeability of plant cell membranes, *Plant Physiol.,* **39**:1043–1050.

Goldsmith, M. H. M., 1967a. Movement of pulses of labeled auxin in corn coleoptiles, *Plant Physiol.,* **42**:258–263.

———, 1967b. Separation of transit of auxin from uptake: average velocity and reversible inhibition by anaerobic conditions, *Science,* **156**:661–663.

———, 1968. The transport of auxin, *Ann. Rev. Plant Physiol.,* **19**:347–360.

Greenidge, K. N. H., 1952. An approach to the study of vessel length in hardwood species, *Amer. J. Botany,* **39:**570–574.

Gutknecht, J., 1967. Membranes of *Valonia ventricosa:* Apparent absence of water-filled pores, *Science,* **158:**787–788.

Hartt, C. E., and H. P. Kortschak, 1964. Sugar gradients and translocation of sucrose in detached blades of sugarcane, *Plant Physiol.,* **39:**460–474.

Hatch, M. D., and K. T. Glasziou, 1964. Direct evidence for translocation of sucrose in sugarcane leaves and stems, *Plant Physiol.,* **39:**180–184.

Heath, O. V. S., 1959. The water relations of stomatal cells and the mechanisms of stomatal movement, in F. C. Steward (ed.), "Plant Physiology," vol. II, pp. 193–250, Academic Press, New York.

Huber, B., 1956. Die Gefässleitung, in W. Ruhland (ed.), "Handb. Pflanzenphysiol," vol. III, pp. 511–513, Springer-Verlag, Berlin.

Katchalsky, A., 1961. Membrane permeability and thermodynamics of irreversible processes, in A. Kleinzeller and A. Kotyk (eds.), "Membrane Transport and Metabolism," pp. 69–86, Academic Press, New York.

Kohn, P. G., and J. Dainty, 1966. The measurement of permeability to water in disks of storage tissues, *J. Exp. Botany,* **17:**809–821.

Kramer, P. J., 1940. Root resistance as a cause of decreased water absorption by plants at low temperatures, *Plant Physiol.,* **15:**63–79.

———, 1942. Species differences with respect to water absorption at low soil temperatures, *Am. J. Botany,* **29:**828–837.

———, 1955. Water relations of plant cells and tissues, *Ann. Rev. Plant Physiol.,* **6:**253–272.

———, 1956. Roots as absorbing organs, in W. Ruhland (ed.), "Encyclopedia of Plant Physiology," vol. III, pp. 188–214, Springer-Verlag, Berlin.

———, 1959. Transpiration and the water economy of plants, in F. C. Steward (ed.), "Plant Physiology," vol. II, pp. 607–726, Academic Press, New York.

———, E. B. Knipling, and L. N. Miller, 1966. Terminology of cell-water relations, *Science,* **153:**889–890.

Kreeb, K., 1965. Studies on the osmotic constants: II. An electronic method for measuring the diffusion pressure deficit (DPD) (NTC-Method), *Planta,* **66:**156–164.

Kuiper, P. J. C., 1964. Dependence upon wavelength of stomatal movement in epidermal tissue of *Senecio odoris, Plant Physiol.,* **39:**952–955.

Laties, G. G., and K. Budd, 1964. The development of differential permeability in isolated steles of corn roots, *Proc. Nat. Acad. Sci.,* **62:**462–469.

Longuet-Higgens, H. C., and G. Austin, 1966. The kinetics of osmotic transport through pores of molecular dimensions, *Biophys. J.,* **6:**217–224.

McCready, C. C., 1966. Translocation of growth regulators, *Ann. Rev. Plant Physiol.,* **17:**283–294.

Meyer, B. S., 1956. The hydrodynamic system, in W. Ruhland (ed.), "Encyclopedia of Plant Physiology," vol. III, pp. 596–614, Springer-Verlag, Berlin.

Münch, E., 1930. "Der Stoffbewegungen in der Pflanze," Fisher, Jena.

Nakayama, F. S., and W. L. Ehrler, 1964. Beta ray gauging technique for measuring leaf water content changes and moisture status of plants, *Plant Physiol.,* **39:**95–97.

Ogston, A. G., 1966. On water binding, *Fed. Proc.,* **25:**986–989.

Penman, H. L., and R. K. Schofield, 1951. Some physical aspects of assimilation and transpiration, *Soc. Exp. Biol. Symp.,* **5:**115–129.

Putman, E. W., and W. Z. Hassid, 1954. Sugar transformations in leaves of *Canna indica:* Synthesis and inversion of sucrose, *J. Biol. Chem.,* **207:**885–902.

Raney, F., and Y. Vaadia, 1965a. Movement of tritiated water in the root system of *Helianthus annuus* in the presence and absence of transpiration, *Plant Physiol.,* **40:**378–382.

———— and ————, 1965b. Movement and distribution of THO in tissue water and vapor transpirated by shoots of *Helianthus* and *Nicotiana, Plant Physiol.,* **40:**383–388.

Ray, P. M., 1960. On the theory of osmotic water movement, *Plant Physiol.,* **35:**783–795.

Richards, L. A., and C. H. Wadleigh, 1952. Soil water and plant growth, in B. T. Shaw (ed.), "Soil Physical Conditions and Plant Growth," pp. 73–224, Academic Press, New York.

Robertson, J. D., 1960. The molecular structure and contact relationships of cell membranes, in J. A. V. Butler and B. Katz (eds.), "Progress in Biophysics and Biophysical Chemistry," vol. 10, pp. 343–418, Pergamon Press, New York.

————, 1966. The organization of cellular membranes, in J. M. Allen (ed.), "Molecular Organization and Biological Function," pp. 65–106, Harper & Row, New York.

Rothfield, L., and A. Finkelstein, 1968. Membrane biochemistry, *Ann. Rev. Biochem.,* **37:**463–496.

Russell, M. B., 1959. Water and its relation to soils and crops, *Adv. Agron.,* **11:**1–131.

Scholander, P. F., B. Ruud, and H. Leivestad, 1957. The rise of sap in a tropical liana, *Plant Physiol.,* **32:**1–6.

————, H. T. Hammel, E. D. Bradstreet, and E. A. Hemingsen, 1965. Sap pressure in vascular plants: negative hydrostatic pressure can be measured in plants, *Science,* **148:**339–346.

Shimshi, D., 1963. Effect of soil moisture and phenylmercuric acetate upon stomatal aperture, transpiration, and photosynthesis, *Plant Physiol.,* **38:**713–721.

Ståhfelt, M. G., 1956a. Die stomatäre Transpiration und die Physiologie der Spaltöffnungen, in W. Ruhland (ed.), "Handb. Pflanzenphysiol," vol. III, pp. 351–426, Springer-Verlag, Berlin.

————, 1956b. Die cuticuläre Transpiration, *ibid.,* pp. 342–350.

Stein, W. D., 1967. "The Movement of Molecules across Cell Membranes," 309 pp., Academic Press, New York.

Stiles, W., 1960. The composition of the atmosphere (oxygen content of air, water, soil, intercellular spaces, diffusion, carbon dioxide and oxygen tension), in W. Ruhland (ed.), "Encyclopedia of Plant Physiology," vol. XII/2, pp. 114–148, Springer-Verlag, Berlin.

Stocking, C. R., 1956. The state of water in cells and tissues, *ibid.,* pp. 15–21.

Thimann, K. V., and E. W. Samuel, 1955. The permeability of potato tissue to water, *Proc. Natl. Acad. Sci.,* **41**:1029–1033.

Ting, I. P., and W. E. Loomis, 1965. Further studies concerning stomatal diffusion, *Plant Physiol.,* **40**:220–228.

Ussing, H. H., 1954. Membrane structure as revealed by permeability studies, in Kitching (ed.), "Recent Developments in Cell Physiology," Colson Papers, 7th Symposium.

Veihmeyer, F. J., 1956. Soil moisture, in W. Ruhland (ed.), "Encyclopedia of Plant Physiology," vol. III, pp. 64–123, Springer-Verlag, Berlin.

Whiteman, P. C., and K. Koller, 1964. Saturation deficit of the mesophyll evaporating surfaces in a desert halophyte, *Science,* **146**:1320–1321.

Wiebe, H. W., 1966. Matric potential of several plant tissues and biocolloids, *Plant Physiol.,* **41**:1439–1442.

Wilson, T. H., 1962. "Intestinal Absorption," 263 pp., W. B. Saunders Co., Philadelphia.

Woolley, J. T., 1965. Radial exchange of labeled water in intact maize roots, *Plant Physiol.,* **40**:711–717.

Zelitch, I., 1959. The relationship of glycolic acid to respiration and photosynthesis in tobacco leaves, *J. Biol. Chem.,* **234**:3077–3081.

————, 1961. Biochemical control of stomatal opening in leaves, *Proc. Natl. Acad. Sci.,* **47**:1423–1433.

————, (ed.), 1963. Stomata and water relations in plants, *Bull.* #664, Conn. Agr. Exp. Sta., New Haven.

————, 1965b. Biochemical control of stomatal opening and the synthesis of glycolic acid in leaves, *Fed. Proc.,* **24**:868–872.

————, and P. E. Waggoner, 1962a. Effect of chemical stomata on transpiration and photosynthesis, *Proc. Nat. Acad. Sci.,* **48**:1101–1108.

———— and ————, 1962b. Effect of chemical control of stomata on transpiration of intact plants, *Proc. Nat. Acad. Sci.,* **48**:1297–1299.

———— and D. A. Walker, 1963. Some effects of metabolic inhibitors, temperature, and anaerobic conditions on stomatal movement, *Plant Physiol.,* **38**:390–396.

———— and ————, 1964. The role of glycolic acid metabolism in opening of leaf stomata, *Plant Physiol.*, **39**:856–862.

Zimmermann, M. H., 1958. Translocation of organic substances in trees: III. The removal of sugars from sieve tubes in the white ash (*Fraxinus americana* L.), *Plant Physiol.*, **33**:213–217.

————, 1963. How sap moves in trees, *Sci. Am.*, **208**(March):132–142.

————, 1964a. The effect of low temperature on ascent of sap in trees, *Plant Physiol.*, **39**:568–572.

————, 1964b. The relation of transport to growth in dicotyledonous trees, in M. H. Zimmermann (ed.), "The Formation of Wood in Forest Trees," pp. 289–301, Academic Press, New York.

# 4
# Inorganic Nutrition

## 4.1   INTRODUCTION

The study of the inorganic nutrient requirements of green plants and fungi is enormously easier than that of other kinds of organisms. The reason is that green plants and fungi typically require very little in the way of organic growth factors, so that growth media may be prepared in high states of purity. Consequently the phenomenology of inorganic plant nutrition was already well advanced in the nineteenth century. It became such a popular field that by the mid-twentieth century, there were few agriculturally significant plants whose macronutrient requirements had not been described and no major crop that had escaped at least cursory examination of micronutrient requirements as well. We shall try to define "macronutrient" and "micronutrient" below.

Despite the luxuriant crops of "N-P-K" studies and the persistent search for more efficient means of supplying "trace metals," we have faltered in exploiting plants for the pursuit of theoretical information on the roles of nutrient elements and their dynamism in the life of the plant. As we shall learn, there are fascinating examples of the deductive use of chemistry in tracing metal ions in plant metabolism. There are also important discoveries in plants that relate the action of inorganic nutrients to the most fundamental biological processes. But curiously, much of what is known of the generalities of inorganic nutrients has been derived from studies on mammals, even though such discoveries could often have been done more directly, more easily, more elegantly, and earlier with plants as test organisms (McCollum, 1957).

### Essentiality

In general we shall concern ourselves only with the nutrients that are *essential*. As with most notions in science, the idea of essentiality is one which seems at first to be an uncomplicated idea, but becomes a little difficult when we try to define it.

Arnon and Stout (1939) proposed a set of necessary and sufficient criteria for the essentiality of nutrients:

"*a*) A deficiency of the element makes it impossible for the plant to complete the vegetative or reproductive stage of its life cycle; *b*) such deficiency is specific to the element in question, and can be prevented or corrected only by supplying this element; and *c*) the element is directly involved in the nutrition of the plant quite apart from its possible effects in correcting some unfavorable microbiological or chemical condition of the soil or other culture medium. From that standpoint a favorable response from adding a given element to the culture medium does not constitute conclusive evidence of its indispensability in plant nutrition."

The criteria of Arnon and Stout are certainly sufficient, but I am not sure they are exclusive and therefore necessary. The requirement that a complete life cycle should not occur in the absence of a nutrient works a hardship on our friends in animal physiology. Until recently, there were few half-way satisfactory demonstrations of the Arnon-type essentiality of zinc in animals. In Elvejem's famous experiments with rats (cf. Underwood, 1962; McCollum, 1957), the diet was so modified that the control rats given zinc grew rather poorly. Again, "porcine keratosis" has been identified as a zinc deficiency, but one that is induced by an unnaturally high level of calcium in the rations fed to pigs. But because zinc is an indispensable part of certain NAD-linked dehydrogenases, we need have no doubt that zinc is required for life (and therefore is essential). Again, in the digestive tract of the horse, grasses are fermented to a number of primary and secondary alcohols. A horse without liver alcohol dehydrogenase would have no means of processing these alcohols and would consequently never have become man's sober and faithful servant. Therefore, if zinc is an essential component of liver alcohol dehydrogenase, then zinc is essential to the horse.

It seems only reasonable to add therefore that *a nutrient is also essential if it is an indispensable part of an indispensable component of the organism.*

### Quantitative Requirements

The necessary evidence to prove the absence of a requirement for an element is simple enough to specify: that the organism will grow normally with less than one atom per organism. This requirement has of course been impossible to meet, even if we were to modify it to read "one atom per cell."[1] Rather than conclude that an element is nonessential, we have to state that the plant requires "less than $x$ $\mu$moles per gram dry weight." Thus the notion of *nonessentiality* is strictly operational and quantitative.

### Quantitation—Macronutrients and Micronutrients

Information on the quantities of nutrients required by plants is of enormous theoretical and practical importance. It is clearly not enough to know how much of a nutrient is necessary for a plant merely to survive. We usually specify that there is a sufficiency of a given nutrient when growth is no longer limited by the nutrient, but we may wish to vary this definition from time to time. For example, we may inquire of the *quantity of nutrient* required for normal chlorophyll synthesis, for an optimum root-to-shoot ratio, etc.

The analytical base on which we calculate quantities is also important. Stiles (1916) clearly stated that we must distinguish between the effects of *concentrations* and of *amounts* of nutrients (cf. also Nobbé, 1865).

In the simplest view, a plant responds to changes in the concentrations of nutrients by changes in the rates of plant processes, and to changes in the amounts of nutrients by changes in the amounts of chemicals transformed.

[1] Each cell of the alga *Scenedesmus* requires on the order of 15 atoms of molybdenum.

A mathematical model for responses to changes in the concentrations of nutrients in soil or culture solutions, all other factors held constant, is

$$\frac{dp}{dt} = f_e(c_e)$$                                Eq. 4-1

$$p = \int_{t_1}^{t_2} f_e(c_e)\, dt$$                        Eq. 4-2

where $p$ is the plant variable in question, $c_e$ is the external concentration of nutrient in the external medium, and $t_2 - t_1$ is the interval of observation. Unless continuous-flow procedures are employed (Johnston and Hoagland, 1929; Asher et al., 1965), $c_e$ is neither constant nor a known function of $t$ over the course of the experiment, and it becomes impossible to deduce $f_e(c_e)$ from merely observing $p$. For example, the yield of oats in relation to different amounts of phosphate in the nutrient medium, even with the medium replenished from time to time, provides negligible information on the rate of growth of oats as a function of phosphate concentration.

For nutrients that can be accumulated or translocated, the responses of plants are equally dependent on the history of the plant. For example, luxury consumption of zinc (Knauss and Porter, 1954; Scharrer, 1941) permits a plant to grow normally, with levels of zinc in the medium which would otherwise produce deficiency. Marine algae can store up enough phosphate to prosper for several generations in the absence of detectable phosphate in the surrounding sea water. Equations 4-1 and 4-2, therefore, are meaningful for such nutrients only under conditions of continuous culture as well as continuous flow; that is, when a constant rate of growth is maintained by a constant limiting concentration of the nutrient, such as in the chemostat (T. W. James, 1961).

A second approach to quantitating the response of a plant to a nutrient is to determine the rate of a plant process (for example, growth) as a function of the concentration of the nutrient within the plant. This approach is the basis for *foliar analysis*.

The equation, analogous to Eq. 4-1, that relates the rate of a process to the internal concentration of a nutrient is then

$$\frac{dp}{dt} = f_i(c_i)$$                                Eq. 4-3

where $c_i$ is the concentration of the nutrient within the plant.

It has been useful to distinguish between those nutrients required in large quantity—such as nitrogen, phosphate, potassium, and calcium—from those required in much smaller quantity—manganese, zinc, copper, and boron. Following the suggestion of Hoagland (1944), we speak of *macronutrients* and *micronutrients*. But, if only to keep our concepts straight, it is important to specify the analytical base, which is the amount of nutrient required per gram of plant material. The reason that we use per unit of cell material (rather than per unit of culture medium) is that these amounts dictate directly or indirectly the composition of nutrient media in all but the most exquisitely designed experiments. Also, we simply do not know what are the true limiting concentrations in media for most nutrients.

A list of macronutrients and micronutrients is presented in Table 4-1.

Table 4-1   Critical Amounts Required for the Growth of Multicellular Plants*

The amounts shown are averaged values for many plants and ignore special requirements or requirements exaggerated by competition from other nutrient species. *Critical* amounts denote the amounts that are barely adequate for normal growth and development. The upper set are usually referred to as *micronutrients,* the middle set as *macronutrients.* The lower set might not be regarded as nutrients in the usual sense since they are present in the atmosphere or as water, but they are included for sake of completeness.

| Element | Concentration in Dry Matter† | | Relative Number of Atoms with Respect to Molybdenum |
| --- | --- | --- | --- |
| | µatoms/g | µg/g | |
| Molybdenum | 0.001 | 0.1 | 1 |
| Copper | 0.1 | 6 | 100 |
| Zinc | 0.3 | 20 | 300 |
| Manganese | 1.0 | 50 | 1,000 |
| Iron | 2.0 | 100 | 2,000 |
| Boron | 2.0 | 20 | 2,000 |
| Chlorine | 3.0 | 100 | 3,000 |
| Sulfur | 30 | 1,000 | 30,000 |
| Phosphorus | 60 | 2,000 | 60,000 |
| Magnesium | 80 | 2,000 | 80,000 |
| Calcium | 125 | 5,000 | 125,000 |
| Potassium | 250 | 10,000 | 250,000 |
| Nitrogen | 1,000 | 15,000 | 1,000,000 |
| Oxygen | 30,000 | 450,000 | 30,000,000 |
| Carbon | 35,000 | 450,000 | 35,000,000 |
| Hydrogen | 60,000 | 60,000 | 60,000,000 |

* After Stout, 1961.
† Because of variable amounts of cell walls and other inert components, it would probably be better to express requirements on the basis of protein.

## 4.2   PLANTS AS UNIQUE ORGANISMS FOR THE STUDY OF INORGANIC NUTRITION

The first advantage of green plants (including algae) and fungi for studying inorganic nutrition is purely methodological. Compare the nutrient media for growing typical green plants or algae with those for culturing a protozoan (Table 4-2). The protozoan *Tetrahymena pyriformis* is of special interest as the first animal grown in axenic culture, that is, without any other organism being present. The presence of large amounts of organic materials in the nutrition of an organism interferes both analytically and functionally when one

Table 4-2  Nutrient Media for a Typical Higher Plant, an Alga, and a Protozoan

Amounts of components in $\mu$moles per liter

| Higher Plant* | | *Peridinium* sp. (an algal dinoflagellate)† | | *Tetrahymena pyriformis* (a ciliate protozoan)‡ | |
|---|---|---|---|---|---|
| KNO₃ | 4,000 | | 500 | | |
| Ca(NO₃)₂ | 5,000 | CaCO₃ | 1,000 | | |
| | | CaCl₂ | 360 | 200 | |
| KH₂PO₄ | 1,000 | | 147 | 3,700 | |
| K₂HPO₄ | | | | 2,900 | |
| MgSO₄ | 2,000 | | 407 | 570 | |
| Fe | 9 | | 36 | 160 | |
| B | 23 | | | | |
| Mn | 4.6 | | 14.5 | 6.3 | |
| Zn | 3.8 | | 12 | 1.9 | |
| Cu | 0.3 | | 0.016 | 12 | |
| Mo | 0.1 | | 5.2 | | |
| Co | | | 0.17 | | |
| EDTA | | | 1,000 | | |
| | | NaH Glutamate§ | 6,900 | Amino acids | Alanine, arginine, aspartic acid, glycine, glutamic acid, histidine, isoleucine, leucine, lysine, methionine, phenylalanine, proline, serine, threonine, tryptophan, valine | approx. 10,000 each |
| | | Glycine§ | 9,400 | | | |
| | | Alanine§ | 7,800 | | | |
| | | Glucose§ | 3,400 | | | |
| | | | | Vitamins | Pantothenic acid, nicotinamide, pyridoxine, pyridoxal, pyridoxamine, riboflavin, pteroylglutamic acid, thiamine, biotin, choline, lipoic acid | 0.001–50 ecah |
| | | | | Nucleic acid precursors | Guanylic, adenylic, cytidylic acids, and uracil | approx. 20 each |
| | | | | | Glucose | 17,000 |
| | | | | | Sodium acetate | 12,000 |
| | | | | | Tween 80 | 24 g |

* Robbins, 1946 (cf. Hewitt, 1966).
† Provasoli and Pintner, 1953.
‡ Kidder and Dewey, 1951.
§ Promotes growth, but not essential.

wishes to study inorganic nutrients. How does one trace the path of ammonia in the presence of amino acids? Of phosphate in the presence of nucleotides? In addition, amino acids complex metals with great tenacity. It is extremely difficult to free amino acids of minute amounts of iron, manganese, copper, and zinc, so that the mere demonstration of deficiencies for these metals in animals becomes a monumental task, usually involving the use of highly unnatural diets. Moreover in the presence of a mixture of amino acids it is extremely difficult to calculate the effective concentration of a metal ion (see Sec. 4.3).

The requirements for substantial amounts of organic materials appear to have been confusing to organisms as well as to their analytical chemistry. Animals typically ingest inorganic nutrients which either are integral parts of organic nutrients or are tightly complexed with them. Although active transport can occur across the walls of the gut, inorganic nutrients are largely an unsought dividend of the animal's quest for carbohydrates, fats, and protein. Children are occasionally deficient in iron and then only from subsisting almost exclusively on cow's milk. Nutrient deficiency in a higher animal therefore is usually the result of an abnormal diet or some other pathological stress. Land plants and both phyto- and zoö-plankton, on the other hand, live in rather dilute environments. Since soil and water environments may vary enormously in the availability of essential ions, these forms have acquired highly efficient mechanisms for accumulating ions from their surroundings. In some cases the transport mechanisms are specific not only for selecting a class of ions but in accepting an essential ion (e.g., calcium) in preference to a nonessential one (e.g., strontium) (cf. Drescher-Kaden and Schwanitz, 1956). In other cases, plants that live among peculiarly high concentrations of an ion may exclude it (R. B. Walker, 1954) or show unusual accumulations of it (cf. Bollmann and Schwanitz, 1957). Higher plants, in addition, have developed the means of moving metals and other inorganic nutrients from one place to another *within the plant*. This phenomenon is discussed in each case study as *mobility*.

We have said that fungi, algae, and higher plants have advantages in studying inorganic nutrition. We did not include bacteria. Obviously, the microbial habit brings convenience and precision within the grasp of the investigator. But bacteria have certain curious disadvantages. For one thing, unless one goes to a rich brew of amino acids, the amount of protoplasm obtainable per unit volume of culture medium is usually less in the case of bacteria than in such luxuriant bloomers as the mold *Aspergillus* or the alga *Euglena gracilis*. The importance of this is that trace amounts of contaminants in the water or in medium components may completely satisfy a bacterial growth requirement for a micronutrient. Secondly, and quite unaccountably, the requirements of bacteria for many metals may be an order of magnitude less than the corresponding requirements of algae and fungi. The consequence of both these circumstances is that the growth requirements of bacteria for iron, manganese, zinc, and copper, are usually satisfied by the level of metal contamination in media and glassware that remains after all but the most exhaustive purification.

Specific organisms and groups of organisms are often advantageous for the study of individual nutrients (Fig. 4-1). Fungi typically need unusually high concentrations (ca. $10^{-5}$ $M$) of copper (for phenol oxidase?), whereas the requirement for *Chlorella* is lower (ca. $10^{-6}$ $M$) and that for *Euglena* is undetectable (less than $10^{-8}$ $M$). Requirements for calcium are at least ten-

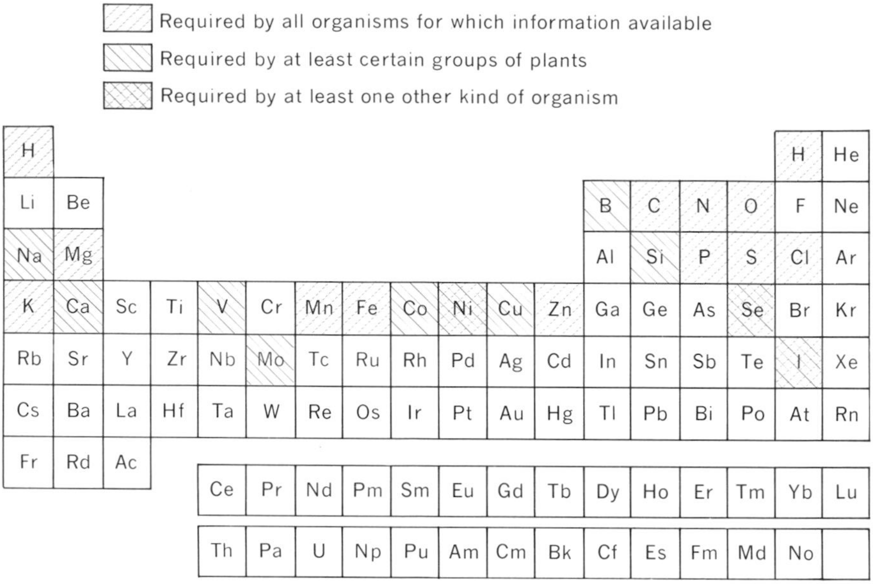

Fig. 4-1   A typical essentiality set: patterns of essentiality among organisms.

fold greater in multicellular plants than in fungi or unicellular forms. A requirement for boron has been detected only in multicellular plants and in diatoms. Molybdenum requirements have been demonstrated principally in plants that reduce nitrate or fix atmospheric nitrogen (cf. Evans, 1956). Inorganic cobalt is required only in plants that fix atmospheric nitrogen. There are many unicellular algae that require cobalt in the form of vitamin $B_{12}$, but no such requirement has ever been demonstrated for multicellular plants.

## 4.3   SOME CONCLUSIONS FROM CHELATE CHEMISTRY: HOW TO BE REACTIVE THOUGH COMPLEXED

The metals required for the growth of plants exist for the most part in cells and tissues as *complexes* and especially as *chelates*. Chelation is a well-developed branch of chemistry, and we shall discover that the fruits from this branch carry the seed of prophecy. That is, we can predict much of the composition and function of metals in biological systems from a knowledge of their chelation chemistry.

A *complex* is a combination of a metal ion and an electron donor, called a *ligand*. Bonds between metal and ligand are called coordination bonds and may be almost purely ionic, or high ligand-field, or purely covalent, or low ligand-field (Orgel, 1960). (Bond types used to be distinguished as "ionic" and "covalent" in coordination chemistry. These terms have been replaced by a concern for the identification of the orbitals involved.)

A chelating ligand is one containing two or more functional groups with each able to donate electrons to a single metal ion. The fact that the several functional groups of a chelating ligand are more or less fixed in space greatly increases the probability that the metal will form complexes with all of these groups simultaneously. This greatly increases the stability of chelating or multidentate over monodendate ligands (Fig. 4-2 and Table 4-3).

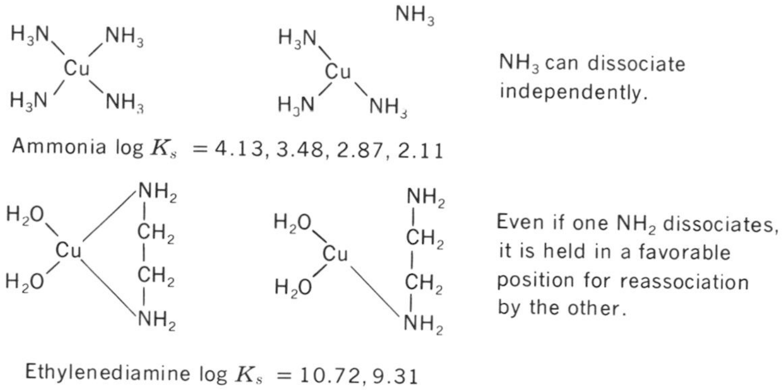

$$H_3N \diagdown \diagup NH_3$$
$$Cu$$
$$H_3N \diagup \diagdown NH_3$$

$$NH_3$$
$$H_3N \diagdown$$
$$Cu$$
$$H_3N \diagup \diagdown NH_3$$

NH₃ can dissociate independently.

Ammonia log $K_s$ = 4.13, 3.48, 2.87, 2.11

Ethylenediamine log $K_s$ = 10.72, 9.31

Even if one NH₂ dissociates, it is held in a favorable position for reassociation by the other.

Fig. 4-2   The stability of multidentate ligands (after Chaberek and Martell, 1959). When the number of possible chelate rings is increased, the stability of the chelate increases. An explanation of this effect is that the fixation in space of one end of the molecule increases the probability of another end forming a bond with the metal. The effect is illustrated with cupric amines.

Table 4-3   Stabilities of Mono- and Multidentate Ligands*

Increasing the number of rings increases the stability constants.

| | log $K_{MA}$ | | |
|---|---|---|---|
| Ligand | Cu(II) | Ni(II) | Co(II) |
| H₂NCH₂COOH | 8.6 | 6.2 | 5.2 |
| HN with CH₂COOH, CH₂COOH | 10.6 | 8.2 | 7.0 |
| N—CH₂COOH, CH₂COOH, CH₂COOH | 12.7 | 11.3 | 10.6 |

* After Chaberek and Martell, 1959.

Given equilibrium between metal ions and ligands

$$M^{+m} + L^{-n} \rightleftharpoons [ML^{m-n}]$$

the *stability* constant $K$ of a complex is simply defined as

$$K \equiv \frac{[ML^{m-n}]}{[M^{+m}][L^{-n}]} \qquad\qquad \text{Eq. 4-4}$$

Wyman (1965) has proposed a general mathematical formulation for complexation to include the binding of a variety of groups to macromolecules.

### Properties of Chelates

The following properties of metals in chelates are of importance in biological systems: coordination number, stability, redox potential, exchangeability, chemical reactivity, and color (of the whole chelate). We shall also be concerned with the relative stability or specificity of a chelating ligand for a series of metal ions.

A list of the functional groups typically found in ligands is shown in Table 4-4.

Table  4-4  Types  of  Functional Groups  in  Ligands*

| | |
|---|---|
| —O— | Enolate |
| —NH$_2$ | Amino |
| —N=N— | Azo |
| N | Ring N |
| $-C{\Large\diagup}^{\textstyle O}_{\diagdown O}$ | Carboxylate |
| —O— | Ether |
| $_\diagdown C{=}O$ | Carbonyl |
| —OH | Alcohol |
| —SH | Sulfhydryl |
| —SO$_2$O$^{\ominus}$ | Sulfonate |
| —PO$_3{}^{\ominus}$ | Phosphate |

*After  Chaberek  and  Martell, 1959.

### Coordination Number

The maximum number of functional groups with which a metal will complex is said to be the coordination number (Table 4-5). These numbers are obtained by comparing the stability constants for complexes of a metal with increasing

Table 4-5 Coordination Numbers of Some Biologically Important Metals*

| Coordination Number | Configuration | Metal Ions |
|---|---|---|
| 2 | Linear | $Cu^+$, $Ag^+$, $Hg^+$, $Hg^{++}$ |
| 4 | Tetrahedral | $Mg^{++}(?)$, $B^{3+}$, $Zn^{++}$, $Cd^{++}$, $Hg^{++}$ |
| 4 | Planar | $Co^{++}$, $Ni^{++}$, $Cu^{++}$, $Ag^{++}$ |
| 6 | Octahedral | $Mg^{++}(?)$, $Ca^{++}$, $Sr^{++}$, $Ba^{++}$, $Ti^{4+}$, $V^{4+}$ $Cr^{3+}$, $Mn^{++}$, $Mn^{3+}$, $Fe^{++}$, $Fe^{3+}$, $Co^{++}$, $Co^{3+}$ $Ni^{++}$, $Ni^{4+}$, $Cd^{++}$, $Zn^{++}(?)$, $Al^{3+}$, $Si^{4+}$ |
| 8 | Configuration in doubt | $Mo^{4+}$ |

* After Chaberek and Martell, 1959.

numbers of ligands found. Coordination numbers of a metal typically equal the number of electrons required to fill a noble gas shell. They may differ for different oxidation states, as in the case of copper.

## Stability

The stability constants for EDTA-metal chelates are shown in Table 4-6. The values for alkali metals are low and those for ions with occupied $d$-orbitals are high. This means that, except for alkali metals, the concentrations of free metal ions in the presence of equivalent EDTA would be vanishingly small. We say that EDTA binds metals very tightly.

For any given ligand, there is a predictable relation of stability constants for different metals. The log $K$ values among benzoylmethane chelates, for example, are a linear function of the ionization potentials of the metals (Fig. 4-3). On the other hand, thiol-containing ligands such as cysteine bind in relation to the solubility products of the corresponding metal sulfides:

$$Mg < Mn(II) < Fe(II) < Co(II) < Zn < Cu < Fe(III) < Hg$$

The structure and composition of the chelating ligand is obviously crucial to the stability of the chelate. The following factors contribute:

1. Number of chelate rings. The notion that entropy decreases when ligands are tied together led Schwarzenbach and Adamson (cf. Charberek and Martell, 1959, p. 144) to predict an increment of approximately 2 log $K$ units for each chelate ring formed compared with the corresponding monodentate complex. The additional stability conferred by the additional rings has been ascribed by R. J. P. Williams (1959) to the overcoming of repulsive polar forces among the complexing ligands.

2. The size of the chelate rings is critical. It must be long enough to accommodate the directed angles of electronic orbitals of the metal, but it must be short enough to contribute to the negative entropy and heat of reaction

Table 4-6  Stability Constants of Some Metal-EDTA Complexes*

$$^-OOCCH_2 \qquad\qquad CH_2COO^-$$
$$H-\overset{+}{N}CH_2CH_2\overset{+}{N}-H$$
$$HOOCCH_2 \qquad\qquad CH_2COOH$$

EDTA expressed in the table as $H_4A$

| Metal Ion | Metal Chelate Formed | Equilibrium Constant | |
|---|---|---|---|
| | | Formulation | Value (log) |
| H(I) | HA | [HA]/[H][A] | 10.26 |
| | $H_2A$ | [$H_2A$]/[H][HA] | 6.16 |
| | $H_3A$ | [$H_3A$]/[H][$H_2A$] | 2.67 |
| | $H_4A$ | [$H_4A$]/[H][$H_3A$] | 2.00 |
| Li(I) | MA | [LiA]/[Li][A] | 2.79 |
| Na(I) | MA | [NaA]/[Na][A] | 1.66 |
| Mg(II) | MA | [MgA]/[Mg][A] | 8.69 |
| | MHA | [MgHA]/[Mg][HA] | 2.28 |
| | $M_2A$ | [$Mg_2A$]/[MgA][Mg] | 0.70 |
| Ca(II) | MA | [CaA]/[Ca][A] | 10.59 |
| | MHA | [CaHA]/[Ca][HA] | 3.51 |
| Sr(II) | MA | [SrA]/[Sr][A] | 8.63 |
| | MHA | [SrHA]/[Sr][HA] | 2.30 |
| Ba(II) | MA | [BaA]/[Ba][A] | 7.76 |
| | MHA | [BaHA]/[Ba][HA] | 2.07 |
| Mn(II) | MA | [MnA]/[Mn][A] | 14.04 |
| | MHA | [MnHA]/[MnA][H] | 0.47 |
| Fe(II) | MA | [FeA]/[Fe][A] | 14.33 |
| | MHA | [FeHA] [FeA][H] | 1.31 |
| | MAOH | [FeAOH]/[FeA][OH] | 4.9 |
| | MA(OH) | [FeA(OH)$_2$]/[FeAOH][OH] | 4.1 |
| Fe(III) | MA | [FeA]/[Fe][A] | 25.1 |
| | MAOH | [FeAOH]/[FeA][OH] | 6.45 |
| | MA(OH) | [FeA(OH)$_2$]/[FeAOH][OH] | 4.53 |
| Co(II) | MA | [CoA]/[Co][A] | 16.31 |
| | MHA | [CoHA]/[CoA][H] | 3.09 |
| Ni(II) | MA | [NiA]/[Ni][A] | 18.62 |
| | MHA | [NiHA]/[NiA][H] | 5.20 |
| Cu(II) | MA | [CuA]/[Cu][A] | 18.80 |
| | MHA | [CuHA]/[CuA][H] | 5.58 |
| Zn(II) | MA | [ZnA]/[Zn][A] | 16.50 |
| | MHA | [ZnHA]/[ZnA][H] | 3.28 |
| Cd(II) | MA | [CdA]/[Cd][A] | 16.46 |
| | MHA | [CdHA]/[CdA][H] | 3.34 |
| Hg(II) | MA | [HgA]/[Hg][A] | 21.8 |
| | MHA | [HgHA]/[HgA][H] | 8.48 |
| Pb(II) | MA | [PbA]/[Pb][A] | 18.04 |
| | MHA | [PbHA]/[PbA][H] | 5.02 |
| Al(III) | MA | [AlA]/[Al][A] | 16.13 |
| | MHA | [AlHA]/[AlA][H] | 3.41 |
| | MAOH | [AlA]/[H][AlAOH] | 5.16 |

* After Chaberek and Martell, 1959.

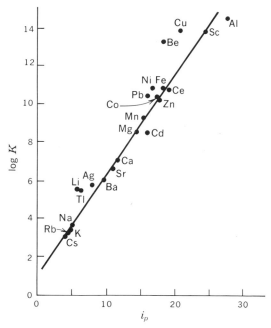

*Fig. 4-3* Stability constants of dibenzoylmethane chelates as a function of the ionization potentials of the metal ions (from Chaberek and Martell, 1959).

described in the paragraph above. Five-membered rings are typically the most stable.

3. The composition of the chelating ligand can have highly specific effects. For example, the addition of alcoholic hydroxyl groups usually adds little or strongly decreases stability, except for the Fe(III) chelates.[1] Dihydroxyethyl glycine binds most metals about as strongly as glycine, but Fe(III) is bound so tightly ($\log K = 30.1$) that the ferric complex is stable to pH 10.

Chelate chemistry has only whetted our appetite for understanding the specificity of natural chelates. Why, for example, do chlorophylls contain magnesium; cytochromes *c*, iron; carboxypeptidase, zinc; nitrate reductase, molybdenum; ascorbic acid oxidase, copper; and cyanocobalamin, cobalt?

Some of these specificities are doubtless due to uniquely high stabilities of the ligands for the metals. In the case of protoporphyrins however, either iron or magnesium is inserted, leading to heme and ultimately chlorophyll, respectively. Specific enzymes are responsible, but what protein structure would select $Mg^{++}$ out of a soup containing $Fe^{++}$?

---

[1] Ferrichrome (cf. Fig. 4-18) is an iron-storing polypeptide produced by the smut fungus *Ustilago sphaerogena* and is stable at pH 12!

In the case of carboxypeptidase, the zinc chelate is stronger than the iron or manganese chelate, but copper and mercury chelates are tighter still. We can imagine that zinc wins on the grounds of simple mass action, there being normally very low concentrations of copper or mercury ion in the pancreas, which is the site of synthesis.

Natural chelates containing invariably one kind of metal may result therefore from high stability constants for the metal, from the exclusion of competing ions, or from negligible exchangeability subsequent to synthesis (see below).

Bjerrum et al. (1957) and Chaberek and Martell (1959) have provided very useful tabulations of stability constants for natural and synthetic ligands.

## Oxidation Potential

Consider an inorganic oxidation-reduction pair, such as $Fe^{++}/Fe^{3+}$. The oxidation potential is an abstraction from the equilibrium of the oxidation and reduction of a second oxidation reduction pair:

$$Fe^{++} + X^+ \rightleftharpoons Fe^{3+} + X$$

We normally express the oxidation potential with respect to $H/H^+$.

$$Fe^{++} + H^+ \rightleftharpoons Fe^{3+} + \tfrac{1}{2}H_2 \qquad\qquad \text{Eq. 4-5}$$

But since not all oxidation-reduction systems do react with hydrogen, we employ the notion of a reaction with free electrons, so that Eq. 4-5 is the sum of two "half reactions."

$$\tfrac{1}{2}H \rightleftharpoons H^+ + e^- \qquad E^\circ = 0.000 \text{ volt} \qquad\qquad \text{Eq. 4-6}$$
$$Fe^{++} \rightleftharpoons Fe^{3+} + e^- \qquad E^\circ = 0.771 \text{ volt} \qquad\qquad \text{Eq. 4-7}$$

The potential of a cell is determined from the familiar Nernst equation:

$$E = E^\circ + \frac{RT}{NE} \log \frac{[\text{oxidized form}]}{[\text{reduced form}]} \qquad\qquad \text{Eq. 4-8}$$

where $E$ is the voltage between two given half cells, $E^\circ$ is that for half cells at unit concentration, $N$ is the number of electrons required for each oxidation, and $F$ is the Faraday constant.

In these expressions the electrons may be thought of as participants in the equilibria and that the "electron pressure" or electrical potential is analogous to the chemical potential of the reactants, specifically.

$$zFE^\circ = \mu_{Fe^{++}} + \mu_{Fe^{3+}} \qquad\qquad \text{Eq. 4-9}$$

where $z$ is the electron charge per atom and $F$ is the Faraday.

The essential point is that an oxidation-reduction pair is expressible as an equilibrium. It follows that anything which affects the equilibrium will also affect the oxidation potential.

A primary factor that affects the oxidation potential is pH. From Eq. 4-8 and our definition of hydrogen as the reference voltage,

$$E = 0.000 + \frac{RT}{F} \log \frac{[H^+]}{[H_2]^{\frac{1}{2}}}$$

At 1 atm of $H_2$ and pH 7

$$E = 0.000 + 0.059 \log \frac{10^{-7}}{1}$$

$$= -0.413 \text{ volt}$$

Then the oxidation potential of the hydrogen electrode at pH 7 is —0.413 volt.

Since physiological processes often occur near pH 7 and may involve one or more hydrogen ions, oxidation potentials are often calculated for pH 7 and are then reported as $E'$ with the corresponding standard potential as $E^{\circ\prime}$.

A second major factor affecting oxidation potentials is the presence of chelating agents.

Let us imagine that a ligand L capable of binding $Fe^{3+}$ and $Fe^{++}$ is introduced into the system. In the usual case L will bind $Fe^{3+}$ more tightly than it will $Fe^{++}$. The equilibrium between $Fe^{++}/Fe^{3+}$ will then be altered:

$$Fe^{++} \rightleftharpoons Fe^{3+}$$
$$+ \qquad +$$
$$L \qquad L$$
$$\Updownarrow \qquad \Updownarrow$$
$$[FeL]^{++} \quad [FeL]^{3+}$$

Since the oxidized form ($Fe^{3+}$) is stabilized, the presence of L will clearly raise the oxidation potential.

EXAMPLE 4-1.   In the range of pH 3.5 to 6.5, $Fe^{++}$ and $Fe^{3+}$ react with EDTA as follows:

$$Fe^{++} + EDTA^{4-} \rightleftharpoons Fe(EDTA)^{--} \qquad \log K_s = 14.3$$
$$Fe^{3+} + EDTA^{4-} \rightleftharpoons Fe(EDTA)^{-} \qquad \log K_s = 25.1$$

Calculate the oxidation potential of a mixture of $mM$ $Fe^{++}$, $mM$ $Fe^{3+}$, and 100 $mM$ $EDTA^{4-}$.

If no complexing agent were present the cell voltage would be calculated from Eq. 4-8:

$$E = -0.77 - \frac{0.059}{1} \log \frac{Fe^{3+}}{Fe^{++}}$$

$$= -0.77$$

In the presence of 100 m$M$ EDTA$^{4-}$, the concentration of the metal ion will be

$$Fe^{++} = \frac{[Fe(EDTA)^{--}]}{[EDTA^{4-}] \times 10^{14.3}}$$

and

$$Fe^{3+} = \frac{[Fe(EDTA)^{-}]}{[EDTA^{4-}] \times 10^{25.1}}$$

Equation 4-8 becomes

$$E = E_0 - 0.059 \log \left[ \frac{[Fe(EDTA)^{-}]/[EDTA^{4-}] \times 10^{25.1}}{[Fe(EDTA)^{--}]/[EDTA^{4-}] \times 10^{14.3}} \right]$$

$$= -0.77 - 0.059 \log \frac{10^{-3}}{10^{-1} \times 10^{25.1}} \frac{10^{-1} \times 10^{14.3}}{10^{-3}}$$

$$= -0.77 - 0.059 \log (10^{-10.8})$$

$$= -0.77 + 0.64$$

$$= -0.13 \text{ volt}$$

Ligands typically, but not invariably, stabilize preferentially the form with a higher valence and therefore raise the redox potential (Table 4-7). The

Table 4-7   Oxidation Potentials of Several Ions and Chelates*

| Oxidation-reduction Pair | | $E°$, volts |
|---|---|---|
| Co$^{++}$($aq$) | Co$^{3+}$($aq$) | −1.84 |
| Co(NH$_3$)$_6$$^{++}$ | Co(NH$_3$)$_6$$^{3+}$ | −0.10 |
| Co(CN)$_6$$^{4-}$ | Co(CN)$_6$$^{3-}$ | +0.83 |
| Mn$^{++}$($aq$) | Mn$^{3+}$($aq$) | −1.51 |
| Mn(CN)$_6$$^{4-}$ | Mn(CN)$_6$$^{3-}$ | +0.22 |
| Fe$^{++}$($aq$) | Fe$^{3+}$($aq$) | −0.77 |
| FeEDTA$^{--}$ | FeEDTA$^{-}$ | −0.13 |
| Fe(CN)$_6$$^{4-}$ | Fe(CN)$_6$$^{3-}$ | −0.36 |
| ferredoxin(red) | ferredoxin(ox) | −0.44 |
| Fe($o$-phen)$_3$$^{++}$ | Fe($o$-phen)$_3$$^{3+}$ | −1.14 |

\* After Chaberek and Martell, 1959.

factors that are thought to be involved are discussed by Chaberek and Martell (1959), and by B. R. James et al. (1962). Of special interest in plant physiology is the low oxidation potential of ferredoxin (Sec. 2.5) which enables it to donate electrons to NADP in the light reaction of photosynthesis.

### Exchangeability and Reaction Rate

The most important generalization about the exchangeability and reaction rates of complexed metals is their independence from thermodynamic stability. For example, $Cr^{3+}$ amines are chemically unreactive yet thermodynamically unstable; $Mn^{++}$ complexes, by comparison, are extremely reactive, yet may be extremely stable.

Taube (cf. Orgel, 1960) related the activity of complexed metals to the electronic nature of the complexes:

| Reactive | Unreactive |
|---|---|
| 1, 2, and $3d$ electrons | $3d$ electrons |
| High spin | 6 $t_{sg}$ electrons |
| Octahedral complexes | Low spin |

These generalizatons can be rationalized on the idea that reactivity requires that there be free orbitals available for a replacing ligand whereas substantial activation energy would be required if reaction or exchange necessitated that an electron be driven into a new orbital.

The same reasoning applies to the rates of oxidation-reduction reactions. Here high reactivity requires that an electron be passed from or to a metal without extensive electronic reorganization. This is an application of the Franck-Condon principle and is completely analogous to the rule governing allowedness in photochemistry. A molecule will be reduced rapidly if the oxidized and reduced complexes have the same or similar configuration, or similar ionic radii, or if a bridge atom or conjugated structure can pass the electrons. We can well imagine that the structures of electron transport carriers are conditioned on these requirements.

It is interesting how the catalytic activity of a metal in oxidation-reduction reactions can be enhanced by chelation. For example, ascorbic acid oxidation by cupric ion is 100 times greater in ascorbic acid oxidase than in the free ion (Meiklejohn and Stewart, 1941).

Looked at from the other direction, catalase is one of the most active enzymes known, but its catalytic activity is only 200 times greater than that of ferric tetraamine chelate (Wang, 1955).

There are sometimes advantages to an organism in unreactive, i.e., nonexchangeable, chelates. No exchange of magnesium under physiological conditions can be detected in chlorophyll (Aronoff, 1963). Indeed if the metal were labile, we should expect it to be displaced by thermodynamically more stable ions such as copper, manganese, or iron.

## Color

The absorption spectra of metal chelates have been the principal means of assessing their electronic structure (Orgel, 1960). There is, for example, a number of forbidden transitions (Sec. 2.2) which reveal the extent of $d$-orbital splitting under the influence of ligand fields.

The allowed transitions and therefore intense absorptions (Sec. 2.2) in the visible spectrum of many metal chelates of biological importance correspond to transitions within conjugated ring systems within the ligand itself, such as the Soret band of porphyrins, and to charge-transfer complexes, in which an electron from the ligand (or the metal) is excited into a molecular orbital shared by the metal and ligand. This phenomenon involves then an increase (or decrease) in the valence state of the metal and is of possible importance for photochemical processes (cf. Sec. 2.5). An example of a charge-transfer transition is the absorption in the red of ferrous-*o*-phenanthrolene, in which an electron from $Fe^{++}$ is excited into a molecular orbital in the ligand.

### Conclusions about Chelates

Complexation in general and chelation in particular are important to the plant in the following ways: (1) chelates may stabilize structural elements; (2) the electrical charge on a metal may be neutralized so that it can be brought into lipid phases in concentrations otherwise unattainable; (3) the reactivity of metal ions, and especially their oxidation potentials, may be controlled over wide ranges through changes in ligands; (4) the stability of chelates, and changes in stability through changes in molecules bearing ligands, serve as vehicles for the accumulation of nutrient ions (cf. Sec. 4.4).

### 4.4  THE ACTIVE TRANSPORT OF IONS

As we noted above, plants live in a dilute ionic environment, and it is only by accumulating ions to 10- or even 10,000-fold over the concentrations in their environment that plants are able to survive. This process, by which substances are moved against an overall gradient of chemical potential, is called *active transport* (Fig. 4-4). It is as remarkable as it is essential. One might even say that active transport is an identifying feature of living organisms.[1]

Active transport of ions has been studied in whole higher plants, in algae, and in isolated roots, slices of storage tissues, and leaves. Since we should not go far astray by generalizing from one to all of these systems, we shall focus principally on a generalized hypothesis of active transport of ions in roots.

### The Phenomena

Root segments, cut from barley seedlings, will absorb ions and accumulate them in the xylem sap against large concentration gradients. Since the xylem sap and, of course, the external solutions are relatively free of sites for adsorption or complexation, an increase in *concentration* of an ion necessarily means

---

[1] There are, however, "transport-defective" mutants of many organisms which are unable to move one or another substance against potential gradients.

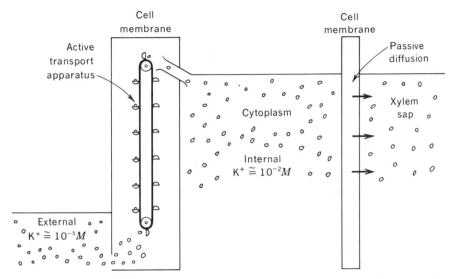

*Fig. 4-4* Diagram of overall process of active ion transport (after van den Honert, 1937).

an increase in chemical potential. The movement of ions into the barley roots is therefore appropriately called "active."

The anatomy of roots (Fig. 3-2) leads one to consider the path that ions must traverse. Ions might move (*a*) through the cell walls only, (*b*) across the epidermal wall and then through the cytoplasm only, (*c*) through the cytoplasm and vacuoles, or (*d*) combinations of the first three. As in the case of water movement, the presence of the Casparian strip at the endodermis has been thought to act as a check valve, preventing ions from diffusing back through the cell walls in this region.

Although the xylem does not extend all the way to the tip meristem, xylem bundles are differentiated well within the regions of principal ion absorption.

### The Symplasm Hypothesis

Crafts and Broyer (1938) proposed that ions are transported actively into the cytoplasm of cortical cells of roots and migrate through cytoplasmic connections (called *plasmadesmata*) from one cell to another toward the stele. The inter-communicating cytoplasm is called the *symplasm*. From the cells of the stele itself, the ions are leaked passively into the xylem.

### The Evidence

Nearly all the evidence for this general hypothesis was obtained before the hypothesis was evolved to its present form. We may note the different kinds of evidence as being consistent (or inconsistent), but only a few items are predictions from, and therefore tests of, the hypothesis.

## Anatomical Differences in Ion Absorption

A prediction from the hypothesis is that cells on the cortical side of the root will accumulate ions whereas those on the stelar side will tend to leak them. This has been demonstrated in elegantly simple experiments from Laties' laboratory (Laties and Budd, 1964). It was found possible to remove the stele and test for ion absorption in the epidermal and stelar side of the cortex separately. Ions were vigorously accumulated from the epidermal side but only weakly from the stelar side (Table 4-8). Curiously the differences were true only of

### Table 4-8  Active Ion Absorption by Isolated Tissues of Maize Roots*

The primary roots of maize seedlings were separated into stele and cortex and the cortex allowed to absorb radioactive chloride ($Cl^-$) from solutions.

| Tissue | Ion Uptake, cpm | |
|---|---|---|
| | per 10 sections | per 100 mg fresh weight |
| Intact roots | 5,180 | 3,550 |
| Stelar side of cortex | 92 | 484 |
| Epidermal side of cortex | 4,880 | 12,230 |

\* From Laties and Budd, 1964.

the freshly prepared cortex. Within a relatively few hours after the stele had been excised, the stelar side of the cortex acquired greatly enhanced power for ion accumulation. Exactly the same behavior is shown by freshly cut slices of storage tissue, such as carrot or potato. Laties and Budd infer that exposure to the "outside" somehow signals a tissue to fabricate ion accumulation apparatus. If this generalization is correct, we should inquire into the nature of the signals and of the response (cf. Sec. 5.4).

## Free Spaces, Outer Space, Inner Space

Water and solutes may penetrate intercellular water and certain adjacent parts of plant tissues by free diffusion. This region is called the *water free space.* Its volume, which may be determined by the movement of substances that do not readily cross cell membranes, such as mannitol or polyglucose, may be as large as 25 percent of the tissue volume. Dissolved ions will also penetrate the water free space by free diffusion, but they may, in addition, be adsorbed by fixed ions on cell walls and membranes in a kind of *Donnan equilibrium* or *exchange adsorption* (Fig. 4-5). That ions are adsorbed to roots, as in ion exchange resins, had been inferred since 1916 (cf. Sutcliffe, 1962), but it is clearly proved by the facts that radioactive species are displaced by unlabeled species and that the rates of adsorption and desorption are what one would

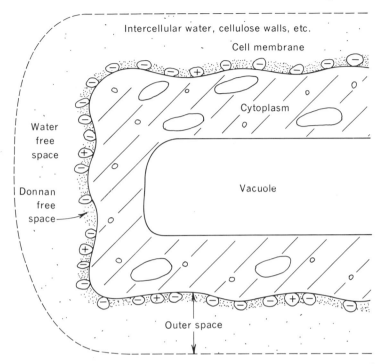

*Fig. 4-5* Water free space, exchange adsorption, and outer space of plant tissues.

predict for diffusion-limited processes (Briggs, et al., 1958). From the observation that cations are adsorbed at higher concentrations than anions, we infer that the fixed ions or ligands are principally anionic.

It is sometimes useful to refer to the region penetrated freely by an ion as the *apparent free space* or *outer space* of that ion. The outer space would include the water free space and the Donnan free space. (All of these terms must be defined operationally in experimental work.)

Since the Donnan free space is not a volume that can be readily calculated, the outer space remains a kinetic and not a physical volume. *Because* the concentration of ions is high in the region of adsorption compared to the rest of the water free space, one can *not* assume that the amount of ions in free diffusion within a tissue occupies the same volume as it would in the surrounding fluid, and thereby calculate the volume for an outer space. Much nonsense has been written about outer spaces through this error. It follows that the volume associated with water free space should be measured with a nonionic and nonmetabolizable solute.

If one follows the time course of ion movements into plant organs, there is typically a rapid first-order absorption, followed by a slow linear uptake (Fig. 4-6). The first phase can be interpreted as corresponding to the entry of the ions into outer space and the second to entry into inner space. The kinetics of ion absorption at constant external ion concentrations can be described by an equation of three terms:

$$c_t = c_i + c_e(1 - \epsilon^{k_d t}) + vt \qquad \text{Eq. 4-10}$$

where

$c_t$ = concentration of ions in tissue at time $t$
$c_i$ = concentration of ions in tissue at zero time
$c_e$ = equilibrium concentration of exchangeable ions in outer space
$k_d$ = diffusion-limited constant for movement into outer space
$v$ = rate of movement into inner space

The values of $c_i$, $c_e$, and $v$ can be determined graphically (Fig. 4-6). If radioactive species are used, $c_i$ disappears. After an interval of absorption,

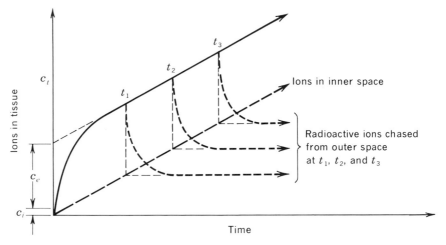

*Fig. 4-6* Kinetics of ions into tissues. Graphical evaluation of parameters of Eq. 4-10.

nonradioactive or "cold" ions may be used to elute or "chase" radioactive ions out of the outer space. Assuming no leakage from inner to outer space, the radioactivity left is then attributable to ions in inner space. In some tissues, especially at low external concentrations, ion absorption first increases exponentially and then becomes linear with time. In others the exponential term of Eq. 4-10 is negligible and $k_d$ is nonlimiting. These effects have been interpreted as supporting opposite hypotheses concerning the location of ion spaces and transport processes (cf. Laties, 1969; Welch and Epstein, 1969).

The size of the outer space is not correlated with active transport, and so our principal concern is with $v$, the rate of movement of ions into inner space. The factors governing the value of $v$ are considered below in the light of the *carrier hypothesis*.

### The Carrier Hypothesis

The idea that ions are transported across biological membranes by carriers was first proposed in 1900 by the plant physiologist Pfeffer (cf. Sutcliffe, 1962).

The carrier hypothesis for the active transport of ions may be generalized as follows: Ions diffuse passively across the cell wall, are adsorbed or otherwise attached to a carrier or mechano-enzyme in the plasma membrane, and then are pumped or secreted into the cytoplasm at a higher concentration (Fig. 4-7). We may elaborate the symplasm hypothesis by noting that the protoplasm

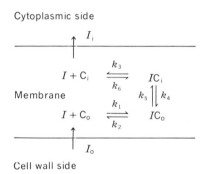

*Fig. 4-7* Diagram of carrier hypothesis.

of cortex cells is in communication through plasmadesmata, so that the entire protoplasm functions as a unit. But on the border between cortex and stele the membranes become inactive or "leaky," and the ions diffuse passively into the xylem vessels.

There are additional pumps into internal pools such as that across the tonoplast into the vacuole. The vacuoles having much larger volumes than the cytoplasm then become reservoirs of high ion concentration. The vacuoles are not, however, along the main pathway to the xylem (see below).

The pumps consist of carriers C that have the property of adsorbing ions reversibly and have a lower affinity for the enzyme on the cytoplasmic side of the plasma membrane than on the cell wall side.

Carriers need not be enzymes; in fact an adsorption model reduces to the same mathematics, but physiologists prefer to deal with as few equations as possible. So by common consent, ion transport kinetics are described in terms of $K_T$ and $V_{max}$ values of hypothetical carriers (Epstein and Hagen, 1952).

Unlike the case of an enzyme-catalyzed transformation of a substrate molecule, there is no reason to think that the transported ion undergoes any net chemical change (other than an increase in chemical potential). We must therefore make certain distinctions between Michaelis-Menten kinetics and the equations describing ion transport. Let

$v \equiv$ rate of irreversible accumulation of ion

$k_1, k_2 \equiv$ rate constants for formation and dissociation of carrier-ion complex on outside of membrane

$k_3 \equiv$ rate constant of dissociation of ion from carrier on cytoplasmic side of membrane. (It is usually assumed that $k_4$ and $k_5$ in Fig. 4-6 are non-limiting and that $k_6$ is negligible.)

$$K_T = \frac{k_2 + k_3}{k_1}$$

$C \equiv$ total concentration of carrier sites (combined and free)

$I \equiv$ external concentration of ion

then

$$v = \frac{k_3 C I}{K_T + I} = \frac{V_{max} I}{K_T + I} \qquad \qquad \text{Eq. 4-11}$$

Analogous to enzyme kinetics, a low $K_T$ would be equivalent to a high affinity between carrier and ion, and a high $k_3 C = V_{max}$ would mean a high transport capacity. $V_{max}$ is typically 2 to 4 meq hr$^{-1}$ (kg fresh weight)$^{-1}$.

We shall also encounter inhibition of ion absorption, some of which is similar but not completely analogous to competitive inhibition by substrate analogs (Sec. 1.13). For example, the addition of Rb to roots absorbing K decreases the rate of K accumulation. Since the Rb ion is also transported, it is inappropriate to speak of this as a competitive inhibition. We should use instead the equations for multiple substrates. On a Lineweaver-Burk plot, the kinetics appear the same as those for competitive inhibitors.

Consider two ions P and Q competing for the same carrier sites. In the absence of Q,

$$v_p = \frac{V_{max} P}{K_p + P}$$

In the presence of Q,

$$v'_p = \frac{V_{max} P}{K'_p + P}$$

where the symbols are analogous to those in Eq. 4-11 and

$$K'_p = K_p \left( 1 + \frac{Q}{K_Q} \right)$$

Actually the $K_T$ values of some ion-tissue systems are not single valued, but may show two or more components. The existence of such complex kinetics might encourage us to discard the whole approach, were it not that the two components can be separated physically. Tori and Laties (1966) have shown that, in the case of transport of alkali ions, meristematic tissues, containing only tiny vacuoles, exhibit only the high-affinity $K_T$ whereas the low-affinity set of carriers appears when the cells develop the large central vacuole typical of mature cells. The kinetically separable carriers also show different sensitivity to ion competition.

The evidence is by no means consistent in pointing toward the tonoplast membranes as sites of the low-affinity carriers. Welch and Epstein (1969) found no delay between the onset of the low-affinity and high-affinity transport. Epstein's alternative hypothesis is of a mosaic of carrier systems on the plasma membrane, with low- and high-affinity transport operating side by side.

### Physiological Significance of Carriers

Let us assume not only that the carrier hypothesis is correct but that there are two or more kinds of carriers, with high and with low affinities for the ions. Which, if any, of these carrier mechanisms are relevant to the absorption of essential nutrients by intact plants?

One would like to see transport kinetics determined on the same set of plants as that used for determining the critical concentrations of nutrients. This has not been done, but a generalization is possible: *the range of concentrations of ions which are critical for growth corresponds to the high-affinity (low $K_T$) transport mechanisms.*

For example, Epstein et al. (1964) found kinetic evidence with barley roots for one carrier for potassium with a $K_T$ of $2 \times 10^{-5}$ $M$ and a multiplicity of other transport processes with $K_T$ values up to $2 \times 10^{-2}$ $M$ (Epstein and Rains, 1965). The critical concentrations for the growth of a variety of pasture plants (Asher and Ozanne, 1967) showed critical concentrations for K at $2 \times 10^{-5}$ $M$. It seems then that only the high-affinity–low $K_T$ transport process is essential to the growth of the plants. Moreover, Fried and Shapiro (1961) have pointed out that the typical concentrations of nutrient ions in soil water correspond only to the high-affinity–low $K_T$ processes.

Since it is unlikely that the plants would often find themselves in soils with high ionic concentrations, the low-affinity–high $K_T$ appears of secondary interest. One can imagine that the low-affinity processes might operate in "luxury consumption," storing nutrients against a dilute day.

From the kinetic data we infer that plant cells possess a number of carrier molecules with specificities for different kinds of ions (Table 4-9). If these carriers were proteins, one would expect to find mutants which lacked or had greatly reduced ability to accumulate certain kinds of ions. Such mutants have in fact been observed (cf. Pardee, 1968).

Table 4-9  Carrier System of Barley Roots as Interpreted from Kinetic Data*

In instances where multiple carriers were inferred, only the high-affinity carriers are tabulated.

| Carrier | $V_{max}$ $\mu$moles hr$^{-1}$(g wet wt)$^{-1}$ | Ions | $K_T$, $M$ | Reference |
|---|---|---|---|---|
| Alkalis | 11.9 | K$^+$ | $3 \times 10^{-6}$ | Epstein et al., 1964 |
|  |  | Rb$^+$ | $4 \times 10^{-6}$ |  |
|  |  | Na$^+$ | $10 \times 10^{-6}$ |  |
|  |  | NH$_4^+$ | Approx. $30 \times 10^{-6}$ |  |
| Alkaline earths |  | Mg$^{++}$ |  |  |
|  |  | Ca$^{++}$ | $3 \times 10^{-6}$ |  |
|  |  | Sr$^{++}$ | $3 \times 10^{-6}$ |  |
| Halides | 4.7 | Cl$^-$ | $10^{-5}$ | Elzam and Epstein, 1965 |
|  |  | Br$^-$ | $2 \times 10^{-5}$ |  |
|  |  | I$^-$ | Negligible |  |
| Sulfate |  | SO$_4^=$ | $10^{-5}$ |  |
|  |  | SeO$_4^=$ | $7 \times 10^{-6}$ |  |
|  |  | Cr$_2$O$_7^=$ |  |  |
| Phosphate |  | H$_2$PO$_4^-$ | $1$–$3 \times 10^{-6}$ |  |
| Nitrate† |  | NO$_3^-$ | $2 \times 10^{-6}$ |  |

\* Compiled by Fried and Shapiro, 1961.
† Corn.

### Accumulation of Ions in Aerial Parts of Plants

Active transport is not the only means by which plants accumulate nutrients. Ions may also be swept in passively with the transpiration stream. It is commonly observed that the amount of ions reaching the leaves of a plant can be strongly affected by transpiration rates (Table 4-10). In general, however, transpiration promotes ion uptake by plants that are already rich in salts and is relatively less important for low-salt plants, where active uptake proceeds more vigorously.

In the xylem the ions are presumably not subject to active transport processes, but living cells along the transpiration stream accumulate ions from it. Studies have been made on accumulation by leaf tissue (Smith and Epstein, 1964).

### The Mechanism of Active Transport

Although there is no shortage of speculations (cf. Sutcliffe, 1962) on the means by which the hypothetical carriers harness metabolic energy to effect local increases in chemical potential, there is little among these speculations

Table 4-10  Ion Absorption in Low and High Rates of Transpiration*

| Plants | Environmental Conditions | Water Absorbed, ml (g fresh weight of shoot)$^{-1}$ | Salt Absorbed, meq (g fresh weight of plant)$^{-1} \times 10^2$ | | Calculated Br Conc. in Xylem Sap, meq/1 |
|---|---|---|---|---|---|
| | | | K$^+$ | Br$^-$ | |
| Low-salt status | Low humidity, light | 9.60 | 10.85 | 9.52 | 9.7 |
| | High humidity, light | 3.60 | 10.40 | 9.65 | 25.1 |
| High-salt status | Low humidity, light | 8.10 | 5.20 | 6.07 | 6.7 |
| | High humidity, light | 2.58 | 3.24 | 4.24 | 14.8 |

* From Broyer and Hoagland, 1943.

which can yet be subjected to rigorous criteria (cf. Rosenberg and Wilbrandt, 1963).

The one common thread among the attempts to identify the mechanisms of active transport is phosphorylation. Inhibitors that have been shown to eliminate oxidative phosphorylation or photophosphorylation inevitably abolish active ion transport. The simplest interpretation is that carriers at some stage in the cycle of transport undergo phosphorylation and subsequently use the free energy of hydrolysis of a phosphate ester either to dissociate ions from the carrier binding sites or to generate new sites (cf. Goldacre, 1952; Mitchell, 1967).

For a time there was speculation and some evidence that mitochondria might be the intracellular sites of ion accumulation (cf. Millard et al., 1964; Mitchell, 1967), but with the possible exceptions of calcium and phosphate, the amounts of ions accumulated and the concentration gradients achieved rule out mitochondria as other than secondary sites (Hodges and Hanson, 1965; Kenefick and Hanson, 1966). Similar transport processes have been observed in chloroplasts (cf. Nobel and Parker, 1965).

Ion absorption can be inhibited by chloramphenicol (MacDonald et al., 1966), an inhibitor of protein synthesis that is specific for mitochondria and chloroplasts.

### Case Studies: Questions and Evidence

With the background discussions of chelate chemistry and carrier-mediated transport in mind, we can summarize inorganic nutrition in three questions:

1. How much (i.e., concentration and amount) of a nutrient element is required?
2. What is the transport system, if any?
3. How does the plant respond to different concentrations (or amounts) of the element?

[Other information concerning nutrient elements may provide a useful background or be of interest in terms of chemical transformations or growth and development, or as agricultural problems, but they are only of oblique relevance to the theory of inorganic nutrition. For example, knowledge of the role of iron in porphyrins (B. R. James et. al., 1962) or of calcium in amylase (Takagi and Isemura, 1965) is obviously relevant to the biochemistry of electron transport and the enzymology of starch hydrolysis, respectively, but such knowledge does not necessarily illuminate our understanding of a plant's responses to the levels of nutrients in the external and internal environments.]

What kinds of evidence will we accept as answers to these questions?

1. The question of quantitative essentiality was developed above (Sec. 4.1). We need add only that essentiality must be defined with respect to specific processes: growth, yield (as of seeds or tubers), or perhaps the formation of some chemical component, e.g., protein, which may be incidentally related to commercial "quality."

2. We should first establish, in the case of transport, whether it is *active,* that is, whether the nutrient is moved from lower to higher chemical potential. A *carrier* should be identifiable as a molecule or macromolecule that is present in the region where transport occurs (presumably in the cell membrane) and that can combine reversibly with the nutrient. The $K_T$ and $V_{max}$ of the carrier should be consistent with the performance of the intact plant.

We should want to understand the biochemical or biophysical changes in the carrier that allow for loading and unloading; implicit in this would be an understanding of the means by which metabolic energy is coupled to the transport process.

A secondary, but still significant, aspect of transport is *mobility or retranslocation,* the movement of nutrient elements after they have been deposited in a tissue. Mobility is presumably related directly to the reversibility of the complexes in which the nutrient occurs. The rates of translocation from organ to organ, as for example, from older to younger leaves, can be readily determined with those nutrients for which radioisotopes are available (cf. Biddulph, 1955). For others, mobility has been inferred from the symptoms of deficiency: the appearance of symptoms in the younger tissues first is consistent with lack of mobility (Table 4-11).

Curiously, all of the nutrients in seed storage organs are readily translocatable. For example, the calcium requirements of seedlings can be supplied from cotyledonary supplies and iron can be translocated during germination (Mayer, 1954).

3. In determining the response of a plant to changes in the concentrations or amounts of a nutrient, we must again define the response (e.g., growth), then the dose-response function. A rational understanding of the plant response would consist of determining compositional changes and especially the identification of limiting factors over the entire dose-response curve. One approach to this is to inventory all the forms of the nutrient element

in the cell and then to determine which ones decrease to limiting concentrations over critical concentration ranges of the nutrient. This approach has been made only in the cases of molybdenum and cobalt nutrition where only one form of each is known to be present.

Table 4-11 Mobility of Inorganic Nutrients in Plants

The designation of an ion as "mobile" or "nonmobile" is far from absolute, but refers to the tendency of the ion in mature plants to move from one tissue to another, once it has left the transpiration stream.

| Mobile Ions | Nonmobile Ions |
| --- | --- |
| Nitrate | Calcium |
| Ammonium | Borate |
| Phosphate | Iron |
| Sulfate | Manganese(?) |
| Potassium | Zinc |
| Magnesium | |

In the case of nutrients that reversibly form chelates, a relatively simple (if somewhat naive) model is available. It is simply that the potential complexing agents (e.g., proteins) compete for a dwindling supply of the nutrient. Those agents with strong stability constants succeed in chelating the ion; the weak are deprived, and ultimately one of them becomes limiting (cf. Price, 1968).

The mathematics of complexation of a limiting factor in an expanding (growing) system have been explored and some simplified treatments have been applied to microbial systems in the chemostat (Moser, 1958), but there has not been sufficient data to apply these equations to higher plant systems.

## 4.5 CASE STUDY: NITROGEN

Quantitatively and qualitatively, nitrogen is easily the most important nutrient for higher plants. In contrast to most other nutrients, the amounts of nitrogen required by plants are intimately related to such variables as temperature, light intensity, and photoperiod (cf. Nightingale, 1937, 1955).

Starting with some ideas of Klebs (1903) on flowering, Kraus and Kraybill (1918) developed a physiological *Weltanschauung* based on ratios of carbohydrate to organic nitrogen. According to this scheme (cf. Nightingale, 1948, 1955), high carbohydrate:nitrogen ratios favored root development and drought resistance, whereas low ratios favored rapid vegetative growth and succulence. The CN rule, as it came to be called, was evolved before the discovery of growth substances or the development of a theory of growth regulation (Chap. 5), but the rule remains a remarkably useful guide for the horticulturist. We shall see that there is, in fact, some rational basis for it.

### Several Sources of Nitrogen

Nitrate is the principal nitrogenous component of soil, aquatic, and marine environments.[1] Ammonia secreted by animals or released from urea is slowly converted to nitrate by nitrifying bacteria, such as the genus *Nitrosomonas*.

In contrast to plants in nature, which only occasionally receive a gift of ammonium or urea from a passing animal, agricultural plants are now fertilized heavily with both of these potentially toxic materials. We shall see that the plant absorbs and responds to nitrate, ammonium, and urea nitrogen quite differently and that there are remarkably different responses among plants to these different forms of nitrogen.

### Quantitative Requirements

Plants generally require about 1 mmole of nitrogen per gram dry weight (Table 4-1). Because of the quantitative and qualitative importance of nitrogen in plant nutrition, nitrogen requirements of most agricultural plants have been studied intensely. Probably the best documented case is that of sugar beet. Ulrich (1964) has found that the measurement of nitrate in the petioles of young sugar beet leaves provide the most consistent indication of a critical internal concentration of nitrate for the growth of roots and tops (Fig. 4-8).

As an example of the necessity for defining the criteria for quantitative requirements, the critical concentration for *growth* of sugar beet is 70 to 350 m$M$ nitrate, whereas, the optimum concentration for *sugar accumulation* is only 35 m$M$.

There are fewer data available on the critical external concentrations of nitrate or ammonium. Pirschle (1931) estimated critical concentrations in flowing nutrient solutions for maize of 2,000 $\mu M$ nitrate and ammonium, but for peas the critical concentrations were 100 $\mu M$ nitrate and less than 20 $\mu M$ ammonium. Lemon, wheat, and *Lemma* are said to grow adequately on 70 $\mu M$ nitrate with flowing solutions (cf. Chapman and Liebig, 1940), but analytical and mechanical problems have clearly prevented workers from maintaining lower concentrations. The most accurate data to date are probably those of van den Honert and Hooymans (1955) who found a $K_T$ for nitrate absorption of 250 $\mu M$. The $K_T$ but not the $V_{max}$ was independent of pH (Fig. 4-9).

---

[1] Organic nitrogen may dominate in heavily polluted water. Ryther (1954) showed that Great South Bay on Long Island, which had been badly polluted by duck farms (and more recently by people), contained large populations of *Nannochloris atomus* and *Stichiococcus* sp. When these algae were cultured in the laboratory, they were found to accept organic and inorganic nitrogen indiscriminately, whereas typical marine diatoms require nitrate or nitrite.

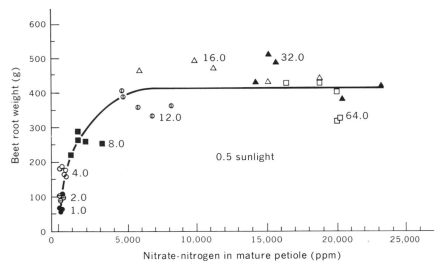

*Fig. 4-8* Growth of sugar beet roots as a function of the concentration of nitrate-nitrogen in mature petioles (Ulrich, 1964). The numbers refer to the initial concentration of nitrogen in milliquivalents per liter of nutrient solution for each set of experimental plants.

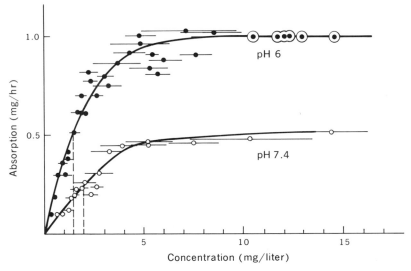

*Fig. 4-9* Absorption of nitrate by sugar beet as a function of nitrate concentration at two pH levels (van den Honert and Hooymans, 1955).

### Transport Processes

The uptake of nitrate and ammonium from vigorously stirred solutions proceeds at concentrations much lower than the critical concentrations observed for growth. Rye plants absorbed nitrate at constant rates down to 3 $\mu M$; half-maximal rates appeared to be about 1 $\mu M$ (Olsen, 1950). Similar results were obtained with the aquatic *Helodea* (Olsen, 1953).

Tromp (1962) obtained detailed and precise data on ammonium absorption by wheat roots. The half-maximal rates ($K_T$) were 7 $\mu M$ (Fig. 4-10).

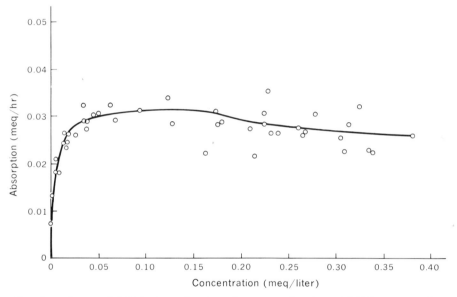

*Fig. 4-10* Rates of NH₄⁺ absorption by wheat as a function of NH₄⁺ concentration (Tromp, 1962). Intact wheat plants absorbed NH₄⁺ from solution maintained with constant concentrations by continuous-flow techniques. Potassium was 0.34 m$M$ throughout.

Since even continuous-flow experiments must inevitably err by overestimating critical concentrations, and growth experiments are especially suspect because of the long intervals required, it seems prudent to accept the lowest values obtained.

The rates of uptake of nitrate and ammonium are affected in opposite directions by the pH of the external solution (Table 4-12) (Clark and Shive, 1934). The strikingly different effects of pH on the two transport systems have at least two alternative explanations. Since it is unlikely that the pH of the cytoplasm is greatly affected by the outside pH, we may infer that the transport system is located at least in part in the plasma membrane. The effects of pH may represent competition by OH⁻ and H⁺ in the case of nitrate and ammonium

Table 4-12 Rates of Absorption of NH$_4$- and NO$_3$-Nitrogen by Tomato Plants at Different pH*

Tomato plants were grown in solution of different pH and the disappearance of nitrogen from the culture solutions measured over a 6-hr period.

| pH of Culture Solution | N Absorbed, mg per 100 g fresh weight in 6 hr | | |
|---|---|---|---|
| | NH$_4$-N | NO$_3$-N | total N |
| 4.0 | 3.4 | 4.8 | 8.2 |
| 5.0 | 4.2 | 5.9 | 10.1 |
| 6.0 | 4.6 | 4.1 | 8.7 |
| 7.0 | 6.6 | 3.0 | 9.6 |

* From Clark and Shive, 1934.

transport, respectively. Alternatively, the effects may be due to titration of a group crucial to the structure of a carrier.

### Mobility

The mobility (or redistribution) of nitrogen is the most clearly demonstrable of all nutrient transport processes. When plants are allowed to become deficient, the meristems and young leaves remain green whereas the older leaves turn yellow and senesce. Most dramatically, a major portion of the nitrogen of grasses can be recovered in the grain, leaving the straw of minimal nutritional value for animals. The mechanism of this translocation is instructive: proteins constitute most of the nitrogen of plants with less than adequate nitrogen; since amino radicals cannot dissociate from proteins, the only conceivable mechanism is the breakdown of proteins and liberation of amino acids to the phloem.

We know, in fact (Sec. 5.2), that many of the proteins of plants are in a "steady state," undergoing continuous decomposition and synthesis. Such a steady state, together with some means of favoring transport into the meristem, might be sufficient to account for the loss of nitrogen by older organs. It is also possible that proteolysis is under metabolic control (cf. Sec. 5.2) and is actively promoted in older organs during nitrogen deficiency.

### Response of Plants to Nitrogen

Thanks to the work of Palladine, Schultze, Prianishnikov, and other pioneers (cf. Prianishnikov, 1951), we have long enjoyed a fairly complete knowledge of the changes in the nitrogenous constituents of plants in response to different amounts of the several kinds of nitrogen nutrients.

With large doses of ammonium, most plants accumulate large amounts of amides, principally glutamine, together with lesser amounts of glutamate and other free amino acids. In the case of nitrate fertilization, glutamine accumulates only under high light intensity and temperatures of above 20° or so. Asparagine varies to a much lower degree and often in an opposite direction.

The so-called "acid plants," such as *Begonia* or rhubarb, accumulate ammonium salts of organic acids in their vacuoles, sometimes to concentrations of as much as 0.1 to 0.2 $M$ and at pH 1.5 (Ruhland and Wetzel, 1927)!

Yemm's group (cf. Folkes, 1959) has provided elegant kinetic evidence on the sequential incorporation of $N^{15}H_3$ into glutamate, glutamine, and protein amino acids (Fig. 4-11).

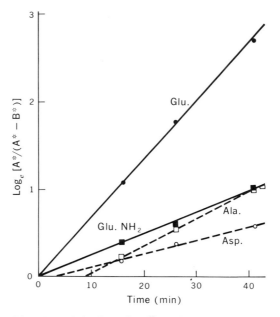

*Fig. 4-11* Kinetics of $N^{15}H_4^+$ incorporation into organic nitrogen in the yeast *Torulopsis utilis* (Folkes, 1959). A kinetic test, described in the text, permits one to identify components which are formed directly from a nutrient as contrasted with those that pass through intermediates (see text).

The incorporation of $N^{15}$ into glutamine, glutamic acid, alanine, and asparagine was analyzed by a simple test of product-precursor kinetics. If $A \rightarrow B$ in exponentially growing cells, it can be shown (cf. Sec. 1.2 and Appendix B) that the incorporation of label follows the relation

$$B^* = A^*(1 - e^{-bt})$$

so that a plot of log [A*/(A* — B*)] against time should give a straight line passing through the origin.

In the present case, a compound that is formed directly from $NH_4^+$, or one that passes through intermediates with negligible pool sizes, will show this relation. One formed by a subsequent reaction, such as a transamination, will cross the $t$ axis at some positive value. We see that glutamic acid and glutamine are formed directly from $NH_4^+$ whereas alanine and asparagine must pass through some N-containing intermediate.

The early workers recognized a reciprocal relation between assimilated nitrogen (i.e., forms other than nitrate) and carbohydrate reserves.

These phenomena are in part understandable from the metabolic pathways of nitrogen (Fig. 4-12). According to this scheme, nitrate is reduced to ammo-

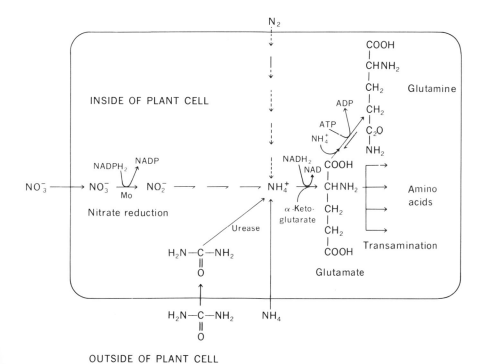

OUTSIDE OF PLANT CELL

*Fig. 4-12* Metabolic pathways involved in the nitrogen nutrition of plants.

nium by a series of steps involving reductases, with molybdenum as an essential component. Alternatively ammonium may enter the cell as such, arise from the hydrolysis of urea, or in the case of a gifted few, by reduction of gaseous nitrogen (Burris, 1966). Ammonium is converted by reductive ammination of α-ketoglutarate to glutamate and then by amidation to glutamine.

Free ammonia, it turns out, is toxic; so it is understandable (teleologically at least) that once ammonium is formed, plants are obliged to store it, which they do as amino acids, amides, or ammonium salts. The reciprocity between carbohydrates and organic nitrogen observed under the so-called C/N rule is therefore understandable. Prianishnikov (1951) spoke of carbohydrates as "detoxifying" ammonia.

Thus, although a nitrogen-deficient plant in nature has a plentiful store of carbohydrates, a plant on an extended ammonium spree can exhaust its carbohydrate reserves. A plant in the shade fed excessive ammonium may become depleted of carbohydrate to the point of death. Similarly, a plant in the shade fed excessive ammonium may undergo undesirable changes, very familiar to horticulturists, such as soft, succulent stems. Rice and other grasses, fertilized during cloudy weather, are apt to fall over or "lodge." Such plants, moreover, are extremely susceptible to both cold and drought damage. It is obvious that crops such as potatoes and sugar beets will produce less if carbohydrates are continuously drained into glutamine and amino acids.

We can also imagine that plants deprived of carbohydrates might have less enthusiasm for such adventures as tuberization or sexual reproduction, but these phenomena may well be related only indirectly (cf. Sec. 5.2).

Teleology aside, we should not have predicted from a first inspection of Fig. 4-12 that nitrate could accumulate under conditions where ammonium is metabolized. The delicious fact is that nitrate reductase, the first enzyme in the reduction of nitrate, is under *metabolic control*. Deferring the question of the *mechanism* of metabolic control to Chap. 5 (cf. Sec. 5.2), we note here that formation of nitrate reductase is *induced* by nitrate and *repressed* by ammonium. The enzyme also disappears in the dark and is resynthesized in the light (cf. Beevers et al., 1965).

Through the vehicle of enzyme control, plants are sometimes able to accumulate large amounts of nitrogen as nitrate under unfavorable growing conditions without exhausting carbohydrate stores.

Urea may be thought of as an indirect means of supplying ammonium (Fig. 4-12) and one which occurs both in nature and in agricultural practice. Urea is often absorbed directly by leaves and hydrolyzed in situ. It is interesting that plants which are especially sensitive to urea toxicity have high urease levels, a phenomenon consistent with our observations on ammonium toxicity.

An unfailing source of nitrogen is gaseous nitrogen in the atmosphere. Some bacteria (e.g., *Azotobacter* spp.), many algae, and certain fungi have learned to reduce $N_2$ (Yocum, 1960). Perhaps the most interesting case is that of *Rhizobium,* a bacterium which forms symbiotic nodules in the roots of certain legumes. Each species of *Rhizobium* is specific for a very few higher plant species. The nodules, but not the isolated bacteria, are capable of reducing $N_2$. Legumes have thus become at least partially independent of soil nitrogen.

## 4.6 CASE STUDY: THE ALKALI METALS

From Table 4-1 we learned that plants typically require about 250 $\mu$moles of potassium per gram dry weight. Evans and Sorger (1966) have compiled K requirements for a number of species and concluded that normal K contents average about twice those of deficient plants. The requirements for potassium are qualitatively absolute, although rubidium exerts a sparing action for some plants. The other alkali metal ions will not substitute at all (cf. Evans and Sorger, 1966). The potassium requirement for simple increase in dry weight may be less than that for other processes. In the tomato, for example, amounts of potassium that are supraoptimal for growth stimulate the production of citric acid, a component that happens to be important for the quality of canning tomatoes.

Concentrations of potassium on the order of $10^{-5}$ $M$ are sufficient to support normal rates of growth (Fig. 4-13). Similarly, the kinetically inferred $K_T$ con-

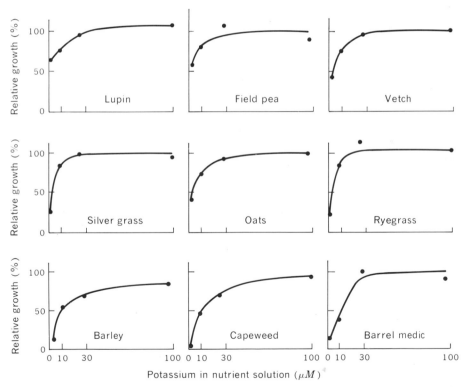

*Fig. 4-13* Growth of plants at different concentrations of potassium (Asher and Ozanne, 1967). Relative yields of dry weights of whole plants, tops, and roots were averaged for 14 plants. Continuous-flow cultures were used and K$^+$ was the only limiting nutrient.

stants for the carriers of alkali ions are on the order of $3 \times 10^{-6}$ $M$ (Table 4-9). This extremely tight binding between potassium and the hypothetical carriers is quite at variance with the stability constants of known K complexes, which are not known to exceed $10^3$ (Bjerrum et al., 1957). Our only conclusion is that plants have again bested organic chemists.

The sites of action of potassium are equally mysterious. Just because the complexes of K are relatively weak and nonspecific, it has been impossible to extract K complexes from plants; hence the inventory approach has not been feasible.

Some thirty or so enzymes require or are activated by alkali ions (cf. Evans and Sorger, 1966). The behavior of pyruvate kinase may be taken as typical: maximal activities are obtained with about 50 m$M$ $K_T$ and the dissociation constant (inferred from the $K_T$ with respect to $K^+$) is on the order of 5 m$M$. Relative activities with other alkali ions are

$$K^+ > Rb^+ > NH_4^+ > Na^+ \gg Li^+ = Tris^+$$

The concentrations of potassium in normal and deficient plants, on the order of 50 and 25 m$M$, respectively, are not inconsistent with the idea that potassium is required for the activation of enzymes similar to pyruvate kinase, although the physiological effects (e.g., decrease of growth rate) seem out of proportion to the predicted decrease in enzyme activities.

ADPG-starch transglucosylase (starch synthetase) is another example of an alkali ion–requiring enzyme; $K^+$ is again the most effective (Murata and Akazawa, 1968). The $K_M$ ($K^+$) of the sweet potato root enzyme is 13.3 m$M$. Starch production in the intact plant also requires $K^+$ to a similar degree.

[The nature of the activation of enzymes by alkali cations is of interest: Sorger et al. (1965) report a correspondence between the influence of an ion on the electrophoretic behavior of pyruvate kinase and on the ability of the ion to activate the enzyme.]

Even though generations of plant physiologists have muttered about the roles of K and Ca in "maintaining the proper colloidal state" (cf. Manery, 1966), such a function has not provided a theoretical understanding of the uniqueness of K as an essential nutrient.

The lability of potassium complexes probably also accounts for the extraordinary mobility of K in translocation. A deficiency of potassium is invariably seen first in older tissue, as the young, growing tissues apparently exercise priorities in the complexation of this metal.

Sodium ion is also required by *Cyanophyceae* and halophytes, such as *Atriplex vesicaria* (Brownell, 1965), and it promotes growth of a number of other plants, notably *Beta vulgaris* (beets, Swiss chard, etc.).

In their favor, the alkali ions have isotopes with convenient half-lives and their emission and absorption spectra provide easy and sensitive means of analysis. Because of the analytical advantages of the alkali ions, a great deal more quantitative and more certain data are available about these ions in soils, and in absorption and transport, than about any others. It is in the interpretation of these data that our understanding is lacking.

## 4.7 CASE STUDY: PHOSPHATE

### Requirements

Asher and Loneragan (1967) have confirmed with elegant techniques earlier estimates of van den Honert and others that the critical concentration of phosphate for plant growth is about 50 $\mu M$. There are however fairly wide variations among plants. For example (Fig. 4-14), although silver grass grows optimally

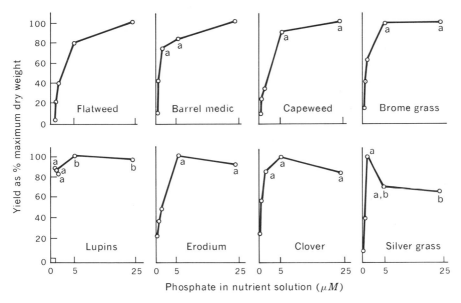

*Fig. 4-14* Growth of a variety of pasture plants in different concentrations of phosphate (Asher and Loneragan, 1967). Continuous-flow cultures were employed.

at 1 $\mu M$ phosphate, several other species grew at less than half their maximum rates at this concentration. Since phosphate is frequently limiting in natural environments, these differences could be of great ecological and agricultural significance.

The critical concentration for phosphate absorption for the stems and leaves of *Helodea,* a water plant, is about 3 $\mu M$ (Olsen, 1953). The changes in absorption rate with pH with a variety of plants are consistent with the idea that $H_2PO_4^-$ is the principal species absorbed.

Algae are able to remove phosphate from natural waters down to extremely low levels. The concentration of inorganic phosphate in equilibrium with the marine diatom *Phaeodactylum tricornutum* (Bohlin) is 1 to $4 \times 10^{-9}$ M (Kuensler and Ketchum, 1962)! The cells can concentrate phosphate internally to a concentration $10^8$ times higher than the external concentration and 30 times over the amounts that would limit growth. The limiting amount per cell is about $2 \times 10^{-15}$ mole (Kuensler and Ketchum, 1962). Similar values have been reported for other algae (cf. review of Krauss, 1958). The limiting internal phosphate concentration for the blue-green alga *Microcystis aeruginosa* is 0.12 percent of dry weight (Gerloff and Skoog, 1954) and that for potato leaves is 0.35 percent (Hill and Cannon, 1948).

A number of cations form very tight complexes with phosphate and especially polyphosphates, but Nature appears not to have been able to use these complexes in the biological accumulation of this essential anions. The mechanisms of phosphate accumulation by intact plants are not known. The usual pathway is through the roots, but phosphate may also be absorbed by leaves.

### Transport

An accumulation of calcium phosphate by isolated plant mitochondria has been described (Hodges and Hanson, 1965; Kenefick and Hanson, 1966), but it is not known whether this phenomenon is of physiological significance in phosphate storage.

*what has this to do with transport?*

### Essentiality

We have no difficulty accounting for the essentiality of phosphate: it is present in nucleic acids and nucleotides, phospholipids, and sugar phosphates, all of which are essential components of the cell.

The identity of the different kinds of phosphate compounds varies greatly over the course of development of a plant. In the seeds of grains, phytin (Ca and Mg salts of inositol phosphates) accumulate spectacularly (Fig. 4-15) mostly at the expense of inorganic phosphate and pyrophosphate. In algae provided with abundant quantities of phosphate, polyphosphates may become the principal form of phosphate.

Low concentrations of phosphate may limit oxidative phosphorylation in the phosphorylation of ADP to ATP by isolated mitochondria. It is not known whether this or any other phosphorylation per se becomes limiting in intact plants (cf. Roux, 1966), but it is evident that a sharp decrease in the growth of *Spirodela* occurs just when the concentration of inorganic phosphate is exhausted (Bieleski, 1968).

In glycolysis and respiration of a few test organisms, the controlling role

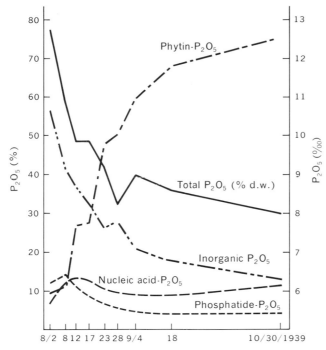

*Fig. 4-15* Distribution of phosphate compounds in ripening corn grain ($P_2O_5$ in percent of total $P_2O_5$; total $P_2O_5$ in parts per thousand (%₀) of dry weight) (Schmalfuss, 1941).

of phosphate is reasonably well documented. Chance and Maitra (1963) define a useful term:

$$\text{Phosphate potential} \equiv \log \frac{\text{ATP}}{\text{ADP} \cdot \text{P}_i}$$

The equilibria of electron transport are such that cytochrome $c$ and NAD are about equally oxidized and reduced at a phosphate potential of 5, which is about the value of $P_i$ calculated for ascites tumor cells.

As an example of the use of phosphate potential, one can say that, if the concentrations of ATP and ADP were equal ($\text{ATP}/\text{ADP} = 1$), inorganic phosphate would be limiting electron transport at internal concentrations of about 10 $\mu M$.

Similar considerations apply to glycolysis (Uyeda and Racker, 1965) and photophosphorylation (Lynn and Brown, 1967).

The signs of phosphate deficiency are very characteristic: in the initial stages, plants as diverse as tomatoes, Merion blue grass, green algae, and marine diatoms show normal rates of photosynthesis coupled with low rates of respiration. As a result, carbohydrates accumulate, the leaves of plants become a deep green, while growth, especially of aerial parts, virtually ceases (cf. review of Pirson, 1955).

Phosphate-deficient plants appear to be like misers, who continue to accumulate capital but are reluctant to spend it. One reason may be that the chloroplasts, the principal sites of carbon fixation, may have a lower critical concentration for phosphate than the mitochondria, which are the presumed catabolic centers.

Almost unique among nutrient deficiencies, phosphate deficiency is often instantly reversible (Fig. 4-16). Plants which have virtually ceased growing resume "normal" rates of respiration within minutes. Clearly the lesion in

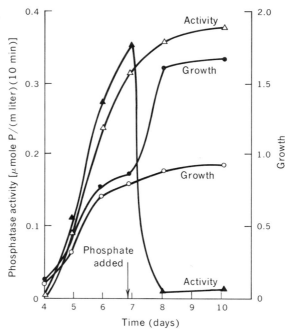

*Fig. 4-16*   Time course of phosphatase activity in phosphate-deficient *Euglena* (Price, 1962). One flask (△, ○) was given 100 μmoles of phosphate per liter of medium at time of inoculation and allowed to exhaust phosphate. The second flask (▲, ●) was treated identically until 6.9 days, at which time (see arrow) 2,000 μmoles of phosphate per liter of medium was added. Growth (○, ●) was measured as optical density of the culture suspension, corrected for self-absorbance. Phosphatase activity (△, ▲). Final specific activity of phosphate-deficient cells was 18 times greater than initial activity and 52 times greater than activity of cells that had received additional phosphate.

early phosphate deficiency cannot involve the degradation of crucial catalysts, except perhaps by the reversible loss of phosphate.

One reasonable hypothesis is that the pace-setting process in phosphate deficiency is a phosphorylation involved in the catabolism of food reserves such as those which might be catalyzed by a kinase.

## Incidental Intelligence

*? wow!!*

Tissues of phosphate-deficient plants are mechanically tough and resistant to desiccation. There is commonly a large root-to-shoot ratio. In fact, a mild phosphate deficiency may be advantageous during drought, transplantation, or similar physiological stresses. It may even be worthwhile for a horticulturist to induce phosphate deficiency if he wishes to minimize growth.

Phosphate deficiency is very common in nature: serpentine soils ($<2$ mg phosphate per 100 g soil, compared with normal values of 5 to 32 mg), woodland soils of New Jersey, the red soils of Georgia, uncontaminated lakes, and large parts of the ocean. The Grand Banks owe their abundance of fish in large part to local currents which force an upwelling of deep ocean water, rich in both nitrate and phosphate.

A curious and extremely widespread response to phosphate deficiency is the formation of phosphatases by the process of de-repression (cf. Sec. 5.2). Stated simply, the presence of inorganic phosphate prevents the formation of an enzyme, phosphatase. As the organism runs out of inorganic phosphate, the repression is removed and enzyme is formed (Fig. 4-16). The process is teleologically reasonable, as this enzyme makes available to the plant an additional "substrate space," namely, phosphate esters.

As much as half the phosphate of natural waters can be organic; hence algae capable of producing phosphatases may enjoy a substantial advantage.

A list of plants showing repressible phosphatases is shown in Table 4-13.

Table 4-13   Repression of Phosphatase Activities among Algae*

| Algal Division | Species Showing Repression | Species Not Showing Repression |
|---|---|---|
| *Chrysophyceae* | 3 spp. | None |
| *Bacillariophyceae* | 12 spp. | None |
| *Cryptophyceae* | None | 3 spp. |
| *Cyanophyceae* | *Oscillatoria* sp. | *Coccochlosis* |
| *Dinophyceae* | 2 spp. | None |
| *Chlorophyceae* | *Chlamydomonas* sp. | 3 spp. |
| *Euglenophyta* | *E. gracilis*† | |

\* After Kuenzler and Perras, 1965.
† Price, 1962.

Since the enzyme (at least in *E. coli*) is a zinc metallo-enzyme (Plocke and Vallee, 1962), interactions between zinc and phosphate nutrition are possible.

## 4.8   CASE STUDY: IRON

Iron in the physiology of plants is a jigsaw puzzle in which most of the intricate background of the landscape has been pieced together but with so many pieces missing in the foreground, we have still not caught the meaning of the whole picture.

That plants require iron was demonstrated by Gris in 1844 when he painted solutions of iron salts on yellowing (chlorotic) grape leaves and thus restored the natural color to the leaves.

Plants typically require about 2 $\mu$moles of iron per gram dry weight (Table 4-1). That the bulk of this requirement is associated with chloroplast formation is shown from studies with *Euglena gracilis* (Fig. 4-17). This alga requires

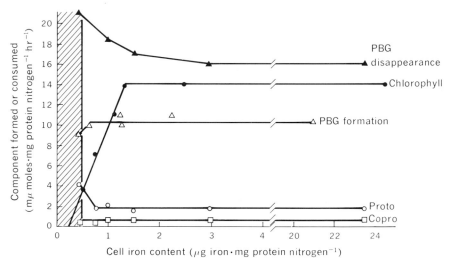

*Fig. 4-17*  Growth and chlorophyll formation in mixotrophic *Euglena gracilis* as a function of the iron contents of the cells (Carell and Price, 1965). Comparison of activities of enzymes associated with porphyrin metabolism with rates of chlorophyll formation. Rates of equivalent PBG disappearance, and Copro, Proto, and chlorophyll formation, and $\delta$-aminolevulinate dehydratase activity ($\triangle$) were determined with frozen and thawed *Euglena* suspensions. All rates are presented as functions of the iron contents of the cells.

only 0.5 $\mu$atom iron per milligram protein for heterotrophic growth in the dark, but almost 1.5 $\mu$atoms to satisfy chloroplast formation.

There is no information on the critical concentrations of iron in the external solution. The problem of maintaining subcritical levels of transition and heavy metal ions in flowing culture solutions may be insuperable. Even with special purification schedules, the contamination levels for iron and zinc are typically in the order of 10 n$M$, which may very well satisfy most plants.

## Mobility

In contrast to the response of plants to low concentrations of nitrogen, phosphate, and the alkali ions in plants, iron-deficiency symptoms appear first in the young, developing leaves. One can have soybean plants, recently deprived of iron, in which the mature leaves are a healthy deep green while the newly expanded leaves are yellow with barely a trace of green. We do not know whether this failure of translocation of iron is due to an absence of turnover of the iron compounds or to a failure of the translocatory machinery of the phloem to transport iron.

## Transport System

A decisive problem for plants in the absorption of iron is the extreme insolubility of ferric hydroxide [$K_{sp}$ of Fe(OH)$_3$ is $1.1 \times 10^{-36}$]. The equilibrium concentration of Fe$^{3+}$ at pH 4 is $1.1 \times 10^{-6}$ $M$ and is, of course, drastically less at higher pH. A consequence is that plants growing in neutral or alkaline soils had to evolve means of complexing or reducing ferric ions in order to extract iron from even iron-rich soils.

J. C. Brown's group (cf. Tiffin, 1966) have reported the occurrence of ferric citrate as the principal iron component of the transpiration stream of soybean plants. They also report that it is missing from varieties that accumulate iron poorly. Although the overall movement of iron from the solution around the roots to the leaves is strongly temperature-dependent (Riekels and Lingle, 1966), the notion that iron transport is "active" remains unproved.

Plants adapted to acidic soil environments (e.g., *Azalea* spp. and the *Ericacae* generally) typically lack any special mechanisms for absorbing iron and depend on low pH in the soil to maintain ferric iron in solution.

A number of circumstances exacerbate iron deficiency in addition to high pH: high concentrations of calcium ("lime-induced chlorosis") and high concentrations of manganese (Table 4-14). Many farm and garden crops have been aided in their quest for iron by the application of synthetic chelating agents, such as EDTA, usually added as the iron chelate (cf. Wallace, 1962). A spectacular case is that of pineapple culture in Hawaii: the plants are grown on soils which are rich in manganese. Although the soils are also rich in iron, the best way of providing the pineapple plant with sufficient iron has been to spray iron chelates directly on the foliage!

Even though we do not know the limiting concentrations of iron for any plant, we can predict that it might be very low. Our reasoning is as follows:

Table 4-14   Conditions that Result in Chloroses Similar to That of Iron-deficiency Chlorosis

The characteristic sign is yellowing of newly expanded leaves.

| Condition | Comments |
| --- | --- |
| Iron deficiency | Prevented by iron |
| High levels of calcium | Prevented by iron or by the addition of chelating agents |
| High levels of manganese and certain other transition and heavy metal ions | Prevented by foliar applications of iron or by the addition of chelating agents |
| Certain genetic mutations affecting the transport of iron | Prevented by high levels of iron in the nutrient solution. Xylem sap of these mutants is low in citrate* |
| High or low temperatures | May be due to excessive absorption of manganese or phosphate |

* Tiffin, 1966.

imagine a carrier for ferric ion with a stability constant similar to that of ferric citrate, $K \simeq 10^{12}$. If a plant could maintain 0.1 m$M$ of the free carrier within the root cell membranes, the concentration of ferric ion in static equilibrium with 0.1 m$M$ of the ferric carrier complex should be $10^{-12}$ $M$! The $K_T$ of such a transport system would be also about $10^{-12}$ $M$.

There are of course ferric chelates that are far more stable than ferric citrate: ferrichrome A (Fig. 4-18), which is a cyclic polypeptide with six oxygens donating electrons to the ferric ion, has an apparent stability constant in excess of $10^{30}$. This substance, thought to be a form of iron storage in the smut fungus *Ustilago sphaerogena,* is stable to pH 12! A carrier containing this structure at a concentration of 1 $\mu M$ could be expected to sop up iron from solutions of less than $10^{-32}$ $M$!

A carrier must, of course, be able to be "turned off" as well; that is, there must be a mechanism for release. Nature may not be able to use some of the most stable chelate structures as carriers simply because she has not discovered the key to reversing their chelation.

Biochemical Basis for Iron Requirements

We can write an impressive catalog of iron-containing compounds in plants (Table 4-15). Some of these are present at lower concentrations in iron-deficient tissues. We can appropriately assign part of the total requirements for iron to these compounds, but we lack sufficient information to construct an inventory of iron compounds in normal or deficient plants. We do not know which, if any, of the known iron compounds become limiting to chlorophyll synthesis and growth.

Several authors agree (Liebich, 1951; Whatley et al., 1951) that most of the iron of spinach leaves is in the chloroplasts, but these experiments need to be repeated with modern methods of intracellular localization (cf. Sec. 1.2).

An early hypothesis was that the iron requirement for chlorophyll synthesis was assignable to the step in porphyrin synthesis at which coproporphyrinogen is converted to protoporphyrin (Sec. 1.9). As attractive as this hypothesis is, the evidence with the alga *Euglena* is that the iron requirement for chlorophyll

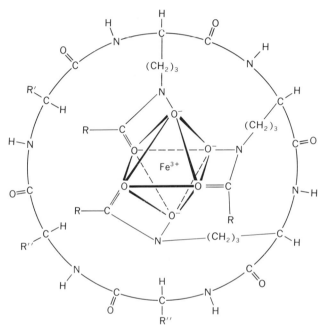

*Fig. 4-18* Structure of ferrichrome A (Neilands, 1964). This cyclic polypeptide forms a cage of oxygen ligands for ferric ion. It serves as an iron-storage compound for the smut fungus *Ustilago sphaerogena* and as a growth factor to certain other microorganisms.

synthesis is still evident at concentrations of cell iron where the coproporphyrinogen step is satisfied (Fig. 4-17). Moreover, in cowpeas (*Vigna sinensis* L.) δ-aminolevulinate is incorporated into chlorophyll at normal rates by iron-deficient tissue (Marsh et al., 1963b), which is inconsistent with a block after this intermediate (cf. Sec. 1.9).

Iron-deficiency chlorosis may, in fact, be due to a requirement for iron in the structural development of the chloroplast. Bogorad et al. (1959) had observed grossly abnormal plastid structures in electron photomicrographs of iron-deficient maize plants. In *Euglena,* not only chlorophyll but all of the

Table 4-15   Behavior of Different Iron-containing Compounds in Iron Deficiency*

| Compound | Response in Iron Deficiency | Comments | References |
|---|---|---|---|
| Catalase | Strongly decreased | Reported for a number of higher plants and *Neurospora* | † |
| Cytochrome *c* | Strongly decreased | Two cases only: cytochrome *c* in *Ustilago* may be aberrant; in *Neurospora*, an indirect assay was employed | Healy et al., 1955; Neilands, 1957 |
| Ferredoxin | Strongly decreased | In *Vigna sinensis* | Marsh et al., 1963a |
| Ferrichrome | Decreased | *Ustilago:* less sensitive than cytochrome *c* | Neilands, 1957, 1964 |
| | Increased production of free ligand in leaves of several plants | Derepression of free ligand proposed | Page, 1966 |
| Hematin‡ | Strongly decreased | Decrease in *Sinapis alba* is proportional to that of chlorophyll | Davenport, 1958; De Kock et al., 1960 |
| Heme | Decreased | In *Vigna sinensis*, decrease was less than that of chlorophyll | Marsh et al., 1963a |
| Cytochrome oxidase | Slightly decreased | In *Neurospora* and *Helianthus*, decrease is approximately 30 percent in extreme deficiency | Healy et al., 1955; Weinstein and Robbins, 1955 |
| Peroxidase | Unchanged | Only slight changes observed in several species | De Kock et al., 1960; Healy et al., 1955; Marsh et al., 1963a |

* From Price, 1968.
† Many authors: Agarwala et al., 1964, 1965; Brown and Hendricks, 1952; De Kock et al., 1960; Healy et al., 1955; Weinstein and Robbins, 1955.
‡ Heme iron constitutes approximately 10 percent of total iron; heme iron is equivalent to about $\frac{1}{60}$ of chlorophyll at all levels of iron (Davenport, 1958).

chloroplast pigments are more or less coordinately deficient at levels of cell iron which were sufficient for heterotrophic growth.

The critical concentration of iron in nutrient solutions (ca. $10^{-6}$ $M$) is near the $K_M$ value ($8 \times 10^{-6}$ $M$) with respect to iron of ferrochelatase (Jones, 1968), the enzyme responsible for the incorporation of $Fe^{++}$ into protoporphyrin IX.

Proto IX $+$ $Fe^{++}$ $\xrightarrow[\text{ferrochelatase}]{}$ protoheme $\rightarrow \rightarrow \rightarrow$ various iron porphyrin enzymes

It follows that this step could be limiting to chloroplast development, through the formation of essential iron porphyrins, such as cytochrome $f$.

## 4.9 CASE STUDY: ZINC

### Phenomena

In the years before 1900, a number of the transition and heavy metal ions were found to promote growth. Among them was zinc, which Raulin reported in 1869 to stimulate the growth of the mold *Aspergillus*. The first promotion by zinc of growth of a higher plant (maize) was not demonstrated until 1914 (Mazé, 1914).

Scientific opinions were divided over whether these stimulations represented true nutritional requirements or nonspecific stimulation. Consequently these metals were often labeled "Reizstoffe."

It was not until the 1930s that sufficient absolute and specific requirements had been demonstrated for plant physiologists to accept the generality of zinc as an essential plant nutrient.

Following the work with higher and lower plants, workers found evidence for the essentiality of zinc for animals, ultimately including humans (cf. Underwood, 1962; Prasad, 1967).

In addition to difficult problems in the analysis of zinc, our understanding of zinc nutrition advanced slowly because the symptoms of zinc deficiency are rather nonspecific: mottling, small leaves, shortened stems (Hewitt, 1963; T. Wallace, 1961; Baumeister, 1955, 1958).

### Quantitative Requirements

Higher plants require on the order of 1 $\mu$mole zinc per gram dry weight (Table 4-1).

We obtained dose-response curves for the growth of *Euglena* by analyzing the growth rates and zinc contents of *Euglena* while the cultures were running out of zinc (Fig. 4-19). We found that, for low concentrations of zinc, the growth rates were linear functions of the internal zinc concentrations.

There are several instances in which the synthesis of a cell constituent requires more zinc than normal growth does: cytochrome $c$ in *Ustilago sphaerogena* (Fig. 4-20), alcohol dehydrogenase, tryptophan synthetase, and hexokinase in *Neurospora,* and alkaline phosphatase in *E. coli* (Table 4-16). These are each advantageous systems in which to study the role of an essential metal, in that striking decreases in the rate of synthesis of the protein can be detected while growth, as measured by bulk protein, continues to increase at normal rates. Under these conditions one can be confident that the effect is not a

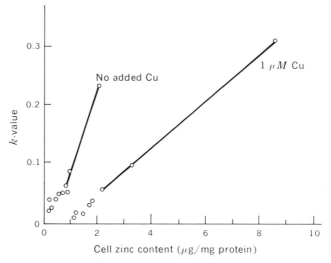

*Fig. 4-19* Growth as a function of limiting zinc content in *Euglena gracilis* (Price and Quigley, 1966). The growth "constant" for binary fission (cf. Eq. 5-4) is plotted as a function of the zinc content of the cells. Copper acts as a competitive inhibitor of growth in the presence of limiting concentrations of zinc.

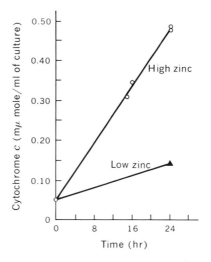

*Fig. 4-20* Requirement of zinc for cytochrome *c* synthesis in the smut fungus *Ustilago sphaerogena* (Grimm and Allen, 1954; D. H. Brown et al., 1966).

consequence of an inhibition of growth, as could occur through negative feed-back inhibition (cf. Sec. 5.2).

The observed requirement of zinc for tryptophan synthetase appears to have a curious consequence to growth physiology: one of the growth substances, indol-3-acetic acid (auxin; cf. Sec. 5.5), is probably derived from tryptophan. Before the zinc requirement for tryptophan synthetase in *Neurospora* was discovered, Skoog (1940) and Tsui (1948) had found low rates of stem elongation, low auxin activities, and low tryptophan contents in zinc-deficient tomato plants.

As was the case with iron, we do not know the critical concentrations for zinc in the external solution. We do know that plants can remove zinc down to less than 10 n$M$. There is no evidence either for or against the active transport of zinc (cf. Epstein, 1966).

Zinc, once settled in a tissue, is virtually immobile (Wood and Sibly, 1950; Milikan and Hanger, 1965).

## Responses of Cells to Zinc

As with iron, one can compile a list of proteins in the cell which contain zinc as essential components (Table 4-16). As we noted earlier, the formation

Table 4-16   Some Zinc-containing Enzymes in Plants

| Enzyme | Behavior in Zinc Deficiency | References |
|---|---|---|
| Many NAD- and NADP-dependent dehydrogenases | Alcohol dehydrogenase greatly decreased in *Neurospora*, yeast; glucose-6-phosphate and triose phosphate dehydrogenases decreased in *Aspergillus* | Nason et al., 1951, 1953 Curdel, 1966 Bertrand and de Wolf, 1957 |
| Certain aldolases | Decreased in clover; zinc not identified in this enzyme | Quinlan-Watson, 1951 |
| Alkaline phosphatase | Decreased in *E. coli* | Torriani et al. (quoted by Price, 1966) |
| NAD-independent D-lactate dehydrogenase | Decreased in *Euglena*, yeast | Price, 1961; Curdel, 1966 |
| Carbonic anhydrase | Decreased in oats | Wood and Sibly, 1952 |

of some of these is more sensitive to low levels of zinc than is growth itself, but none of those identified appear to be the direct instruments for the decreased rates of growth at critical concentrations of zinc.

One possibility is that zinc deficiency interferes with RNA synthesis (Schneider and Price, 1962; Wacker, 1962). Decreased rates of RNA (but not DNA) formation were detected just prior to the onset of growth inhibition

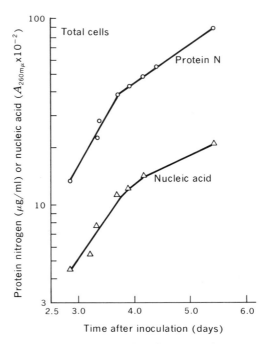

*Fig. 4-21* Nucleic acid and protein formation during the onset of zinc deficiency in *Euglena gracilis* (Schneider and Price, 1962). A decrease in the rate of RNA accumulation occurs just before a failure in protein synthesis.

in zinc deficiency (Fig. 4-21). In late zinc deficiency, a sharp decrease in the concentration of RNA occurs (Kessler and Monselise, 1959; Schneider and Price, 1962; Wacker, 1962; D. H. Brown et al., 1966). RNA (but not DNA) also contains zinc (Wacker and Vallee, 1959; Fuwa et al., 1960), but there is no reason to think that zinc is essential to the action of RNA.

## SUGGESTED CASE STUDIES

### Adaptations in Halophytes

Many plants growing in strongly saline environments either tolerate or actually require sodium chloride. In some cases, salt appears to be actively extruded from such plants through "salt glands."

What adaptations in transport processes, enzyme activation, membrane structure, etc., are associated in the salt-tolerant plant?

*References:* Evans and Sorger, 1966; Kates et al., 1965; Branton, 1969.

### Boron

Borate is required by higher plants and by certain diatoms, but no requirements have ever been detected for green algae, fungi, or animals. What are the peculiarities of multicellular plants that require borate? The element is not translocated, and seems to be unavailable once deposited in a cell.

*References:* Albert, 1965; Bowen and Gauch, 1966; Neales, 1964; Olson and Johnson, 1949; Perkins and Aronoff, 1956; Slack and Whittington, 1964; Yih et al., 1966; Lee and Aronoff, 1966, 1967.

### Manganese

With the discovery that manganese-deficient plants have low Hill reaction activities, we should be close to an understanding of manganese nutrition. But decarboxylases appear to be manganoproteins. What then are the molecular roles of manganese and which are sensitive to manganese deprivation?

*References:* Bergman, 1955; Eyster et al., 1956; Homann and Gaffron, 1964; Hopkins, 1934; Lane et al., 1966; Munns et al., 1963; Pirson, 1937.

### Cobalt

Cobalt is required by plants only when they (or their symbionts) are required to fix atmospheric nitrogen. The requirement for cobalt in some cases can be attributed wholly to a requirement for the synthesis of vitamin $B_{12}$.

*References:* De Hertogh et al., 1964; Johnson et al., 1966.

### Magnesium

Although magnesium is clearly an essential component of chlorophyll, plants deprived of magnesium show various symptoms including decreased growth rates before they become chlorotic. Might the principal requirement of magnesium be to hold ribosomes together?

*References:* Goldberg, 1966; Lindahl et al., 1966; Tempest et al., 1967.

### Nitrate Reduction

We have only outlined the path of nitrate reduction in green plants. The identification of the enzymes involved, the characteristics of metabolic control, and the fact that nitrate reductase itself contains nearly all the functional molybdenum of plant tissues provides subject matter for several detailed case studies. One of the interesting side lights is that the reduction of nitrate liberates oxidizing equivalents, so that "nitrate respiration" can arise as a quantitatively significant factor.

*References:* Beevers et al., 1965; De Renzo et al., 1953; Evans, 1956; Hewitt et al., 1966; Kessler, 1964; Medina and Nicholas, 1957.

## REFERENCES

Agarwala, S. C., C. P. Sharma, and A. Kumar, 1964. Interrelationship of iron and manganese supply in growth, chlorophyll, and iron porphyrin enzymes in barley plants, *Plant Physiol.,* **39**:603–609.

————, ————, and S. Farooq, 1965. Effect of iron supply on growth, chlorophyll, tissue iron and activity of certain enzymes in maize and radish. *Plant Physiol.,* **40**:493–499.

Albert, Luke S., 1965. Ribonucleic acid content, boron deficiency symptoms, and elongation of tomato root tips, *Plant Physiol.,* **40**:649–652.

Arnon, D. I., and P. R. Stout, 1939. The essentiality of certain elements in minute quantity for plants with special reference to copper, *Plant Physiol.,* **14**:371–375.

Aronoff, S., 1963. Introduction of Mg into chlorophylls *a* and *b in vivo, Plant Physiol.,* **38**:628–631.

Asher, C. J., P. G. Ozanne, and J. F. Loneragan, 1965. A method for controlling the ionic environment of plant roots, *Soil Sci.,* **100**:149–156.

———— and J. F. Loneragan, 1967. Response of plants to phosphate concentration in solution cultures: I. Growth and phosphorus content, *Soil Sci.,* **103**:225–233.

———— and P. G. Ozanne, 1967. Growth and potassium content of plants in solution cultures maintained at constant potassium concentrations, *Soil Sci.,* **103**:155–161.

Baumeister, W., 1955. Über den Einfluss des Zinks auf das Pflanzenwachstum, *Protoplasma,* **45**:133–149.

————, 1958. Hauptnährstoffe, in W. Ruhland (ed.), "Handb. Pflanzenphysiol.," vol. IV, pp. 483–557, Springer-Verlag, Berlin.

Beevers, L., L. E. Schrader, D. Flesher, and R. H. Hageman, 1965. The role of light and nitrate in the induction of nitrate reductase in radish cotyledons and maize seedlings, *Plant Physiol.,* **40**:691–698.

Bergman, L., 1955. Stoffwechsel und Mineralsalzernährung Einzelliger Grünalgen: II. Vergleichende Untersuchungen über den Einfluss mineralischer Faktoren bei heterotropher und mixotropher Ernährung, *Flora,* **142**:493–539.

Bertrand, D., and A. de Wolf, 1957. Necessity of zinc as an oligo element, for glucose-6-phosphate dehydrogenase and 6-phosphogluconic acid dehydrogenase of *Aspergillus niger, Compt. Rend.,* **245**:1179–84.

Biddulph, O., 1955. Studies of mineral nutrition by use of tracers, *Botan. Rev.,* **21**:251–95.

Bieleski, R. L., 1968. Effect of phosphorus deficiency on levels of phosphorus compounds in *Spirodela, Plant Physiol.,* **43**:1309–1316.

Bjerrum, J., G. Schwarzenbach, and L. G. Sillen, 1957. "Stability Constants," Part I, "Organic Ligands," pp. 1–105, The Chemical Society, Special Publication #6, London.

Bogorad, L., G. Pires, H. Swift, and W. V. McIlrath, 1959. The structure of chloroplasts in leaf tissue of iron deficient *Xanthium, Brookhaven Symp.,* **11**:132–136.

Bollmann, A., and F. Schwanitz, 1957. Über den Eisengehalt in Pflanzen einer eisenreichen Standorten, *Z. Botan.,* **45**:39–42.

Bowen, J. E., and H. G. Gauch, 1966. Nonessentiality of boron in fungi and the nature of its toxicity, *Plant Physiol.,* **41**:319–324.

Branton, D., 1969. Membrane structure, *Ann. Rev. Plant Physiol.,* **20**:209–238.

Briggs, G. E., A. B. Hope, and M. G. Pitman, 1958. Exchangeable ions in beet disks at low temperature, *J. Exp. Botany,* **9**:128–141.

Brown, D. H., R. A. Cappellini, and C. A. Price, 1966. Actinomycin D inhibition of zinc-induced formation of cytochrome *c* in *Ustilago, Plant Physiol.,* **41**:1543–1546.

Brown, J. C., and S. B. Hendricks, 1952. Enzymatic activities as indications of copper and iron deficiencies in plants, *Plant Physiol.,* **27**:651–660.

Brownell, P. F., 1965. Sodium as an essential micronutrient element for a higher plant (*Atriplex vesicaria*), *Plant Physiol.,* **40**:460–468.

Broyer, T. C., and D. Hoagland, 1943. Metabolic activities of roots and their bearing on the relation of upward movement of salts and water in plants, *Am. J. Botany,* **30**:261–273.

Burris, R. H., 1966. Biological nitrogen fixation, *Ann. Rev. Plant Physiol.,* **17**:155–184.

Carell, E. F., and C. A. Price, 1965. Porphyrins and the iron requirement for chlorophyll formation in *Euglena, Plant Physiol.,* **40**:1–7.

Chaberek, S., and A. E. Martell, 1959. "Organic Sequestering Agents," pp. 1–616, John Wiley & Sons, New York.

Chance, B., and P. K. Maitra, 1963. Determination of the intracellular phosphate potential by reversed electron transfer, *8th Symp. Soc. Gen. Physiol. 1961* (Pub. 1963), pp. 307–332.

Chapman, H. D., and G. F. Liebig, Jr., 1940. Nitrate concentration and ion balance in relation to citrus nutrition, *Hilgardia,* **13**:141–173.

Clark, H. E., and J. W. Shive, 1934. The influence of the pH of a culture solution on the rates of absorption of ammonium and nitrate nitrogen by the tomato plant, *Soil Sci.,* **37**:203–225.

Crafts, A. S., and T. C. Broyer, 1938. Migration of salts and water into xylem of the roots of higher plants, *Am. J. Botany,* **25**:529–535.

Curdel, A., 1966. Etude du rôle du metal dans l'activité de la D-lacticodeshydrogenase de la levure, Ph.D. Thesis, Univ. de Paris.

Davenport, H. E., 1958. The effects of some micronutrient deficiencies on the concentration of haem and chlorophyll in leaves, *Proc. IV Intl. Congr. Biochem.,* pp. 11–51.

De Hertogh, A. A., P. A. Mayeux, and H. J. Evans, 1964. The relationship of cobalt requirement to propionate metabolism in *Rhizobium, J. Biol. Chem.,* **239**:2446–2453.

De Kock, P. C., K. Commisiong, V. C. Farmer, and R. H. E. Inkson, 1960. Interrelationships of catalase, peroxidase, hematin, and chlorophyll, *Plant Physiol.,* **35**:599–604.

De Renzo, E. C., E. Kaleita, P. G. Heytler, J. J. Oleson, B. L. Hutchings, and J. H. Williams, 1953. Identification of the xanthine oxidase factor as molybdenum, *Arch. Biochem. Biophys.,* **45**:247–253.

Drescher-Kaden, F. K., and F. Schwanitz, 1956. Über den Strontium-und Calciumgehalt in Pflanzen einer strontiumhaltigen Standorten, *Z. Botan.,* **44**:501–504.

Elzam, O. E., and E. Epstein, 1965. Absorption of chloride by barley roots: Kinetics and selectivity, *Plant Physiol.,* **40**:620–624.

Epstein, E., 1966. Dual pattern of ion absorption by plant cells and by plants, *Nature,* **212**:1324–1327.

———— and C. E. Hagen, 1952. A kinetic study of the absorption of alkali cations by barley roots, *Plant Physiol.,* **27**:457–474.

————, D. W. Rains, and O. E. Elzam, 1964. Resolution of dual mechanisms of potassium absorption by barley roots, *Proc. Nat. Acad. Sci.,* **49**:684–692.

———— and ————, 1965. Carrier-mediated cation transport in barley roots: kinetic evidence for a spectrum of active sites, *Proc. Nat. Acad. Sci.,* **53**:1320–1324.

Evans, H. J., 1956. Role of molybdenum in plant nutrition, *Soil Sci.,* **81**:199–258.

———— and G. J. Sorger, 1966. Role of mineral elements with emphasis on the univalent cations, *Ann. Rev. Plant Physiol.,* **17**:47–76.

Eyster, H. C., T. E. Brown, and H. A. Tanner, 1956. Manganese requirement with respect to respiration and the Hill reaction in *Chlorella pyrenoidosa, Arch. Biochem. Biophys.,* **64**:240–241.

Folkes, B. F., 1959. The position of amino acids in the assimilation of nitrogen and the synthesis of proteins in plants, *Symp. Soc. Exp. Biol.,* **XIII**:126–147.

Fowden, L., 1962. The non-protein amino acids of plants, *Endeavour,* **21**:35–42.

Fried, M., and R. E. Shapiro, 1961. Soil-plant relationships in ion uptake, *Ann. Rev. Plant Physiol.,* **12**:91–112.

Fuwa, K. I., W. E. C. Wacker, R. Druyan, A. F. Bartholomay, and B. L. Valleee, 1960. Nucleic acids and metals. II: transition metals as determinants of the conformation of ribonucleic acids. *Proc. Nat. Acad. Sci.,* **46**:1298–1307.

Gerloff, G. C., and F. Skoog, 1954. Cell contents of nitrogen and phosphorus as a measure of their availability for growth of *Microcystis aeruginosa, Ecology,* **35**:348–353.

Goldacre, R. J., 1952. The folding and unfolding of protein molecules as a basis of osmotic work, *Intern. Rev. of Cytology,* **1**:135–164.

Goldberg, A., 1966. Magnesium binding by *Escherichia coli* ribosomes, *J. Mol. Biol.,* **15**:663–673.

Grimm, P. W., and P. J. Allen, 1954. Promotion by zinc of the formation of cytochromes in *Ustilago sphaerogena, Plant Physiol.,* **29**:369–377.

Gris, E., 1844. Nouvelles experiences sur l'action de composés ferrugineux solubles appliqué à la végétation et spécialement en traitement de la chlorose et de la débilité des plantes, *C. R. Acad. Sci. (Paris),* **19**:1118–1119.

Healy, W. B., S. Cheng, and W. D. McElroy, 1955. Metal toxicity and iron deficiency effects on enzymes in *Neurospora, Arch. Biochem. Biophys.,* **54**:206–214.

Hendricks, S. B., 1964. Salt transport across cell membranes, *Am. Scientist,* **52**:306–333.

Hewitt, E. J., 1959. The metabolism of micronutrient elements in plants, *Biol. Rev.,* **34**:333–377.

———, 1963. The essential nutrient elements: requirements and interactions in plants, in F. C. Steward (ed.), "Plant Physiology," vol. III, pp. 137–360, Academic Press, New York.

———, 1966. "Sand and Water Culture Methods Used in the Study of Plant Nutrition," Tech. Commun. No. 22, 2d ed., 547 pp. Commonwealth Agric. Bur., Farnham Royal, Bucks, England.

———, D. P. Hucklesby, and G. F. Betts, 1966. Resolution of nitrite and hydroxylamine reductase activities from plants, *Biochem. J.,* **100**:54.

Hill, H., and H. B. Cannon, 1948. Nutritional studies by means of tissue tests with potatoes grown on a muck soil, *Sci. Agr.,* **28**:185–199.

Hoagland, D. R., 1944. "Lectures on the Inorganic Nutrition of Plants," 226 pp., Chronica Botanica Co., Waltham, Mass.

Hodges, T. K., and J. B. Hanson, 1965. Calcium accumulation by maize mitochondria, *Plant Physiol.,* **40**:101–109.

Homann, Peter, and H. Gaffron, 1964. Flavin specific and manganese dependent action of certain herbicides on photoreactions *in vitro, Plant Physiol.,* **39**:xxxiii.

Hopkins, R. C., 1934. Manganese an essential element for green plants, *N.Y. Agr. Exp. Sta. (Cornell) Mem.,* **151**:1–40.

James, B. R., J. R. Lyons, and R. J. P. Williams, 1962. Iron dipyridyl complexes as models for iron-porphyrin proteins, *Biochemistry,* **1**:379–85.

James, T. W., 1961. Continuous culture of microorganisms, *Ann. Rev. Microbiol.,* **15**:27–46.

Johnson, G. V., P. A. Mayeux, and H. J. Evans, 1966. A cobalt requirement for symbiotic growth of *Azolla filiculoides* in the absence of combined nitrogen, *Plant Physiol.,* **41**:852–855.

Johnston, E. S., and D. R. Hoagland, 1929. Minimum potassium level required by tomato plants grown in water cultures, *Soil Sci.,* **27**:89–109.

Jones, O. T. G., 1968. Ferrochelatase of spinach chloroplasts, *Biochem. J.,* **107**:113–119.

Kates, M., L. S. Yengoyan, and P. S. Sastry, 1965. Isolation and characterization of a diether analog of phosphatidyl glycerophosphate from *Halobacterium* cutirubrum, *Biochim. Biophys. Acta,* **98**:252–268.

Kenefick, D. G., and J. B. Hanson, 1966. Contracted state as an energy source for Ca binding and Ca$^+$ inorganic phosphate accumulation by corn mitochondria, *Plant Physiol.,* **41**:1601–1609.

Kessler, E., 1964. Nitrate assimilation by plants, *Ann. Rev. Plant Physiol.,* **15**:57–72.

———— and S. P. Monselise, 1959. Studies on ribonuclease, ribonucleic acid and protein synthesis in healthy and zinc-deficient citrus leaves, *Physiol. Plantarum,* **12**:1–7.

Kidder, G. W., and V. C. Dewey, 1951. The biochemistry of ciliates in pure culture, in A. Lwoff (ed.), "Biochemistry and Physiology of Protozoa," vol. I, pp. 323–400, Academic Press, New York.

Klebs, G., 1903. "Willkürliche Entwicklungsänderung bei Pflanzen," Jena.

Knauss, H. J., and J. W. Porter, 1954. The absorption of inorganic ions by *Chlorella pyrenoidosa, Plant Physiol.,* **29**:229–234.

Kraus, E. J., and H. R. Kraybill, 1918. Vegetation and reproduction with special reference to tomato, *Oregon State College, Agr. Expt. Sta. Bull.,* **149**:1–90.

Krauss, R. W., 1958. Physiology of the fresh-water algae, *Ann. Rev. Plant Physiol.,* **9**:207–244.

Kuensler, E. J., and B. H. Ketchum, 1962. Rate of phosphorus uptake by *Phaeodactylum tricornutum, Biol. Bull.,* **123**:134–145.

———— and J. P. Perras, 1965. Phosphatases of marine algae, *Biol. Bull.,* **128**:27–284.

Lane, M. D., H. Chang, H. Maruyama, R. Miller, and A. Mildvass, 1966. Mitochondrial PEP carboxykinase, *Fed. Proc.,* **25**:585.

Laties, G. G., 1969. Dual mechanisms of salt uptake in relation to compartmentation and long-distance transport, *Ann. Rev. Plant Physiol.,* **20**:89–116.

———— and K. Budd, 1964. The development of differential permeability in isolated steles of corn roots, *Proc. Nat. Acad. Sci.,* **52**:462–469.

Lee, S. G., and S. Aronoff, 1966. Investigations on the role of boron in plants: III. Anatomical observations, *Plant Physiol.,* **41**:1570–1577.

———— and ————, 1967. Boron in plants: A biochemical role, *Science,* **158**:798–799.

Liebich, H., 1951. Quantitativ-chemisch Untersuchungen über Eisen in der

Chloroplasten und übrigen Zellbestandteilen von *Spinacia oleracea, Z. Botan.,* **37:**129–157.

Lindahl, T., A. Adams, and J. R. Fresco, 1966. Renaturation of transfer ribonucleic acids through site binding of magnesium, *Proc. Nat. Acad. Sci.,* **55:**941–948.

Lüttge, U., and G. G. Laties, 1966. Dual mechanisms of ion absorption in relation to long distance transport in plants, *Plant Physiol.,* **41:**1531–1539.

Lynn, W. S., and R. H. Brown, 1967. P/2e⁻ ratios approaching 4 in isolated chloroplasts, *J. Biol. Chem.,* **242:**412–417.

McCollum, E. V., 1957. "A History of Nutrition," 451 pp., Houghton-Mifflin, Cambridge, Mass.

MacDonald, I. R., J. S. D. Bacon, D. Vaughan, and R. J. Ellis, 1966. The relation between ion absorption and protein synthesis in beet disks, *J. Exp. Botany,* **17:**822–837.

Manery, J. F., 1966. Effects of Ca ions on membranes, *Fed. Proc.,* **25:**1804–1810.

Marsh, H. V., Jr., H. J. Evans, and G. Matrone, 1963a. Investigations of the role of iron in chlorophyll metabolism: I. Effect of iron deficiency on chlorophyll and heme content and on the activities of certain enzymes in leaves, *Plant Physiol.,* **38:**632–637.

———, ———, and ———, 1963b. Investigations of the role of iron in chlorophyll metabolism: II. Effect of iron deficiency on chlorophyll synthesis, *Plant Physiol.,* **38:**638–641.

Mayer, A. M., 1954. Some heavy metals in lettuce seeds and changes in them during germination, *Physiol. Plantarum,* **7:**771–781.

Mazé, P., 1914. Influences respectives des éléments de la solution minérale sur le development du maïs, *Ann. Inst. Pasteur,* **28:**21–48.

Medina, A., and D. J. D. Nicholas, 1957. Metallo enzymes in the reduction of nitrite to ammonia in *Neurospora, Biochim. Biophys. Acta,* **25:**138–141.

Meiklejohn, G. T., and C. P. Stewart, 1941. Ascorbic acid oxidase from cucumber, *Biochem. J.,* **35:**755–760.

Milikan, C. R., and B. C. Hanger, 1965. Effects of chelation and of various cations on the mobility of foliar-applied ⁶⁵Zn in subterranean clover, *Aust. J. Biol. Sci.,* **18:**953–957.

Millard, D. L., J. T. Wiskich, and R. N. Robertson, 1964. Ion uptake by plant mitochondria, *Proc. Nat. Acad. Sci.,* **52:**996–1004.

Mitchell, P., 1967. Active transport and ion accumulation, *Comprehensive Biochem.,* **22:**167–197.

Moser, H., 1958. "The Dynamics of Bacterial Populations Maintained in the Chemostat," Carnegie Inst. Publication 614, Washington, D.C.

Munns, D. N., C. M. Johnson, and L. Jacobson, 1963. Uptake and distribution of manganese in oat plants: II. A kinetic model, *Plant and Soil,* **19:**193–204.

Murata, T., and T. Akazawa, 1968. Enzymic mechanism of starch synthesis in sweet potato roots: I. Requirement of potassium ion for starch synthetase, *Arch. Biochem. Biophys.*, **126**:873–879.

Nason, A., 1950. Effect of zinc deficiency on synthesis of tryptophan by *Neurospora* Extracts, *Science*, **112**:111.

————, N. O. Kaplan, and S. P. Colowick, 1951. Changes in enzymatic constitution in zinc-deficient *Neurospora*, *J. Biol. Chem.*, **188**:397–406.

————, ————, and H. A. Oldewurtel, 1953. Further studies of nutritional conditions affecting enzymatic constitution in *Neurospora*, *J. Biol. Chem.*, **201**:435–444.

National Academy of Sciences-National Research Council, 1966. "The Plant Sciences," 167 pp.

Neales, T. F., 1964. A comparison of the boron requirements of intact tomato plants and excised tomato roots grown in sterile culture, *J. Exp. Botany*, **15**:647–653.

Neilands, J. B., 1957. Some aspects of microbial iron metabolism, *Bacteriol. Rev.*, **21**:101–111.

————, 1964. Biochemistry of the ferrichrome compounds, *Experientia (Supplement 9)*, pp. 222–232.

Nicholas, D. J. D., and K. Commissiong, 1957. Effect of molybdenum, copper, and iron on some enzymes in *Neurospora crassa*, *J. Gen. Microbiol.*, **17**:699–707.

Nightingale, G. T., 1937. The nitrogen nutrition of green plants, *Botan. Rev.*, **3**:85–174.

————, 1948. The nitrogen nutrition of green plants: II, *Botan, Rev.*, **14**:185–221.

————, 1955. "Horticultural Science," 111 pp., Rutgers University Press, New Brunswick, N.J.

Nobbé, Fr., 1865. Die Züchtung der Landpflanzen im Wasser betreffend, *Ebenda*, **7**:68–73.

Nobel, P. S., and L. Parker, 1965. Light-dependent ion translocation in spinach chloroplasts, *Plant Physiol.*, **49**:633–640.

Olsen, C., 1950. The significance of concentration for the rate of ion absorption by higher plants in water culture, *Physiol. Plantarum*, **3**:152–164.

————, 1953. The significance of concentration for the rate of ion absorption by higher plants in water culture: II. Experiments with aquatic plants, *Physiol. Plantarum*, **6**:837–843.

Olson, B. H., and M. J. Johnson, 1949. Factors producing high yeast yields in synthetic media, *J. Bacteriol.*, **57**:235–246.

Orgel, L. E., 1960. "An Introduction to Transition-metal Chemistry," 180 pp., Methuen, London.

Page, E. R., 1966. Sideramines in plants and their possible role in iron metabolism, *Biochem. J.*, **100**:34P.

Pardee, A. B., 1968. Membrane transport proteins, *Science,* **162**:632–637.

Perkins, J. H., and S. Aronoff, 1956. Identification of the blue-fluorescent compounds in boron-deficient plants, *Arch. Biochem. Biophys.,* **64**:506–507.

Pirschle, K., 1931. Nitrate und Ammonsalze als Stickstoffquellen für höhere Pflanzen bei konstanter Wasserstoffionenkonzentration: III, *Planta,* **14**:583–676.

Pirson, A., 1937. Ernährungs-und stoffwechselphysiologische Untersuchungen an *Fontinalis* und *Chlorella, Z. Botanik,* **31**:193–267.

———, 1955. Functional aspects in mineral nutrition of green plants, *Ann. Rev. Plant Physiol.,* **6**:71–114.

Plocke, Donald J., and B. L. Vallee, 1962. Interaction of alkaline phosphatase of *E. coli* with metal ions and chelating agents, *Biochemistry,* **1**:1039–1043.

Prasad, A. S., 1967. Nutritional metabolic role of zinc, *Fed. Proc.,* **26**:172–185.

Prianishnikov, D. N., 1951. "Nitrogen in the Life of Plants," S. A. Wilde (trans.), 109 pp., Kramer Business Service, Inc., Madison, Wis.

Price, C. A., 1961. A zinc-dependent lactate dehydrogenase in *Euglena gracilis, Biochem. J.,* **82**:61–66.

———, 1962. Repression of acid phosphatase synthesis in *Euglena gracilis, Science,* **135**:46.

———, 1966. Control of processes sensitive to zinc in plants and microorganisms, in A. S. Prasad (ed.), "Zinc Metabolism," pp. 1–20, C. C. Thomas, Publisher, Springfield, Ill.

———, 1968. Iron compounds and plant nutrition, *Ann. Rev. Plant Physiol.,* **19**:239–248.

——— and E. Millar, 1962. Zinc, growth, and respiration in *Euglena, Plant Physiol.,* **37**:423–427.

——— and B. L. Vallee, 1962. *Euglena gracilis:* A test organism for study of zinc, *Plant Physiol.,* **37**:428–433.

——— and J. W. Quigley, 1966. A method for determining quantitative zinc requirements for growth, *Soil Sci.,* **101**:11–16.

Provasoli, L., and I. J. Pintner, 1953. Ecological implications of *in vitro* nutritional requirements of algal flagellates, *Ann. N.Y. Acad. Sci.,* **56**:839–851.

——— and S. H. Hutner, 1964. Nutrition of algae, *Ann. Rev. Plant Physiol.,* **15**:37–56.

Quinlan-Watson, T. A. F., 1951. Aldolase activity in zinc-deficient plants, *Nature,* **167**:1033–1034.

Raulin, J., 1869. Études chimiques sur la végétation, *Ann. Sci. Naturelles Botanique,* **11**:93–299.

Riekels, J. W., and J. C. Lingle, 1966. Iron uptake and translocation by tomato plants as influenced by root temperature and manganese nutrition, *Plant Physiol.,* **41**:1095–1101.

Rosenberg, T., and W. Wilbrandt, 1963. Carrier transport uphill: I. General, *J. Theoret. Biol.,* **5:**288–305.

Roux, L., 1966. Influence de la deficience en phosphore sur la composition des fractions phosphorées de la plante de tomate, *Ann. Physiol. Veg.,* **8:**137- 145.

Ruhland, W., and U. Wetzel, 1927. Zur Physiologie der organischen Säuren in grünen Pflanzen: III. *Rheum hybridum, Planta,* **3:**765–769.

Ryther, J. R., 1954. The ecology of phytoplankton blooms in Moriches Bay and Great South Bay, Long Island, New York, *Biol. Bull.,* **106:**198–209.

Scharrer, K., 1941. "Biochemie der Spurenelemente," 272 pp., Paul Parey, Berlin.

Schmalfuss, K., 1941. Über die Wandlungen der Phosphorverbindungen in der reifenden Maisfrucht, insonderheit bei verschiedenen Ernährung der Pflanze, *Bodenkunde u. Pflazenernährung,* **20:**51–177.

Schmid, W. E., H. P. Haag, and E. Epstein, 1965. Absorption of zinc by excised barley roots, *Physiol. Plantarum,* **18:**860–868.

Schneider, E., and C. A. Price, 1962. Decreased ribonucleic acid levels: A possible cause of growth inhibition in zinc deficiency, *Biochim. Biophys. Acta,* **55:**406–408.

Skoog, F., 1940. Relationships between zinc and auxin in the growth of higher plants, *Amer. J. Botany,* **27:**939–951.

Slack, C. R., and W. J. Whittington, 1964. The role of boron in plant growth: The effects of differentiation and deficiency on radicle metabolism, *J. Exp. Botany,* **15:**495–514.

Smith, R. C., and E. Epstein, 1964. Ion absorption by shoot tissue: Kinetics of potassium and rubidium absorption by corn leaf tissue, *Plant Physiol.,* **39:**992–996.

Sorger, G. J., E. Ford, and H. J. Evans, 1965. Effects of univalent cations on the immunoelectrophoretic behavior of pyruvic kinase, *Proc. Nat. Acad. Sci.,* **54:**1614–1621.

Stiles, W., 1916. On the interpretation of the results of water culture experiments, *Ann. Bot.,* **30:**427–436.

Stout, P. R., 1961. "Proceedings 9th Annual California Fertilizer Conference," pp. 21–23.

Sutcliffe, J. F., 1962. "Mineral Salts Absorption in Plants," vol. 1, pp. 1–194, International Series of Monographs on Pure and Applied Biology, Pergamon Press—The Macmillan Company, New York.

Takagi, T., and T. Isemura, 1965. Necessity of calcium for the renaturation of reduced taka-amylase A, *J. Biochem.,* **57:**89–95.

Tempest, D. W., J. W. Dicks, and J. L. Meers, 1967. Magnesium and the growth of *Bacillus subtilis, Biochem. J.,* **102:**36.

Tiffin, L. O., 1966. Iron translocation: II. Citrate/iron ratios in plant stem exudates, *Plant Physiol.,* **41**:515–518.

Tori, K., and G. G. Laties, 1966. Dual mechanisms of ion uptake in relation to vacuolation in corn roots, *Plant Physiol.,* **41**:863–870.

Tromp, J., 1962. Interactions in the absorption of ammonium, potassium, and sodium ions by wheat roots, *Acta Botan. Neerland,* **11**:147–192.

Tsui, C., 1948. The role of zinc in auxin synthesis in the tomato plant, *Amer. J. Botany,* **35**:172–179.

Ulrich, A., 1964. The relative constancy of the critical nitrogen concentration of sugar beet plants, in C. Bould, P. Prevet, and J. R. Magness (eds.), "Plant Analysis and Fertilizer Problems," vol. IV, pp. 371–391, Amer. Soc. Hort. Sci., Humprey Press, Geneva, N.Y.

Underwood, E. J., 1962. "Trace Elements in Human and Animal Nutrition," 2d ed., 492 pp., Academic Press, New York.

Uyeda, K., and E. Racker, 1965. Regulatory mechanisms in carbohydrate metabolism: VIII. The regulatory function of phosphate in glycolysis, *J. Biol. Chem.,* **240**:4689–4693.

van den Honert, T. H., 1937. Over de eigenschappen van plantenwortels, welke een rol spelen bij de opname van voedingzoaten, *Natur. Tijdschr. Ned. Indië,* **97**:150–163.

——— and J. J. M. Hooymans, 1955. On the absorption of nitrate by maize in water culture, *Acta. Botan. Neerland,* **4**:376–384.

Wacker, W. E. C., 1962. Nucleic acids and metals: III. Changes in nucleic acid, protein, and metal content as a consequence of zinc deficiency in *Euglena gracilis, Biochemistry,* **1**:859–865.

——— and B. L. Vallee, 1959. Nucleic acids and metals: I. Chromium, manganese, nickel, iron, and other metals in ribonucleic acid from diverse biological sources, *J. Biol. Chem.,* **234**:3257–3262.

Walker, J. B., 1954. Inorganic micronutrient requirements of *Chlorella:* II. Quantitative requirements for iron, manganese, and zinc, *Arch. Biochem. Biophys.,* **53**:1–8.

Walker, R. B., 1954. The ecology of serpentine soils: II. Factors affecting plant growth on serpentine soils, *Ecology,* **35**:259–266.

Wallace, A., 1962. "A Decade of Synthetic Chelating Agents in Plant Nutrition," 195 pp., Arthur Wallace, Los Angeles, Calif. (no publisher listed).

———, 1966. "Current Topics in Plant Nutrition," 224 pp., published by the author, 2278 Parnell Ave., Los Angeles, Calif., 90064.

Wallace, T., 1961. "The Diagnosis of Mineral Deficiencies in Plants by Visual Symptoms," 3d ed., 125 pp., Chemical Publishing Co., New York.

Wang, J. J., 1955. On the detailed mechanism of a new type of catalase-like action, *J. Am. Chem. Soc.,* **77**:822–823.

Waring, W. S., and C. H. Werkman, 1944. Iron deficiency in bacterial metabolism, *Arch. Biochem.,* **4**:75–87.

Weinstein, L. H., and W. R. Robbins, 1955. The effect of different iron and manganese nutrient levels on the catalase and cytochrome oxidase activities of green and albino sunflower leaf tissues, *Plant Physiol.,* **30**:27–32.

Welch, R. M., and E. Epstein, 1969. The plasmalemma: Seat of the type 2 mechanisms of ion absorption, *Plant Physiol.,* **44**:301–304.

Whatley, F. R., L. Ordin, and D. I. Arnon, 1951. Distribution of micronutrient metals in leaves and chloroplast fragments, *Plant Physiol.,* **26**:414–418.

Williams, R. J. P., 1959. Coordination, chelation, and catalysis, in P. D. Boyer, H. Lardy, and K. Myrback (eds.), "The Enzymes," 2d ed., vol. 1, pp. 391–441, Academic Press, New York.

Williams, R. F., 1955. Redistribution of mineral elements during development, *Ann. Rev. Plant Physiol.,* **6**:25–42.

Wood, J. G., and P. M. Sibly, 1950. The distribution of zinc in oat plants, *Aust. J. Sci. Res.,* **3**:14–27.

——— and ———, 1952. Carbonic anhydrase activity in plants in relation to zinc content, *Aust. J. Sci. Res.,* **5**:244–255.

Wyman, J., 1965. The binding potential, a neglected linkage concept, *J. Mol. Biol.,* **11**:631–644.

Yih, R. Y., F. K. Hille, and H. E. Clark, 1966. Requirement of Ginkgo pollen-derived tissue cultures for boron and effects of boron deficiency, *Plant Physiol.,* **41**:815–820.

Yocum, C. S., 1960. Nitrogen fixation, *Ann. Rev. Plant Physiol.,* **11**:25–36.

# 5
# Growth
# and
# Development

*"It has not escaped our notice that the specific
pairing we have postulated immediately suggests a
possible copying mechanism for the genetic material."*
——Watson and Crick, 1953

## 5.1    THE DOGMA[1]

The phenomena of growth, differentiation, development, and reproduction remain among the most mysterious and fascinating of all biological phenomena. To scientists of earlier times these phenomena appeared to be the product of exclusively biological forces of unfathomable subtlety and complexity. Biologists dominated by the doctrine of vitalism despaired of achieving other than a descriptive understanding of the living world. Even today one finds in such previously impregnable areas as morphogenesis some of the last fitfully defended outposts of vitalism.

Out of the proposals by Watson and Crick (1953) for a structure for deoxyribonucleic acids (DNA), an intellectual structure of awesome proportions has arisen. This set of abstractions not only comprehends much of genetics and biochemistry but is beginning to include growth and reproduction as well. The Dogma, as it is jocularly known, accounts for some phenomena in detail; in other areas it is only vague surmise, but throughout its expanding range, predictions from The Dogma have proved so accurate that it has passed from hypothesis to theory, and now bids for canonization as law.

The notion, originally proposed by Francis Crick as a dogma, was that information flows irreversibly from DNA to protein. As with other dogmas, this one has grown. We now would represent information flow as

$$\overset{\curvearrowleft}{\text{DNA}} \rightarrow \text{RNA} \rightarrow \text{protein}$$

In words, we say that DNA may specify itself or RNA, and RNA in turn specifies protein. Let us review some generalities of The Dogma, and add interpretations that are especially applicable to green plants.

### Duplication: The Structure of DNA and Reproduction at the Molecular Level

The significant event in reproduction is the transfer of genetic information as one DNA molecule is copied from another. DNA is a long[2] heteropolymer of deoxyribonucleotides (Fig. 5-1), in which four kinds of purine and pyrimidine bases are joined along an invariant backbone of deoxyribose phosphates. Genetic information is coded into DNA by the sequence of bases (see The Code, Genes, and Cistrons below).

Most DNA molecules exist as a double-stranded helix (Fig. 5-1) in which each base is hydrogen-bonded to its complement: adenine to thymine (and thymine to adenine), cytosine to guanine (and guanine to cytosine). Since each base has a fixed complement, exactly the same information is coded into each complementary strand of a DNA pair, just as a right-hand glove contains the same information as a left-hand glove.

With the benefit of hindsight, one could predict that the relatively rigid

---

[1] References: Watson, 1965; Bonner, 1965; Ingram, 1965; Khorana, 1965.
[2] An *E. coli* cell, 2 $\mu$ long, contains 2 cm of DNA.

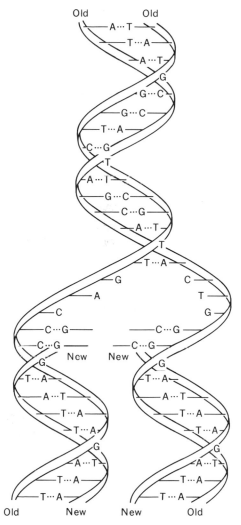

*Fig. 5-1*  Structure and reproduction of DNA
(Watson, 1965). In this revolutionary model,
the molecular structure of DNA provides not
only a means of coding genetic information but
the means of reproducing itself.

double helix of double-stranded DNA would "melt" into two random coils
at temperatures where the thermal kinetic energy exceeds the hydrogen bond
energies. (One sees this experimentally as a small increase in the light absorption
at 260 m$\mu$ as the two strands separate.) In fact, thermal "melting profiles"
provide a criterion for determining if nucleic acids are double-stranded.

During DNA duplication the two strands of each DNA begin to separate (Fig. 5-1). Free deoxyribonucleotides line up opposite their complementary nucleotide residues in the free ends of each chain. An enzyme, DNA polymerase, then zips together the two new chains.

DNA duplication is termed "semiconservative," because each new double helix contains one old strand and one newly synthesized one. The crucial point is that all of the information contained in each of the two old strands is conveyed without error to each new strand. In fact from natural mutation rates one can deduce that either nucleotides align themselves in the correct order with an error of only 1 in $10^6$ or that some equally precise mechanism exists for eliminating errors subsequent to the synthesis of a new DNA.

### The Code, Genes, and Cistrons

We have stated that genetic information is coded into the sequence of nucleotides in DNA. Any kind of *information* including genetic information is a means of selecting among alternatives. The greater the number of alternatives or choices, the greater the information. The alternatives open to a living organism, according to The Dogma, are the selection of amino acids to be incorporated into proteins. The biochemical expression of genetic information is therefore in the amino acid sequence of proteins. The sequence of three nucleotides in DNA specifies one amino acid. For example, the triplet AAA in a molecule of DNA ultimately specifies phenylalanine. A version of the amino acid code is shown in Table 5-1. Since the code contains information and must be *transcribed,* we can gain some insights into the nature of the code by using the metaphors of language. We speak of the nucleotides as an alphabet, of the triplets as words or codons and anticodons, of combinations as having grammar, ambiguity, and redundancy. Thus the grammar of the code is thought to be universal (Basilio et al., 1966), unambiguous, and redundant (or degenerate); it probably does not contain commas and is read from right to left, that is, from the 5' end of the nucleotide (Thatch et al., 1965)!

One of the urgent objectives of molecular biology is to be able to "read" a given sample of DNA—that is, to determine the actual sequence of nucleotides. (The sequence of certain tRNAs has been determined.) There is a possibility that DNA may be read in the electron microscope (Moudrianakis and Beer, 1965).

We can think of a *gene* as that section of a DNA chain required to specify one protein. One can have as many single (point) mutations within a gene as there are nucleotides. Each point mutation results therefore in the alteration of a triplet. An altered gene is an *allele* of the original. An alteration usually results in a triplet specifying a wrong amino acid (Fig. 5-2). The resulting protein may have normal biological activity, altered activity, or no activity. Enzymes from different organisms with similar but not identical properties may differ by a few amino acids (cf. Fitch and Margoliash, 1967; Zucker-

Table 5-1   Codons for the Incorporation of Amino Acids into Proteins*

| First Letter | Second Letter | | | | Third Letter |
|---|---|---|---|---|---|
| | U | C | A | G | |
| | phe | ser | tyr | cys | G |
| U | phe | ser | tyr | cys | C |
| | leu | ser | ochre† | opalescent† | A |
| | leu | ser | amber† | try | G |
| | leu | pro | his | arg | U |
| C | leu | pro | his | arg | C |
| | leu | pro | gln | arg | A |
| | leu | pro | gln | arg | G |
| | ileu | thr | asp | ser | U |
| A | ileu | thr | asp | ser | C |
| | ileu | thr | lys | arg | A |
| | met‡ | thr | lys | arg | G |
| | val | ala | asp | gly | U |
| G | val | ala | asp | gly | C |
| | val | ala | glu | gly | A |
| | val | ala | glu | gly | G |

* After Morgan et al., 1966.
† Ochre, amber, and opalescent are names of classes of bacterial mutants. The codons terminate the reading of a message; that is, they code for "stop here."
‡ AUG codes for both met and formylmethionine. The latter is invariably at the beginning of bacterial proteins; so that one can think of AUG as coding for "start here." It is not known, however, how the cell distinguishes between a code for such chain initiation and for methionine in the middle of a chain.

kandl and Pauling, 1965; Heller and Smith, 1965). The genes specifying these different enzymes may be thought of as analogous to alleles in a single organism.

Mutations may also be *deletions,* in which one or many nucleotides are missing. It follows that there can be no back mutations from a deletion.[1]

If a protein consists of two or more different polypeptide chains, the DNA segment specifying each chain is called a *cistron.*

If more than one identical subunit is present, one can have hybrid proteins (cf. Beckman et al., 1964; Schwartz, 1964).

[1] The deletion of one or two nucleotides in a triplet will scramble the remainder of the message, resulting in nonsense (no protein) or a missense (an inactive protein). The result is called a *frame-shift* mutation. The subsequent deletion of the rest of the triplet will restore the sense of the remainder of the message. A near-normal protein may result so that frame-shift deletion may be phenotypically reversible by subsequent deletions. Such a mutation is called a *suppression.*

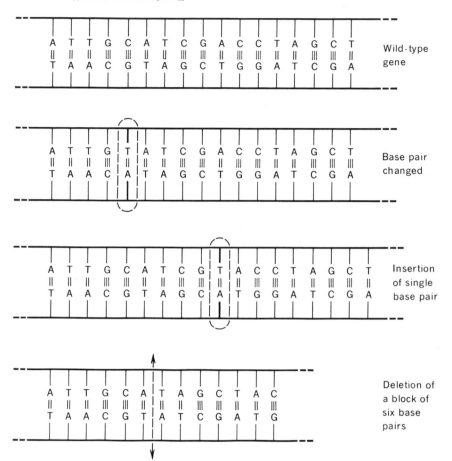

*Fig.* 5-2  Mutation at the molecular level (Watson, 1965). A mutation occurs whenever there is a change in the nucleotide sequence. This can be through the change of one base for another, the addition of a base, or the deletion of a base.

The genes we have been speaking of are *structural genes* in that they specify the structure of functional proteins such as enzymes. There appear also to be *regulator genes* which specify proteins (e.g., repressors) involved in the control of genetic transcription. There may also be *operator genes* which control the production of a series of functionally related proteins.

### Transcription: The Notion of Messenger RNA

We have discussed DNA duplication and the idea that the sequence of nucleotide triplets specifies the sequence of amino acids in the finished protein. We shall now examine the means of transcription, i.e., the first step in the mechanism by which coded information is transformed into the primary structure of proteins.

The information coded into a DNA chain is transcribed into a complementary ribonucleic acid (RNA) chain. Ribonucleotides complementary to the deoxyribonucleotide residues are assembled along a DNA chain and a chain of RNA is formed (Fig. 5-3). DNA-RNA complementarity is identical to DNA-

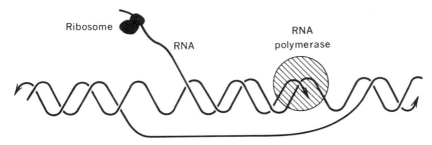

*Fig. 5-3* Transcript of DNA (after Watson, 1965). One strand of the DNA double helix is selected to be "read." RNA precursors—ATP, UTP, GTP, and CTP—align with the complementary bases in the DNA and are connected through phosphate-phospate bonds by the action of RNA polymerase. The resulting "messenger" (mRNA) is then available for protein synthesis. Ribosomes may bind immediately in the case of chloroplasts and mitochondria.

DNA complementarity, except that uracil replaces thymine.[1] It seems likely that the hybrid consists of a double-stranded DNA with a strand of RNA complementary to one of the DNA strands. Moreover it seems that only one of the DNA strands is normally transcribed, or that first a section of one strand, then a section of the opposite strand is transcribed.

Although all RNA appears to be formed in the same way, there are at least three kinds of RNA in cells. *Messenger RNA* (mRNA) carries the information for amino acid sequences of proteins. *Transfer RNA* (tRNA) is attached to activated amino acids. *Ribosomal RNA* (rRNA) provides a structural framework of ribosomes. The formation of activated amino acids proceeds according to Eqs. 5-1 and 5-2 (Fig. 5-4).

$$RCH(NH_2)COOH + ATP \xrightarrow[\text{aa-activating enzyme}]{} RCH(NH_2)C-\textcircled{P}-Ad \qquad Eq.\ 5\text{-}1$$
$$\overset{\|}{O}$$

There is a specific amino acid (aa)-activating enzyme for each of the 20 protein amino acids. Each adenyl-aa is then attached to a molecule of tRNA (Eq. 5-2).

aminoacyl adenylate + tRNA → aminoacyl − tRNA + adenylate        Eq. 5-2

There is one species of aa-activating enzyme and one species of tRNA for each of the several codons for each of the 20 kinds of protein amino acids. These need not all be in the cell at the same time or in the same place (see

[1] For example, the DNA codon for phenylalanine is AAA. The mRNA anticodon is UUU.

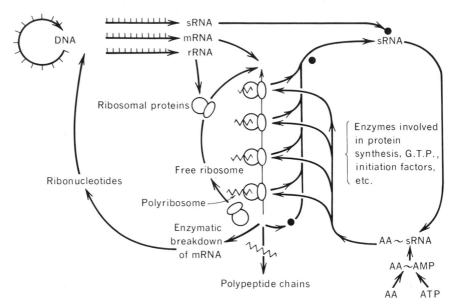

*Fig. 5-4* Translation, including amino acid activation (Watson, 1965). DNA is the source of all genetic information. Most of this information is ultimately expressed in the form of specific proteins, but some functions as ribosomal and transfer RNA.

Intracellular Localization of Informational Macromolecules, below). We must specify *protein* amino acids, since plants possess large numbers of nonprotein amino acids (Fowden, 1958). Each tRNA has at least two functional parts: a reactive group that combines with the amino acyl radical, an "identity card" that is complementary to the mRNA anticodon for the specific aa that it carries, and possibly a "union card" that permits it to be attached to the ribosomes.

*Ribosomal RNA* makes up about half the bulk of ribosomes; its precise function is not known.

As we noted above The Dogma tells us that the direction of flow of information is restricted—

DNA → RNA → protein

DNA may specify itself or RNA. RNA specifies protein but normally[1] not itself. Certain proteins may be instruments in the synthesis of nucleic acids (DNA polymerase, RNA polymerase, enzymes in the synthesis of nucleotides); they do not specify nucleotide sequence and hence the information of nucleic acids. RNA specifies proteins and, through the enzymes mentioned, may control the *rates* of DNA synthesis, but RNA does not influence the information content of DNA.

[1] Enzymes, called RNA replicases, have been found which catalyze the RNA-directed synthesis of RNA. These enzymes occur in the reproduction of RNA viruses. There may be in addition nonpathological processes involving RNA replication.

Taking our definition of information as the selection of alternatives, we may think of a further flow of information from proteins (enzymes) to metabolites. This flow, however, through induction, repression, and end-product inhibition (see below) is at least partially bidirectional.

Both DNA polymerase and DNA-directed RNA synthetase activities have been found in plants (Bonner, 1965).

### Protein Synthesis

TRANSLATION.   RNA or proteins are the functional products of genes; i.e., the characters by which we identify genes are produced by the presence, absence, or alteration of these macromolecules. It is clearly important to discover the factors governing the synthesis of these macromolecules in plants.[1]

After mRNA is synthesized on DNA it moves to the *ribosomes,* where the message is translated into protein (Fig. 5-4). Ribosomes attach to one end of the mRNA chain and move along it. As each ribosome moves along the message tRNA charged with specific amino acids are bound according to the triplet being "read," and the amino acids are linked one by one. Before one ribosome finishes a message, another may attach to the beginning of the mRNA chain, and another, and another so that 6 or 7 (or, in the case of long messages, up to 40) ribosomes may "read" the same mRNA simultaneously. The combination of two or more ribosomes on a single mRNA is called a *polysome.*

In animal systems the time required for the completion of one protein of 150 amino acids on a ribosome is about 3 min.

Ribosomes are small (ca. 100 Å diameter), spherical bodies that can be separated and identified by their sedimentation properties. Taylor and Storck (1964) showed that procaryotes—bacteria and blue-green algae—had 70$s$ ribosomes while the ribosomes of eucaryotes (nucleated organisms) were about 80$s$. This generalization is true for the bulk of eucaryote ribosomes, but a second kind of ribosome can be found in the mitochondria and chloroplasts of eucaryotes which are 70$s$ (Table 5-2). These organelle ribosomes are similar or identical to bacterial ribosomes (see below).

Functional ribosomes—that is, ribosomes capable of synthesizing protein *in vitro*—have been isolated from a few plant tissues: cereal embryos and endosperm (Mans and Novelli, 1964; Marcus and Feeley, 1966; Morton et al., 1964), pea seedlings (Raacke, 1959), and various chloroplasts (cf. Sec. 5.8).

### Intracellular Localization of Informational Macromolecules

DNA.   Most but not all of the DNA of plant cells is in the nucleus. Up to 15 percent of plant DNA is in the chloroplasts (Sec. 5.8) and a much

---

[1] Knowledge of factors governing the composition and quantities of plant proteins will be of peculiar urgency if humans are ultimately required to subsist on leaf proteins (Pirie, 1966).

Table 5-2  Occurrence of 60$s$ and 70$s$ Ribosomes in Plants*

Numbers refer to the sedimentation coefficients as reported by different authors.

| Subcellular Fraction | Plant | Subunits | | | | Ribosomes | |
|---|---|---|---|---|---|---|---|
| | | 30$s$ | 40$s$ | 50$s$ | 60$s$ | 70$s$ | 80$s$ |
| Chloroplasts | Cabbage | | | | | 68 | |
| | Pea | 32 | | 46 | | 62 | |
| | | 31 | | 47 | | 62 | |
| | | 32 | | 45 | | 64 | |
| | | | | | | 70 | |
| | Spinach | 33 | | 47 | | 66 | |
| | | | | | | 66 | |
| | | | | | | 70 | |
| | Tobacco | | | | | 70 | |
| | Euglena† | 30 | | 55 | | 70 | |
| | White clover | | | | | 68–70 | |
| | Chlamydomonas‡ | 28 | | 33 | | 68 | |
| Cytoplasm | Cabbage | | 40 | 55 | | | 83 |
| | Pea | | 38 | 54 | | | 76 |
| | | | 40 | 56 | | | 78 |
| | | | 40 | 56 | | | 77 |
| | | | | | | | 80 |
| | Spinach | | | | | | 83 |
| | | | | | | | 80 |
| | Tobacco | | | | | | 80 |
| | Euglena† | | 46 | 64 | | | 89 |
| | White clover | | | | | | 83 |

* Most of the data compiled by Svetailo et al., 1967.
† Rawson and Stutz, 1968; Mendiola et al., 1969.
‡ Hoober and Blobel, 1969.

smaller fraction (about 1 percent) is in the mitochondria. The amount of DNA per cell is roughly proportional to cell size; with unicellular algae, the proportionality extends over three log decades (Holm-Hansen, 1969). Since the complexity of these algae follows no such trend, we infer that multiple gene copies occur in the larger cells.

The ultrastructure of plant nuclei have been described for several species (cf. Frey-Wyssling and Mühlethaler, 1965; Werz, 1964; Trosko and Wolff, 1965). Nuclear DNA is organized into two or more chromosomes, but apart from the observation that chromosomal DNA is condensed by salt linkage with strongly basic proteins, called *histones* (Sec. 5.2), there is little information about chromosomal structure that has been shown to have predictive capacity (cf. Bonner and Ts'o, 1964; Stern, 1963).

There are several kinds of evidence that the DNA of chloroplasts and mitochondria is in fact specific to these organelles (cf. Sec. 5.8). One of these lines is that organelle DNA, like bacterial DNA, is essentially naked in contrast to nuclear DNA which is condensed with histones.[1]

This generality is consistent with the "ancient symbiont" hypothesis, the notion that mitochondria are derived from bacteria that invaded the ancestors of eucaryotic cells (Ris and Plaut, 1962). Chloroplasts are imagined, according to the same reasoning, to have arisen from blue-green algae. The replication of bacterial and organelle DNA is also similar in that they are both sensitive to nalidixic acid (Lyman, 1967).

Another line of evidence in favor of the specificity of organelle DNA is its equilibrium density. Density profiles of total cell DNA typically show "satellite bands," small quantities of material having a density higher or lower than that of the main band. The DNA of these satellites is typically (but not universally) located in chloroplasts and in mitochondria.

Organelle DNA normally occurs as closed double-stranded loops, again similar to the genophores of many bacteria. These can be seen in electron photomicrographs of DNA preparations (Fig. 5-5) (Sinclair and Stevens, 1966; Avers, 1967). A second, but indirect indication is the property of circular DNA to anneal back into its original density following heat denaturation. This has been shown for chloroplast DNA (Tewari and Wildman, 1966).

Suyama and Bonner (1966) found that mitochondrial DNA from a variety of higher plants had identical densities, despite wide variations in the densities of the main band or nuclear DNA.

A small fraction of metabolically labile DNA has been detected in several plants (Sampson et al., 1963; Cherry, 1964), but its significance has not been discerned.

RNA. Most of the cell RNA is formed in the nucleus, but it accumulates in the ribosomes which are principally in the cytoplasm. Some 80 percent of the RNA of the nucleus is found in the *nucleolus,* and most of this is used in manufacturing ribosomes (McConkey and Hopkins, 1964). From this and other considerations, the nucleolus appears to include that part of the genome which codes for ribosomal RNA, and also for at least some of the histones.

Messenger RNA can presumably be manufactured on any of the DNA and it appears likely that it migrates directly to the ribosomes (Raacke, 1959).

Just as chloroplasts and mitochondria have their own DNA, they also have their own kinds of RNA. There are, as noted above, a multiplicity of codons for each amino acid: some of these codons appear to occur exclusively in organelles as evidenced by specific forms of tRNA and aa-activating enzymes localized in these organelles (Barnett et al., 1967; Fairfield et al., 1969).

---

[1] The dinophyceae are a curious exception: their DNA is localized in nuclei but without histones.

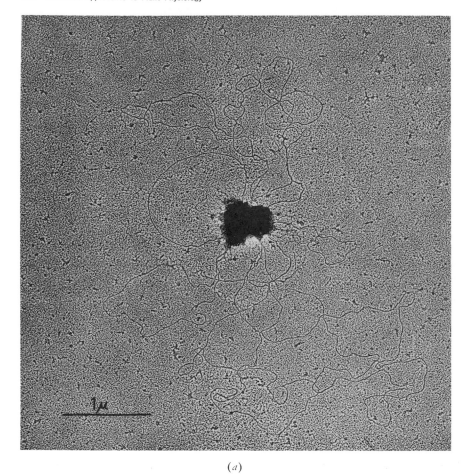

(*a*)

*Fig.* 5-5 Circular DNA from yeast (Avers et al., 1968). Yeast mitochondria were osmotically disrupted, their DNA isolated, and spread on grids for electron microscopy. The total contour length of the circular DNA varies from about 1 to 10 μ. (*a*) DNA filaments extruded upon lysis of a mitochondrion. The total contour length is 48.5 μ; no free ends are visible. (*b*) An isolated DNA filament with a contour length of 9.0 μ.

The organelles also have their own kinds of ribosomes, distinguishable from cytoplasmic ribosomes by their size (see above). Organelle (and bacterial) ribosomes also differ from cytoplasmic ribosomes in their susceptibility to dissociation into subunits at low magnesium concentrations, the size of their subunits, their constituent rRNAs, and their protein components.

There is a feature unique to organelle (and bacterial) ribosomes that is highly significant for the physiologist: 70*s* ribosomes but not 80*s* ribosomes have an affinity for the antibiotic, chloroamphenicol. Protein synthesis by these ribosomes is consequently inhibited specifically by this drug. In contrast, cyto-

(*b*)

*Fig.* 5-5   (Continued)

plasmic ribosomes, but not organelle ribosomes, are inhibited by cycloheximide. These two drugs serve then as diagnostic tools for distinguishing between translation in the cytoplasm and in the semiautonomous organelles.

The size of organelle—especially mitochondrial—genomes presents a serious problem: Mitochondrial DNA in *Euglena* (Edelman et al., 1966) has a molecular weight of only $2.6 \times 10^6$. This is insufficient to code for more than about 10 proteins of 15,000 MW each. If there is only one kind of DNA molecule per mitochondrion, the information is clearly insufficient for the different kinds of tRNA, aa-activating enzymes, rRNA, and ribosomal protein present. Even assuming no redundancy—that is, several different copies of mitochondrial DNA—the total amount of DNA per mitochondrion is sufficient for no more than about 70 different kinds of macromolecules. The evidence of a range of sizes of circular DNA in yeast mitochondria (Avers, 1967) is consistent with a multiplicity of genophores.

## 5.2 CELLULAR CONTROL MECHANISMS

### Induction and Repression[1]

Enzyme induction is the formation of an enzyme in response to the presence of a substrate or substrate analog of the enzyme (Fig. 5-6). If an organism

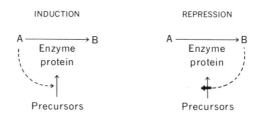

INDUCTION                    REPRESSION

Synthesis is accelerated by        Synthesis is inhibited by
the presence of the reaction       the presence of the reaction
*substrate.*                       *product.*

*Fig. 5-6* Enzyme induction and repression. The terms *induction* and *repression* refer to control over the synthesis of the enzyme protein; that is, control is exerted over the amount of new enzyme available to the cell. Induction and repression is typically controlled by enzyme products or reactants.

may be faced with a multiplicity of possible substrates, there is a clear selective advantage in being able to reserve the costly business of protein synthesis for the occasion when that particular protein is needed.

Bacteria, fungi, and saprophytes generally go in for enzyme induction on a massive scale. *E. coli* can produce a single enzyme to the extent of 5 percent of the total cell protein. Substrate-induced synthesis of enzymes in plants appears neither to be so widespread nor to occur on such a scale as in bacteria (Filner et al., 1969).

The synthesis of nitrate reductase (Beevers and Hageman, 1969—see Chap. 4; Alfridi and Hewitt, 1964) (cf. Sec. 4.5), certain isoenzymes of peroxidase, and glyoxalate cycle enzymes (Syrett, 1966; Hock and Beevers, 1966) are a few cases where proof of enzyme induction in plants has been provided through evidence of *de novo* protein synthesis (cf. Filner et al., 1969).

One would normally encounter enzyme induction only when an organism was in the habit of feeding on a variety of external substrates, but, with the exception of the "acid" flagellates, green plants typically enjoy a rather monotonous diet of inorganic salts, carbon dioxide, oxygen, and water.

[1] References: Wilson and Pardee, 1964; Vogel, 1957, 1961.

Enzyme induction conceivably plays roles in the regulation of enzyme levels in metabolic pathways, so that enzyme levels would respond to increases in the steady-state level of intermediates over a fairly long period of time (hours or days). In this way enzyme induction may play important roles in growth and development.

The complement of induction by enzyme substrates is repression by enzyme products (Fig. 5-6) (Vogel, 1957, 1961; Vogel and Vogel, 1967). The selection value is again obvious: enzymes are not manufactured if the product of the enzyme-catalyzed reaction is present. The synthesis of enzyme upon removal of the product is called *derepression*.

The derepression of acid and alkaline phosphatases (cf. Fig. 4-16) by algae (Price, 1962; Kuenzler and Parras, 1965) and probably by higher plants (Hewitt, 1963) may be of enormous ecological significance. In natural waters, phosphate is frequently a crucial limiting factor in "primary productivity" (i.e., the growth of algae). Since half the total phosphate of sea water can be organic (Kuenzler, 1965) a marine organism can enjoy a great advantage if it can respond to a low-phosphate environment by harvesting phosphate from organic phosphate. Thus the progression of species in plankton blooms might easily be related to the ability of species to utilize organic phosphate.

The presence of organic substrates typically represses the formation of photosynthetic apparatus in plants and conversely autotrophy appears to repress certain heterotrophic pathways (Cook and Carver, 1966; Syrett, 1966). The presence of fermentable substrates, especially glucose, represses the formation of oxidative enzymes in fungi (Slonimski, 1953, Strittmatter, 1957).

Several environmental variables, e.g., light, water during germination, and temperature, routinely stimulate the synthesis or activities of specific enzymes in plants (cf. Filner et al., 1969), but we should be cautious in labeling such phenomena as "enzyme induction" since the molecular mechanisms of such stimulation are not understood.

As with induction, we may consider enzyme repression as a possible factor in regulating the level of enzymes in metabolic pathways as for example in the synthesis of chloroplast pigments (Sec. 5.8) and hexose synthesis (cf. review of Filner et al., 1969).

The detailed mechanism or mechanisms of induction and repression are not understood. A widely favored hypothesis is that of Jacob and Monod (cf. discussions of Watson, 1965; Vogel and Vogel, 1967) in which a substance called a *repressor* can bind to a section of DNA and prevent its transcription. In the presence of the inducer, the repressor molecule is altered as by an allosteric transition (see below), so that it detaches from the DNA. A variant on this model is that the DNA double helix must unwind part way in order to be transcribed and that the repressor holds the two strands together. Repressors were predicted to be proteins, capable of binding specifically to exogenous inducers and to a specific stretch of DNA—structural or operator gene. A protein with the predicted properties of the repressor for $\beta$-galactosidase induction has been isolated from *E. coli* by Gilbert and Müller-Hill (1966). In the case of *repression* the repressor would behave in just the opposite way: binding by the enzyme product converts the molecule into a form that binds to DNA.

Other models for the control of enzyme synthesis are described by Vogel and Vogel (1967). A model for the modulation of amino acid synthesis involves the use of two different species of tRNA for the same amino acid with different affinities for the ribosome. In this model, then, control is exerted at the level of translation.

Another model envisages the uncharged tRNA for an amino acid serving as a repressor for the transcription of enzymes leading to the synthesis of the amino acids (Eidlic and Neidhardt, 1965).

There is also some evidence for messages having different "priorities" (Haselkorn and Fried, 1964).

We should reflect here on some fundamental differences between typical bacteria and typical green plants. As we noted in the discussion of induction, heterotrophy is a very different mode of life from autotrophy. Non-green parts of plants—roots, stem, fruits—are usually heterotrophic organs, but leaves supply a relatively constant mixture of substrates to the dependent organs.

The mean life spans of cells in relation to the life spans of macromolecules reveal other important differences between bacteria and other organisms. The generation times of typical bacteria are less than 1 hr. In a few days a bacterial culture will have completely filled its "substrate space." For the species of bacterial mRNA thus far identified, the mean half-lives are only a few minutes, depending on conditions (Fan et al., 1964), which means that the cell can start and stop making enzymes in response to changed environments many times in a generation. This mode of control would seem to be available to all living cells, but in higher organisms, messengers may be much longer lived (Pitot et al., 1965). For example, in reticulocytes, hemoglobin messenger lasts for days.

THE HISTONE HYPOTHESIS.   Most of what we have learned about cellular control mechanisms has been learned from studies with bacteria. Before we apply unreservedly all of these principles to the questions of control in plants, let us remember that in contrast to the situation in bacteria, most of the DNA of plants and animals is localized in nuclei and, except in the case of dinoflagellates, is complexed with histones which are highly basic and extremely conservative proteins. A pea histone has been found to differ by only a single amino acid residue from the corresponding calf histone (E. L. Smith et al., 1969).

Huang and Bonner (1962, cf. Bonner and Ts'o, 1964) with pea embryo chromatin and Allfrey and Mirsky (1962) with chromatin from animal sources have shown that the addition to DNA of histone fractions strongly and reversibly inhibits the DNA-directed synthesis of RNA. It now appears that a special species of RNA, associated with the histones, may be responsible for the control (Huang and Bonner, 1965; Benjamin et al., 1966).

The *histone hypothesis* generalizes on these and other observations to the proposition that most of the genetic potential of a cell is suppressed at any one time by the complexation of histone-specific RNA with the genes. The sequence of differentiation in a developing cell is determined by the temporary removal of histone RNA from a group of genes, thus permitting transcription.

A blow to the histone hypothesis has come with the discovery that a mutant of the toad *Xenopus laevis* lacking two of the four major classes of histones is capable of phenotypically normal development as far as the tadpole stage (Berlowitz and Birnstiel, 1967). It is difficult to see how histones generally could play roles as specific gene repressors in this organism.

In many developing organisms localized puffing of chromosomes has been observed. The chromosome puffs have been identified as sites of intense RNA synthesis. The location of the puffs is specific to the tissue and developmental stage of the organism. We may speculate that the puffs are chromosome sites released from histone-RNA inhibition and that mRNA is produced in these puffs (Clever, 1964).

Allfrey (cf.; Kleinsmith et al., 1966; Pogo et al., 1966) has an alternate proposal that DNA is released from binding when the histones are esterified with acetate or phosphate.

PROTEIN TURNOVER AND DECOMPOSITION.    Are proteins immortal? What happens to a protein that is no longer needed? Can the cell tear it down to build another?

The mean turnover rate of proteins is about 3 percent/hr in a variety of organisms (Hellebust and Bidwell, 1964; Steward and Bidwell, 1966). This means that on the average 3 protein molecules out of every 100 are decomposed in an hour. This rate is normally balanced by an equal or greater rate of synthesis. Of course, the rates of turnover of individual protein species vary widely.[1]

If the synthesis of a protein species is shut off, we can imagine that the protein will now disappear as a first-order process,

$$-\frac{dp}{dt} = kp \qquad\qquad \text{Eq. 5-3}$$

[1] Clearly a protein turnover rate measured in a whole organism would be only an average value, with many individual proteins deviating from the average by large factors. Moreover, the turnover rate may vary with the physiological state: the turnover rate of growing *E. coli* is 0.6 percent and that of nongrowing cells is 5 percent (Willetts, 1967).

Assuming that $k$ continues to be 3 percent/hr,

$$\ln \frac{p_0}{p_t} = 0.03t$$

where $p_0$ and $p_t$ are the amounts of protein at zero time and $t$ times, respectively. If these kinetics are obeyed, 50'percent of the original protein would disappear in 20 hr and 90 percent in 80 hr.

Clearly in relation to the lifetimes of individual bacterial cells or even cultures, proteins are virtually indestructible. In long-lived cells on the other hand, quite substantial changes in the galaxy of proteins in a cell may occur through decomposition, especially with proteins whose rates of decomposition are greater than the mean. There is evidence that several plant enzymes subject to repressive or inductive control are also subject to accelerated degradation when the need for them vanishes (cf. Fig. 4-16; Alfridi and Hewitt, 1964; Glasziou and Waldron, 1964; Zucker, 1969). Pollock (1959) refers to this as Type I induction (or derepression). We might also call it the Pooh-Bah effect.[1]

In assessing the possible role of protein decomposition we should be concerned not with the absolute life span of the cell but with that interval over which a cell may respond to its internal or external environment. We might call this the "developmental life span," the time between cell divisions or the interval from cell division to when differentiation is complete and irreversible.

The generation times of phototrophic algae typically vary from 4 hr to about one day. In roots the time from cell division to maximum elongation may be as little as 10 hr, in oat coleoptiles 60 hr. Root cells older than about 7 days apparently cease to function in ion and water absorption (Sec. 3.5). At the other extreme the level of total protein in leaf parenchyma will respond to a varying nitrogen supply for several weeks. The ovary wall cells of the sour cherry continue to enlarge for three months (Leopold, 1964). Thus the "developmental life spans" of higher plant cells may range from an order of hours to an order of months.

Thus, if we may generalize, it seems that bacteria meet their needs for different kinds of proteins (within, of course, the capability of their genome) by exercising controls over the rates of essentially immortal proteins. Higher organisms may combine these controls with systems for degrading proteins. Flexibility and nitrogen economy are served at the expense of the degradation and resynthesis of still useful proteins.

### End-product Inhibition, Quaternary Structure of Proteins and the Allosteric Hypothesis

A particularly powerful form of metabolic control occurs when the product of a metabolic pathway inhibits its own synthesis by shutting off the flow

[1] Pooh-Bah: ". . . I'd volunteer to quit this sphere/Instead of you in a minute or two."

of metabolites at the origin of the pathway. Because of the formal similarity between this kind of control and the familiar case of negative feedback in amplifier circuits, this phenomenon is called *negative feedback inhibition* or *end-product inhibition* (Fig. 5-7).

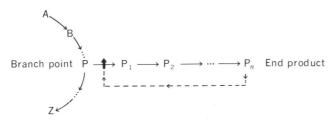

*Fig.* 5-7 End-product inhibition. A form of feedback control, called "end-product inhibition," occurs when the metabolic product of a biosynthetic pathway inhibits an enzyme reaction earlier in the pathway. The step inhibited is typically at a branch point.

End-product inhibition has been studied in detail in metabolic pathways leading to the formation of isoleucine, threonine, lysine, histidine, and cytidine triphosphate (CTP) (Monod et al., 1963; Vogel and Vogel, 1967). In each case the inhibition is exerted on an enzyme operating at a branch point.[1] The inhibition is characteristically restricted to the enzyme at the branch point and is specifically exerted by the end product of the pathway, not by intermediates.

The notion of end-product inhibition contains some principles that not only may account for the mechanisms of enzyme induction and repression but may be relevant to the action of certain hormones.

Let us consider the end-product inhibition of the CTP pathway (Fig. 5-8) as worked out for bacteria by Gerhart and Pardee (1962). A similar system appears to operate in plants (Neumann and Jones, 1964). The enzyme at the branch point is aspartyl transcarbamylase. CTP and to a lesser extent CMP strongly inhibit aspartyl transcarbamylase. This inhibition is contrary to nearly everything we have learned about enzyme kinetics. To begin with there is nothing about CTP chemically that would lead us to expect it to inhibit the enzyme. It does not look like nor behave as a substrate analog or as a noncompetitive inhibitor. Moreover the presence of CTP changes the apparent affinity of the enzyme for its substrates. One has the baffling impression of an enzyme whose fundamental catalytic properties are continuously altered with increasing amounts of substrate. Indeed this appears to be precisely what happens; for CTP alters the quaternary structure of the protein by controlling

[1] A branch point is a metabolic intermediate which can be transformed along either of two pathways; for example, pyruvate can be converted to acetyl CoA or reduced to ethanol and $CO_2$.

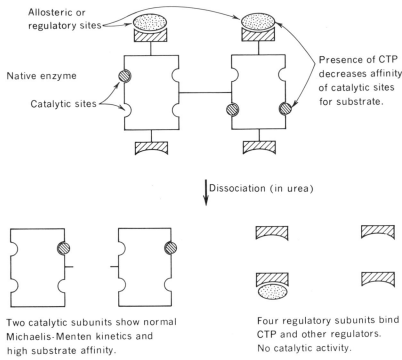

Allosteric or regulatory sites

Native enzyme

Catalytic sites

Presence of CTP decreases affinity of catalytic sites for substrate.

Dissociation (in urea)

Two catalytic subunits show normal Michaelis-Menten kinetics and high substrate affinity.

Four regulatory subunits bind CTP and other regulators. No catalytic activity.

*Fig.* 5-8 End-product inhibition and allosteric behavior of an enzyme. Aspartyl transcarbamylase was the first enzyme in which the molecular mechanism of end-product inhibition was determined (after Gerhart and Schachman, 1965).

the binding of an inhibitor fragment. An internally consistent hypothesis has been proposed by Gerhart and Schachman (1965); aspartyl transcarbamylase has a substrate binding site on part of the protein and a completely independent binding site for CTP on a dissociable part of the protein. When CTP is present, conformational changes occur in the protein that increase the association of the control component. The two forms of the enzyme have different substrate affinities (Fig. 5-9).

Obviously such a system is a fascinating arena for the protein chemist and enzymologist. The important fact for us is that the form and function of a protein may be profoundly modified by small molecules other than its normal substrate to the extent of abolishing the function of the protein.

This phenomenon appears to be general and extends beyond end-product inhibition: The oxygen loading curves of hemoglobin are controlled by sulfhydryl bonding among the monomers (Koshland and Neet, 1968). The activity of phosphorylase *a* and its conformation is controlled by AMP and its conversion is reported to be controlled by insulin (Larner et al., 1963). The activity and substrate specificity of glutamic dehydrogenase is conditioned by the presence of a variety of small molecules including estrogens (Tomkins et al., 1965; Talal and Tomkins, 1964). Animal (Gregolin et al., 1966) but not plant (Burton and Stumpf, 1966) acetyl CoA carboxylase is converted from inactive subunits to an active giant polymer of $6.8 \times 10^6$ daltons by citrate or phosphate.

The structural changes that are evoked by small molecules other than substrates, have been called *allosteric transitions* by Monod et al. (1963, 1965). The site occupied by the nonsubstrate molecule or *allosteric effector* is called the *allosteric site*. Under the umbrella of this allosteric hypothesis, Monod et al. propose a further generalization. *Repressor proteins* are thought to bind

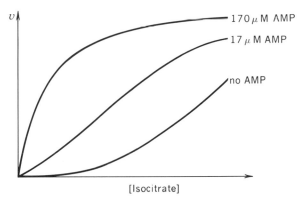

*Fig 5-9* Enzyme kinetics of an enzyme under allosteric control (Atkinson, 1966). The enzyme is NAD-isocitrate dehydrogenase from yeast. *Note:* Although the 170 $\mu$M curve appears to have normal Michaelis-Menten kinetics, careful analysis shows that all three are sigmoid and show the same degree of "cooperativity." The only effect of AMP, the allosteric effector in this case, has been to change the apparent substrate affinity.

to genes, thus preventing transcription of DNA (Fig. 5-10). Repressor proteins possess one or more allosteric sites. In the case of a repressible enzyme, the repressor protein "fits" the gene only when a repressor substance is bound to the repressor. In the case of an inducible enzyme the repressor substance fits the gene only in the absence of an inducer in the allosteric site. Competition between repressors and inducers can easily be imagined.

We should be very cautious in assuming that the several mechanisms of control described for bacterial systems thus far represent the limit of nature's ingenuity. More likely, control is exerted not only at the levels of transcription and at metabolic branch points, but at every nexus in the informational network: duplication (e.g., in cell synchrony), translation (Kenney and Albritton, 1965), and, in the stability of mRNA (Fan et al., 1964), competition for mRNA (Haselkorn and Fried, 1964), and perhaps in the permeabilty of cell membranes to informational macromolecules.

INDUCTION

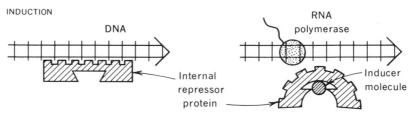

Repressor protein normally binds to a portion of the DNA. An *inducer
molecule* attaches to the repressor protein, decreasing its affinity
for the DNA and thus permitting transcription.

REPRESSION

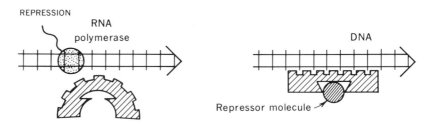

Internal repressor protein does *not* bind to DNA. Transcription proceeds.
A *repressor molecule* attaches to internal repressor protein, increasing
its affinity for DNA and preventing transcription.

*Fig. 5-10*  Model for induction and repression of protein synthesis.

## 5.3  EVIDENCE FOR MECHANISMS OF CELLULAR CONTROL

Section 5.2 is a summary of the mechanisms of control that have been described
thus far for isolated cells, such as bacteria, yeasts, and unicellular algae. We
shall consider the kinds of evidence by which we can identify these different
control processes.

(Additional controls, such as hormones, which operate in tissues and intact
multicellular organisms, are discussed in Sec. 5.5.)

The mechanisms described above are all instances in which an organism
responds to some change in its internal or external environment by changing
the activity of an enzyme or other functional protein. In the case of induction
or repression, whatever the mechanism, the *rate of synthesis* of an enzyme
is affected. In the case of end-product inhibition, the specific activity of the
enzyme is affected.

In assessing a control process, we want first to know whether the *amount*
of enzyme protein is changed. Suppose that an alga responds to phosphate
starvation by showing increased phosphatase activities. We should determine

if the increased activity represents more molecules of phosphatase or increased specific activity of the existing molecules.

We normally distinguish between the two alternatives by testing for *de novo* synthesis, the incorporation of labeled amino acids into the enzyme protein, but this requires physical separation of pure protein. An ingenious variant (Hu et al., 1962) employs incorporation of heavy isotopes (e.g., $O^{18}$) and separation of the proteins in a density gradient. Any new protein can then be distinguished from old protein, even in impure preparations, by a skewed distribution of enzyme activities when centrifuged in a density gradient (Fig. 5-11).

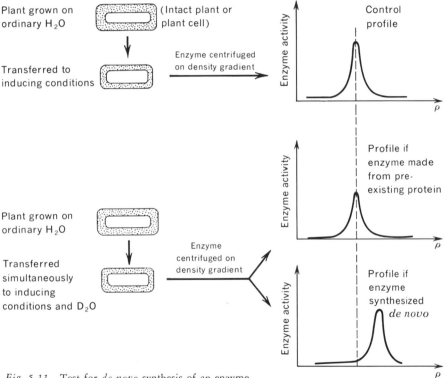

*Fig. 5-11*    Test for *de novo* synthesis of an enzyme.

If substrates or substrate analogs stimulate the formation of new enzyme protein, as in the case of nitrate on nitrate reductase (cf. Beevers and Hageman, 1969; cf. also Chap. 4), we may identify the process as *induction*. Similarly, if the addition of an enzyme product or analog turns off synthesis of the enzyme, as in the case of phosphatases, we may call it *repression*.

On the other hand, if a metabolite or substrate causes a change in the specific activity of an enzyme, we may think in terms of change at the level of protein chemistry: allosteric transitions, inhibitor subunits, and oligomer-polymer equilibria.

One of the characteristics of this latter type of control is nonhyperbolic enzyme kinetics (Fig. 5-9), as shown by aspartyl transcarbamylase. Such kinetics can be taken as very preliminary evidence that the enzyme may be subject to allosteric control.

## 5.4   MORPHOGENESIS: SOME DEFINITIONS, SOME GEOMETRY, AND SOME KINETICS

*"If one looks at a section through a young and growing plant structure, such as an ovary primordium, he sees a mass of cells of various shapes and dividing in many planes. Here chaos seems to reign. When he observes how such a structure develops, however, and finds that it is growing in a very precise fashion, each dimension in step with all the others, he comes to realize that this is not the seat of chaos but of an organizing control so orderly that a specific organic form is produced. This realization is one of the most revealing experiences a biologist can have and poses for him the major problems that his science has to face."*

——Edmund W. Sinnott, 1960

The excitement that The Dogma generates in biologists is rooted in the almost paradoxical contrast between chemistry and traditional biology. Up to now the predictions from The Dogma have concerned phenomena subject to physical and chemical criteria. Yet the Dogmatists foresee it as prevailing over areas of biology which have remained largely closed to those criteria. For example, induction and repression have been called "biochemical differentiation."

Growth, differentiation, and   development are areas in which biologists have sought chemical and  physical criteria almost wholly in vain. The paradox is that we can dimly see The Dogma, derived as it is from chemistry and physics, penetrating morphogenesis, from which the notions of chemistry and physics have been largely excluded. The drama of this paradox is heightened by the recognition that morphogenesis includes, as Sinnott observed, the major problems of biology.

Although the wholesale expansion of The Dogma into morphogenesis remains for the future, we can in the meanwhile gain some insights from the few incursions made thus far, and we can accelerate these events by learning to frame morphogenetic questions in a context from which we can hope to gain molecular biological answers.

We shall define in this and the next several sections some morphogenetic terms in forms calculated to be Dogmatically intelligible. We shall especially strive to point out the kinds of criteria which are needed but cannot yet be supplied in the required form. We shall try to acknowledge where our definitions conflict substantially with others commonly employed. These terms and definitions are of heuristic value only and should be discarded when others with greater predictive capacity are developed.

## Growth

We shall define *growth* with Thimann (cf. Sinnott, 1960) as *irreversible increase in volume*. Thus changes in volume due to turgor changes are not included. Growth may conceivably occur as a net increase in only water. Generally there is concomitant increase in dry matter, especially of cell wall material.

In contrast to the term as used here, many physiologists regard "growth by cell enlargement" or *Streckungswachstum* as only one aspect of growth. They would add "growth by cell division" or *Teilungswachstum*. We shall regard *cell division* as separate from growth, although cell divisions may increase the possibilities for subsequent growth. The events that occur with cell division in the conversion of a male gametophyte or microspores or in sea urchin embryos are *development* and not, within our definition, growth.

The kinetics of growth of living organisms are exceptionally interesting and intensely practical (Whaley, 1961). Biologists of a quantitative turn have often become absorbed in growth kinetics as one of the few frankly biological topics to which one could apply mathematics. With certain exceptions the studies of growth kinetics have not advanced our understanding of growth itself. As D'Arcy Thompson pointed out, ". . . a mere coincidence of numbers may be of little use or none, unless it goes some way to depict and explain the *modus operandi* of growth." The problem is completely analogous to the significance of kinetics in chemical processes (Sec. 1.2): there is always a multiplicity of processes that reduce to the same equation. In addition one can approximate one equation with a number of others. It has not been uncommon for the curve-fitting fraternity to employ the old principle that any curve can be approximated by a power series

$$y = a_0x^0 + a_1x^1 + a_2x^2 + a_3x^3 + \cdots + a_nx^n$$

The parameters of such an equation need not have any functional significance.

There is one general principle of growth kinetics: microbial systems tend to increase in proportion to the amount of living material $X$:

$$\frac{dX}{dt} = kX \qquad\qquad \text{Eq. 5-4}$$

The constant $k$ is characteristic of the organism, temperature, and nutritional state.

Rearranging Eq. 5-4,

$$\frac{dX}{X} = k\,dt \qquad\qquad \text{Eq. 5-4}a$$

and integrating, we obtain

$$\ln X = kt + C$$

where $C$ is a constant of integration.

If the amount of living material at $t = 0$ is $X_0$, the equation reduces to

$$\ln \frac{X}{X_0} = kt \qquad\qquad \text{Eq. 5-5}a$$

or

$$\frac{X}{X_0} = \epsilon^{kt} \qquad\qquad \text{Eq. 5-5}b$$

Growth that obeys the kinetics of Eq. 5-5 is said to be *exponential*. If the length, weight, volume, etc., of the system are plotted against time on semilog paper, one obtains a straight line (Fig. 5-12). As a consequence many writers have perversely referred to exponential growth as *logarithmic!*

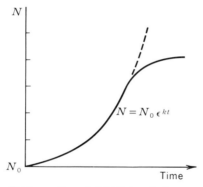
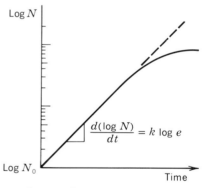

Total growth or "yield." In a finite environment the number of cells will be limited by some nutrient factor or condition.

On a semilog plot ($\log_{10}$) the slope of the growth curve will be proportional to the growth constant $k$.

*Fig. 5-12* Kinetics of microbial (autocatalytic) growth. A culture starts with $N_0$ cells at zero time. Assuming no lag period and nonsynchronous growth, the cell number will increase exponentially. In a finite environment (solid line), the number of cells will be limited by some nutrient factor or condition. Dotted line shows growth curve in an infinite environment. On a semilog plot ($\log_{10}$), the slope of the growth curve will be proportional to the growth constant $k$.

About this time the chemists among us will realize that Eq. 5-5 is not an error, that it is *not* the familiar first-order process so dear to chemical kineticists, $dX/dt = -kX$. The more quick-witted amongst the chemists will also realize that Eq. 5-5 is the equation for autocatalysis and is characteristic of explosive reactions.

It is quite true that growth that follows Eq. 5-5 will be explosive, but the growth of living systems does not follow Eq. 5-5 indefinitely. Later, if not sooner, the system runs out of some essential growth factor, the organisms become inhibited by their own toxic products, or growth is halted by genetic factors.

Growth may be "determinate" or "indeterminate." The growth of stems and roots of many plants can continue indefinitely within broad limits. Such growth, proceeding from *meristems,* is said to be *indeterminate.* The growth rates of individual roots and stems is typically *linear* with time.

$$\frac{dX}{dt} = a$$

Linear growth of roots and stems may proceed even when the entire biomass is increasing exponentially. This seeming paradox is understood when we recognize that periodic branching of roots and stems results in an increasing number of organs, each growing linearly. One aspect of the growth of stems is considered below in Sec. 5.6.

To deal with *determinate* growth, we should consider the expression $dX/X$ which we shall call the *relative growth.* Thus exponential growth is also *linear relative growth* (cf. Eq. 5-4a).

Many determinate organs, e.g., leaves and fruits, after an interval of apparently exponential growth, actually decelerate their growth rate according to a rather precise schedule.

The curve of *specific growth* $= \log (dX/dt)$ has the form of Eq. 5-6 (Fig. 5-13).

$$\log \frac{dX}{dt} = \log b - kt \qquad \text{Eq. 5-6}$$

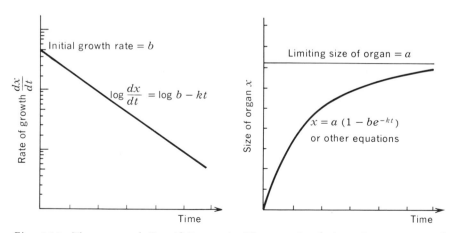

*Fig. 5-13* The curve of "specific" growth. The growth of determinate organs and organisms frequently follows these kinetics. Note that the left-hand ordinate is on a logarithmic scale.

There are several equations which reduce to Eq. 5-6

$$X = a(1 - be^{-kt}) \quad \text{"Monomolecular"}$$

$$X = \frac{a}{1 + be^{-kt}} \quad \text{Logistic}$$

$$X = ae^{-be^{-kt}} \quad \text{Gompertz}$$

These three have in common the notion of $X$ tending to a limiting value:

$$\lim_{t \to \infty} X = a$$

With the exception of the constant $a$, none of the parameters of any of these three equations can be assigned any biological significance, so that there is no rational basis for choosing one equation over the others. In the words of Medawar (1945), "The universal growth equation is . . . a fiction."

Where a single substrate is limiting, as in the case of yeast consuming sucrose, a reasonably rational equation with the same properties can be constructed (Monod, 1942; cf. Fujimoto, 1963).

$$\frac{dX}{dt} = \mu_m X \frac{S/X}{K + S/X} \qquad \text{Eq. 5-7}$$

where $X$ is the concentration of cell material, $\mu_m$ the growth rate constant, $S$ the concentration of rate-limiting substrate, and $K$ a constant analogous to the Michaelis constant but including the specific rate constants for collision of the substrate with the cell, transport into the cell, and finally combination and reaction with the rate-limiting enzyme.

A rigorous approach to the interaction of several rate-limiting substrates or processes has been presented by Lockhart (1965a,b). Empirically, the interactions of day length, light intensity, and temperature have been determined for phototrophic growth in the alga, *Chlorella* (Tamiya et al., 1955).

*Differentiation* clearly refers to differences within a plant. The cells of different tissues of an organism are typically distinctive, even though all (or nearly all) possess the same genome. Since a cell can only vary within the limitations imposed by its genome, the differences among cells or among different parts of a coenocytic organism represent the expression of different genes. *Differentiation, therefore, is gene expression.*

A completely differentiated cell would be one in which all of the genes are expressed. *Reductio ad absurdum,* a completely undifferentiated cell is one in which none are expressed, in which case the cell would be dead! For convenience, we might accept as undifferentiated, a cell which had produced only those proteins required to operate the common core subset of reactions (Sec. 1.1).

According to our view, a leaf parenchyma cell is differentiated because it has produced chloroplast material; a fiber cell on the other hand is differentiated because it has produced massive amounts of secondary cell wall material.

There seems to be no *a priori* reason for restricting ourselves to structural or compositional differentiation; we should include functional differentiation. Meristematic cells, for example, may well be differentiated in the sense that they are concentrating on reproduction to the virtual exclusion of manufacturing pigments, polysaccharides, etc.

The notion that a meristematic cell might be differentiated is not completely heretical. Sinnott (1960) says that there is probably no really undifferentiated structure in a plant. Outside of viruses we at present lack the criteria to determine which genes are expressed, not to mention the problem of quantitative expression. Even though a completely differentiated cell can be imagined, it is a little bewildering to visualize all ten or fifteen thousand genes clamoring simultaneously for expression!

Obviously differentiation may be affected by the internal and external environment of the organism, by such mechanisms as those discussed in the preceding two sections, by growth substances (Sec. 5.5), by the availability of nutrients and other metabolites, and by physical factors such as temperature, light (especially in its control of phytochrome), and diffusion barriers. Moreover we clearly recognize cell differences due to the duration over which genes have been expressed. It seems nonetheless that the common focus of these factors is on gene expression, that is, the production and activities of specific proteins.

One of the factors that cells respond to is infection by symbiotic, pathogenic, or parasitic organisms. Infection can produce strong and specific responses, such as the formation of root nodules in legumes and the massive production of the lactone umbelliferone by sweet potato roots in the vicinity of a fungus infection (Minamikawa et al., 1963; cf. also Weber and Stahmann, 1964; Stahmann et al., 1966; Bhattacharya et al., 1965; Schaffer and Alexander, 1966; and Asahi et al., 1966). Perhaps the most illuminating kind of response occurs in the case of virus infections, where informational macromolecules from the virus may intervene directly in the information transfer system of the host (cf. Reddi, 1966).

### Development

As a low-order approximation, we may think of development as differentiation at the level of organ or organism. The homology is an unsatisfactory one because we have only the most feeble insights into possible mechanisms of gene expression at the organ level. Hormones are one such mechanism (Sec. 5.5).

The development of function, such as the development of photosynthetic capacity in leaves, occurs for the most part as the differentiation of individual cells and subcellular organelles (cf. Sec. 5.8). As a type of problem, therefore, the development of *function* is something we can grasp.

The development of the *form* of organs is far more perplexing. A resort to analytical geometry will illustrate our problem. A leaf blade begins life

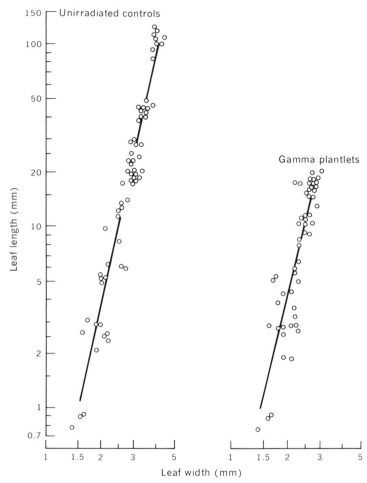

*Fig. 5-14*  Shape of embryonic wheat leaves during development with and without cell division (Haber, 1962). Wheat plants were allowed to grow normally with cell division (unirradiated controls) or when cell division was abolished by irradiation with gamma rays ("gamma plant-lets"). Although the *amount* of growth was less without cell division, the *shape* of the leaves, measured as the ratio of length to width, followed in both cases the same allometric equation.

as a leaf primordium, a roughly hemispherical ball. As it grows it typically grows faster in length than in width. The ratio of length to width continuously increases, but the *relative growth rates of the two dimensions remain constant* (Eq. 5-8). This principle of *constant relative growth* or *allometric growth* was described by Huxley (1932).

$$\frac{1}{l}\frac{dl}{dt} = ax\frac{1}{w}\frac{dw}{dt}$$

Eq. 5-8

If we eliminate time and integrate, we obtain

$$\ln l = a \ln w + C_i \qquad\qquad\text{Eq. 5-9}$$

Equation 5-9 tells us that the logarithm of one dimension is a linear function of the logarithm of another dimension.

The example of the development of the first foliar leaf of wheat is a case of allometric growth which is most instructive to the study of development. One might imagine that each cell in a developing leaf has specific allometric

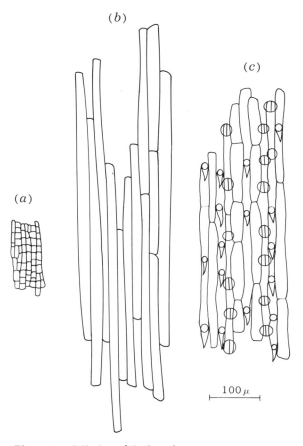

*Fig. 5-15* Cell size of leaf epidermis grown with and without cell division (Haber, 1962). Camera lucida drawings of epidermal cells of (*a*) embryo before germination, (*b*) "gamma plantlets" (no cell division), and (*c*) unirradiated control. The leaves of the "gamma plantlets," in which cell division was suppressed, achieved the same length as the controls through much greater cell elongation. *Note:* stomatal complexes and leaf hairs (small cells in *c*) are absent from the "gamma plantlets."

instructions. Although this may be the case, the plant behaves as if instructions were given to organs and not cells. Haber (1962) subjected wheat seeds to sufficient doses of gamma rays to prevent any subsequent cell divisions. The seeds nonetheless could germinate and grow. Even though the "gamma plantlets" did not reach the same size as normal seedlings, the *form* of the first

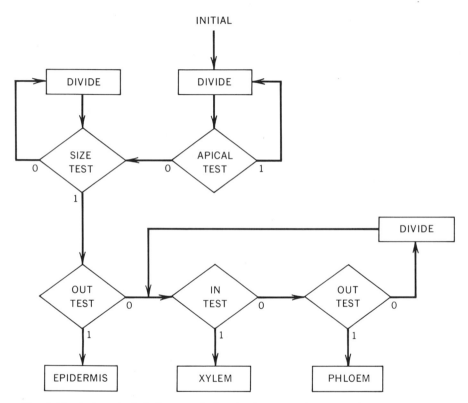

*Fig. 5-16* A hypothetical "program" for development of a plant from a single cell. Bonner (1965) has imagined how a plant might direct its morphogenesis if it were organized along the lines of a computer. For the plant to "choose" between alternatives, the individual cells must be able to sense changes in their immediate environment and to possess some type of a switching network that would permit essentially all-or-none responses.

foliar leaf was completely normal (Fig. 5-14). Since no cell division could occur in the gamma plantlets, each of the irradiated cells must have elongated more than the controls (Fig. 5-15). This simple experiment clearly shows that the morphogenesis of the first foliar leaf of wheat is *expressed at the organ level*. Individual cells expand to fill "organ space."

Although the development of organs and organisms is an expression of their genomes, we cannot yet discern mechanisms that permit genes to be expressed on the organ or organism level. We are forced to ask our questions in allegory.

One such allegory has been put forward by Bonner (1965) in the form of a set of programmed instructions, as to a computer (Fig. 5-16). Bonner's development program is a model in the sense that we can think of development

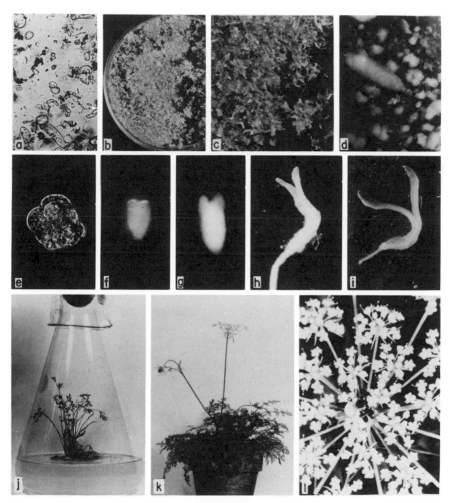

*Fig. 5-17* Embryoids from wild carrot cultures (Steward et al., 1961). The original photograph was kindly supplied by F. C. Steward. (*a*) Free cells obtained by culturing carrot embryos in liquid medium. (*b*) Colonies of cells after plating out on agar. (*c*) Higher magnification of (*b*). (*d*) Free cells after further growth in liquid medium aggregate in various forms; some embryolike forms seen in addition to amorphous masses. (*e* to *i*) Selected stages of development of embryos: (*e*) globular stage, (*f*) heart-shaped stage, (*g*) torpedo stage, (*h*) and (*i*) cotyledonary stages, (*j*) plant produced after embryoid planted on agar, (*k*) plant bearing inflorescences after 6-months growth, (*l*) detail of inflorescence.

happening as a result of a series of "if, then" responses programmed into individual cells.

## Totipotency

If all the information needed to specify an organism is contained in its DNA and if DNA is replicated in each cell of an organism, it follows that one should be able to reproduce any organism from any one of its nucleated cells. This doctrine of totipotentiality—that all cells are totally potent to reproduce—has been vindicated in a few higher plant species. The cells of lower animals also show a great "plasticity" with respect to ultimate differentiation.

The behavior of plant cells is more spectacular. Tissue cultures from roots of the wild carrot (*Daucus carota*) frequently slough off cells and clumps of cells which then develop into "embryoids," curious heart-shaped, cellular structures that look and behave exactly like embryos found in sexual reproduction (Fig. 5-17) (Steward et al., 1964). This discovery is enormously significant as proof of totipotency for these cells, and by implication, cells generally.[1] Of possibly equal significance is the discovery that pollen cells can be made to divide and ultimately form complete, haploid plants (Nitsch and Nitsch, 1969).

It is also extraordinary that workers in the field of animal tissue culture, which appears to exist in another universe, continue to speak of differentiation as an irreversible process.

## 5.5  SUBSTANCES CONTROLLING MORPHOGENESIS: GROWTH REGULATORS, HORMONES, GROWTH SUBSTANCES

There exist classes of low-molecular-weight substances which evoke morphogenetic responses in plants. Loosely termed growth regulators or growth substances, these morphogenetic substances include *auxins, gibberellins,* and *cytokinins.* The study of them has led to a body of information with substantial predictive capacity at the phenomenological level, although probably not to underlying events in development and differentiation.

## Epistemology

The notion of a growth substance presupposes that we can detect it. The criteria for a growth substance are therefore operational: when a plant or plant part is treated with the substance—by spraying, injection, translocation, etc.—we observe a morphogenetic response. The response might be growth or cell division, or indeed any phenomenon of cell differentiation or organ development.

[1] Orchids are among the plants which have yielded embryoids from tissue culture, with a consequent decimation of propagation time and costs. The consequences to the multi-million dollar orchid industry have started to be felt.

Implicit then in the detection of a growth substance is the condition that we can test it on a plant or plant part in which it is either absent or in short supply.

The classic strategy of studying growth substances derives from the work on auxins by Went, his predecessors, and associates (cf. Sec. 5.6 and Went and Thimann, 1937). A test object (oat coleoptile in the example) is selected which does something (grows) only when supplied with a substance (auxin). The same strategy has been applied to several kinds of growth, cell division, flowering, vascular differentiation, branching, rooting, sex expression, etc.

The strategy requires a test object, a plant or plant part which can be persuaded to do something when supplied with a "growth substance," and a source of the growth substance. For example, cultured radish roots or tissues of soybean will grow fairly well with low levels of cytokinin and auxin but produce no xylem elements; when the concentrations of these substances are increased, the numbers of tracheary elements increase enormously (Torrey and Loomis, 1967; Fosket and Torrey, 1969). Investigators have employed all manner of plant extracts including yeast extract, coconut milk, urine, and not infrequently the pages of Beilstein (figuratively, if not literally).[1]

Partly as a consequence of the methodology of growth substances, physiologists have asked two sets of questions: how much of what kinds of growth substances evoke morphogenetic responses? and how do growth substances evoke morphogenetic responses? Some answers and attempted answers will be presented in the case studies of Secs. 5.6 and 5.7.

### Definitions

Growth regulators and its subclasses were defined by a committee of plant scientists (van Overbeek et al., 1954) as follows:

"(Plant) Regulators are organic compounds, other than nutrients, which in small amounts promote, inhibit or otherwise modify any physiological process in plants.

"(Plant) Hormones (Synonym: Phytohormones) are regulators produced by plants, which in low concentrations regulate plant physiological processes. Hormones usually move within the plant from a site of production to a site of action."

The term "growth substance" according to the Committee definition includes synthetic substances which may or may not promote growth but by acting along similar lines may abolish growth or kill plants at very low concentrations.

Since hormones are about the only specific examples we have of *inter*cellular morphogenetic influences, the notion of hormones is crucial to our understanding of development. Unlike van Overbeek's Committee we shall restrict the term

[1] The pages of *The New York Times,* but not those of *The London Times*, were used successfully as a source of juvenile hormone for insects (Slama and Williams, 1966).

to substances that have their action at a site removed from the region of forma-tion. With varying degrees of certainty, we may regard *auxins, gibberellins, cytokinins, abscisins,* and *ethylene* as hormones.

Certain of the vitamins are a special class of growth substances superficially similar to hormones. It is very common for heterotrophic organs, e.g., roots, to be dependent on other organs for a supply of thiamine, nicotinic acid, or pyridoxine (Fig. 5-18). These vitamins are, of course, required for synthesis

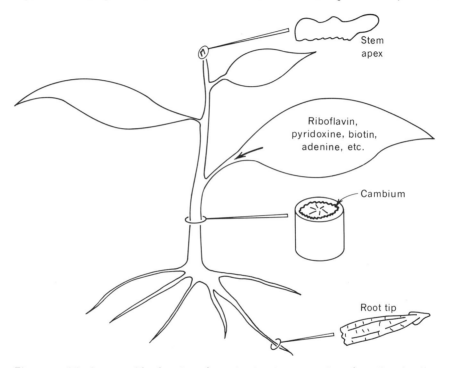

**Fig. 5-18**   The hormone-like function of certain vitamins. A number of small molecules, including many of the vitamins required by animals, are produced in leaf tissue and con-sumed in part in the stem apex, cambium, roots, etc. These latter organs and tissues are therefore dependent on the transport of these small molecules from the leaves. In this limited sense, such molecules may be thought of as hormones.

into coenzymes, such as NAD and pyridoxal phosphate; they are required, therefore, to catalyze the common core subset of chemical processes (Sec. 1.1). Outside of these workaday assignments, vitamins seem not to be able to control growth and development. It seems very unlikely that hormones function as coenzymes.

For the sake of completeness, we should also mention *phermones*, chemical substances by which different individuals communicate with one another. There are certainly examples of chemical substances produced by one plant that inhibit the germination or growth of other plants in the vicinity (Went, 1955). This extreme kind of communication might be better identified as warfare. There is also an instance of a volatile material from oak leaves that turns on sexual activity in the *Polyphemus* moth (Riddiford and Williams, 1967).

Auxins are the oldest and best known among plant growth regulators. They were first discovered to be chemical substances and as hormones by F. W. Went in 1928 (cf. Went and Thimann, 1937, and below). Verified chemical identification of indole-3-acetic acid (IAA) as a naturally occurring auxin followed a few years after.

Auxins were defined by the van Overbeek Committee (van Overbeek et al., 1954) as follows:

"Auxin is a generic term for compounds characterized by their capacity to induce elongation in shoot cells. They resemble indole-3-acetic acid in physiological action. Auxins may, and generally do, affect other processes besides elongation, but elongation is considered critical. Auxins are generally acids with an unsaturated cyclic nucleus or their derivatives.

"Auxin precursors are compounds which in the plant can be converted into auxins.

"Anti-auxins are compounds which inhibit competitively the action of auxins."

This straightforward, operational definition was satisfactory and unambiguous in 1954, but following the recognition of gibberellins (see below), which also promote elongation in shoots, we should be inclined to restrict the operational frame of reference to compounds which induce elongation in etiolated oat coleoptiles, a test system which responds exclusively to auxins.

### Chemical Structure

The range of molecules possessing auxin activities seems at first glance to present a bewildering variety. There are however some rules.

Auxins fall into five chemical classes (Fig. 5-19): indole auxins are the only identified natural auxins, although nonindolic auxins cannot be excluded on chemical grounds; phenoxycarboxylic acids (including the famous 2,4-D); benzoic acids; certain dithiocarbamates; and napthalene carboxylic acids. Within and among these classes there are more rules.

All indole auxins can be transformed in the plant to IAA (cf. Fig. 5-24). Similarly all of the side chains of active auxins can be degraded to acetic acid[1] (cf. Wain, 1967). The side chains of all or nearly all auxins thus terminate

---

[1] Phenoxyaliphatic acids (Fig. 5-19) show strong auxin activity, but only if the plant can degrade the aliphatic side chain to acetic acid (Wain, 1967). The degradation normally occurs by $\beta$-oxidation (Appendix A), and plants vary in their ability to attack higher members of the series. For example, 2,4-dichlorophenoxyoctanoic acid will kill wild carrots in a field of white clover because the wild carrots degrade the molecule to 2,4-dichlorophenoxyacetic acid, while clover does not act upon the molecule.

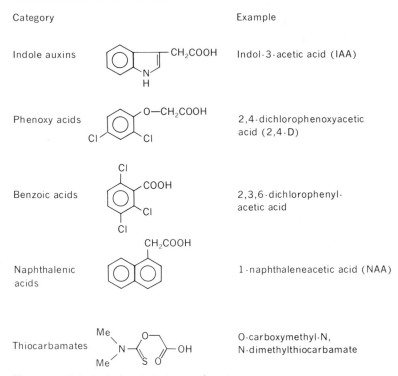

Category

Example

Indole auxins

Indol-3-acetic acid (IAA)

Phenoxy acids

2,4-dichlorophenoxyacetic acid (2,4-D)

Benzoic acids

2,3,6-dichlorophenyl-acetic acid

Naphthalenic acids

1-naphthaleneacetic acid (NAA)

Thiocarbamates

O-carboxymethyl-N, N-dimethylthiocarbamate

*Fig. 5-19*   Principal chemical classes of auxins.

in a carboxylic acid. If optically active, it should be of the D-form. Finally the most active auxins can form structures in which a positive charge is 5.5 Å away from a negatively charged carboxylate oxygen. The molecular site or sites of auxin action presumably have an electrical and spatial complementarity to the structures of active auxins.

There is negligible absorption of visible light by auxins, so that light cannot affect auxin directly under natural conditions.

We learned in Sec. 1.3 that there are enzyme-substrate analogs, antivitamins, and even that certain antibiotics act as specific inhibitors of nucleic acid functions. In the vast efforts expended on testing structure-activity relations, it is surprising that so few antiauxins have been uncovered. One of the best of them is L(—)-2,4,6-trichlorophenoxyisobutyric acid (see Biological Activity below). This substance antagonizes both the positive actions of normal auxins and in addition relieves their inhibitions.

### Biological Activity

The operational definition of auxins (above) includes the biological activity by which these substances were first recognized: promotion of elongation of stem sections of a variety of plants (Table 5-3).

Auxins also promote cambial activity, initiation of roots on cuttings, par-

Table 5-3 Some Biological Activities of Auxins*

The following include the "major" effects of auxins; many other responses of plants have been reported and are often in the way of interactions with other growth substances. Many of the processes that are listed as stimulations may turn into inhibitions at higher concentrations of the growth substance.

Stimulation of stem and coleoptile elongation
Stimulation of cell division in cambia
Stimulation, often as an essential factor, of growth of callus and other types of plant tissue
    cultures
Stimulation of protoplasmic streaming
Initiation of development of root primordia
Inhibition of root elongation
Inhibition of development of bud primordia

* After Thimann, 1937; Leopold, 1964.

thenocarpy of certain fruits, inhibition of root growth, and inhibition of bud development. Auxins are required for the continued growth of most tissue cultures. Auxins are also involved in both the formation and the inhibition of formation of abscission layers, the initiation of vascular differentiation, and the promotion of cell division depending on concentration. Indeed it is characteristic of auxin to promote certain responses at low concentrations and inhibit at high concentrations (Fig. 5-20). This is even true of bud and root growth

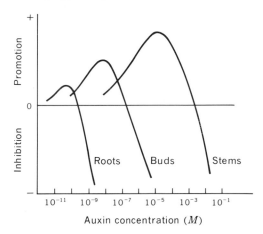

Fig. 5-20 Concentration ranges of stimulation and inhibition of growth by auxin (Thimann, 1937). In general, auxin promotes growth at lower concentrations and inhibits at higher concentrations. Organs differ in their sensitivity, typically in the order, roots > buds > stems.

although the concentration ranges for promotion are narrow and variable. It has been reported, for example, that IAA promotes root hair growth at $10^{-13}$ M (Devlin and Jackson, 1961)!

The rate of growth of coleoptiles as a function of concentration has been analyzed with the Lineweaver-Burk model (Housley, 1961, and cf. Appendix B and Sec. 1.2). Different auxins have a similar "$V_{max}$" but different "affinities" and one substance, 2,4-dichlorophenoxyisobutyric acid, shows good competitive inhibition (Fig. 5-21).

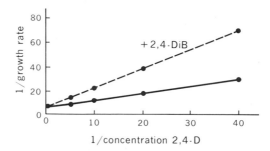

*Fig. 5-21* Lineweaver-Burk analysis of auxin-induced growth and inhibition by a structural analog (McRae and Bonner, 1953). The *rate* of auxin-induced growth is a hyperbolic function of the concentration of the auxin 2,4-D, so that (growth rate)$^{-1}$ is a linear function of (auxin)$^{-1}$. When the growth inhibitor 2,4-dichlorophenoxy-isobutyric acid (2,4-DiB) is added, the kinetics are those of a competitive enzyme inhibitor. Does this prove that 2,4-D and 2,4-DiB act as substrates for an enzyme? (Cf. Sec. 1.2.)

As is evident from Fig. 5-20, auxins are active at extremely low concentrations (typically $10^{-9}$ to $10^{-6}$ $M$). We are dealing with substances that operate at concentrations substantially lower than those of conventional substrates. The actual concentrations of auxins recoverable from tissues that are natural sources of auxin are similarly low. They are produced continuously as can be shown by allowing auxins to diffuse from a tissue into an agar block.

### Auxin as a Hormone

F. W. Went (cf. Went and Thimann, 1937), working in his father's laboratory in Utrecht,[1] showed in 1928 that the growth of subapical tissues of oat coleoptiles was dependent on a substance that diffused from the tips. Hence, his famous dictum, *"Ohne Wuchstoff, kein Wachstum!"* Countless aspiring students of plant physiology of later generations have used Went's famous *Avena* test to demonstrate the hormonal nature of auxin (see Sec. 5.6).

Variations and refinements of the *Avena* test remain today among the most sensitive means of measuring small amounts of IAA. Whether Went's

[1] Went relates that part of this work was done nights while he was ostensibly performing his required military service. This was possible, according to Went, because his military duty did not require being awake.

dictum is absolutely true is not known—whether or not all growth of all plant tissues requires auxin, it is evident that for tissues such as the oat coleoptile, which are dependent on outside sources of auxin for growth, the tissue remains very much alive in the absence of added auxin. In this case we can say only that auxin induces cell differentiation and organ development.

Over longer intervals auxins may be essential to life. This is clearly demonstrated in most tissue cultures, which turn brown and die in the absence of auxin. Similarly horticulturists know that in pruning, one should always leave the shortest possible stub above an internode, else the stub will die and serve as an invitation to infection. Auxins are known to be produced in internodes and the artificial application of auxins to a long stub will keep it alive.

A characteristic of hormones is that they are transported. The basipetal (downward) transport of auxin in the oat coleoptile was clearly demonstrated and measured with remarkable precision by Went and his collaborators in the early 1930s (cf. Went and Thimann, 1937). They showed that the naturally occurring auxin IAA diffused from an agar block into an oat coleoptile and that it was carried physiologically downward, contrary to apparent concentration gradients. If on the other hand an agar block was placed on the basal end of a coleoptile section, diffusion occurred but no transport (Fig. 5-22). This

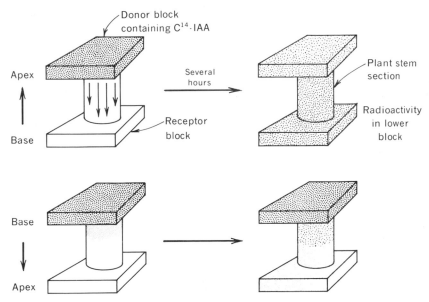

*Fig. 5-22* Polar transport of auxin. The type-experiment employed to measure polar transport uses an excised stem section and two agar blocks, one with C¹⁴-IAA. After the radioactive auxin has been allowed to move through the tissue for a period of time, the donor block is removed, the IAA separated if necessary, and counted. It is easily shown that the differences between basal and acropetal transport are not due to gravity. Arrows show the original orientation of the tissue in the plant.

property of *polar transport* of auxin has been shown repeatedly with a variety of tissue and organ systems (cf. Goldsmith, 1967; Wilkins and Scott, 1968).

Polar transport of auxin is associated, perhaps causally, with *correlative inhibition,* a phenomenon in which auxin that is transported from apical buds inhibits the development of lateral buds.

Many workers have worried about auxin destruction at cut surfaces (see Auxin Metabolism). It was not until $C^{14}$-labeled IAA became available that the early measurements of transport were confirmed and improved upon (Pickard and Thimann, 1964). In fact, auxins are transported in a direction strictly from apex to base in both stems and roots.

The basis for auxin polarities, as with other tissue polarities, is among the mysteries of biology.

### Light

As we noted above, auxin promotes shoot elongation in many plants. Most, or at least the most striking, auxin effects are observed with etiolated plants. We also observe that plants growing in dim light are elongated and typically spindly. The reason for both phenomena is that auxin is rapidly destroyed in the light even at moderately low intensities (cf. Thimann, 1967). The mechanism of auxin destruction in light is not clear, save the negative fact

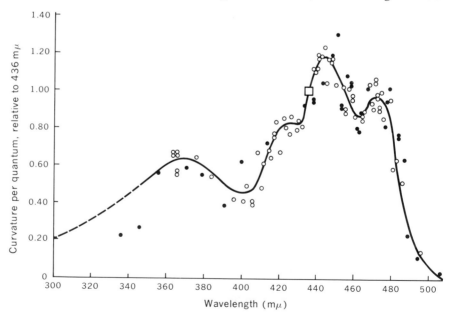

*Fig. 5-23* The action spectrum for the tip curvature of *Avena* coleoptiles (Thimann and Curry, 1960). When etiolated *Avena* coleoptiles are illuminated from one side, they bend toward the light. The curvature is due to differential growth, specifically an inhibition of growth on the illuminated side.

that IAA does not absorb wavelengths which are effective in IAA destruction. Many dyes will mediate auxin inactivation through oxidations of the indole nucleus, and photobiologists have searched for natural pigments that might be operating. The action spectrum for photoinhibition of auxin-induced growth (Fig. 5-23) has led a number of investigators to search for the absorbing pigment. We should infer that it is yellow. It could be a carotenoid or a flavin, but not simply riboflavin.

## Auxin Metabolism

Despite the obvious similarity between IAA and tryptophan and the frequent demonstration of growth promotion by tryptophan, the hypothesis that trypto-phan is a precursor of IAA is vitiated by evidence that the observed conversions of tryptophan to IAA by higher plants is caused by contaminating bacteria (Winter, 1966). Winter has presented an alternate proposal shown in Fig. 5-24,

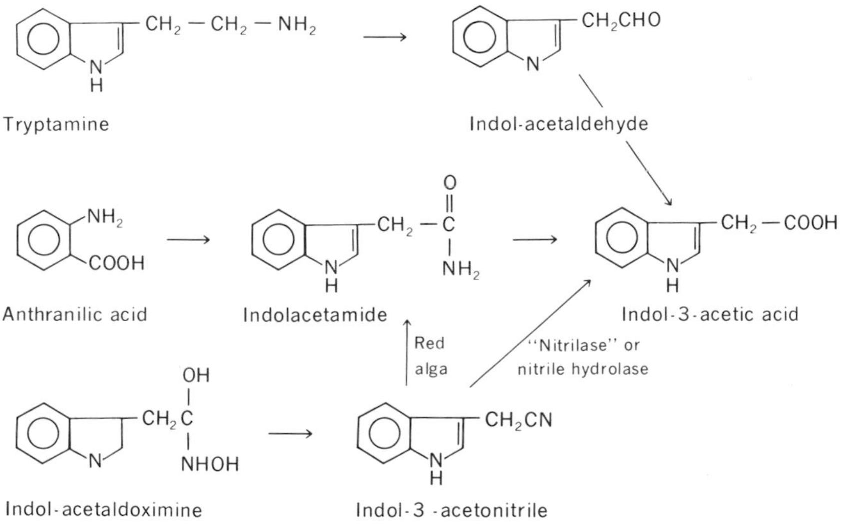

Fig. 5-24  Biosynthesis of indol auxins (cf. Winter, 1966; Mahadevan, 1963; Schiewer and Libbert, 1965, for discussion of alternative hypotheses).

in which tryptamine is oxidized to indolacetaldehyde and then to IAA. The first enzymatic step is strongly inhibited by the growth retardant $\beta$-hydroxyethylhydrazine (Reed, 1965).

The case of maize seeds is a peculiar case. Nutritionists have long known maize to be grossly deficient in tryptophan, but the seeds do contain unusually large amounts of IAA and esters of IAA (Stowe, 1959).

The conversion of IAN to IAA in certain families of plants can be ascribed directly to the action of a nonspecific nitrile hydrolase. The enzyme is found uniquely in tissues which grow when given IAN as well as IAA (Mahadevan and Thimann, 1964).

Numerous enzyme systems have been described which are capable of destroying IAA. They are typically peroxidases and therefore require a divalent metal for optimal activity.

IAA in some tissues is also esterified with several kinds of amino acids and inositols (cf. Zenk and Scherf, 1964; Labarca et al., 1965, 1966). It is not known whether these condensation products are themselves active in growth.

IAA is consumed during the growth of some tissues, and there is good *pari passu* evidence that peroxidases act on IAA *in vivo*. This is especially clear in the case of *Omphalia flavida,* a fungus that infects coffee plants. It secretes a peroxidase into its growth medium and, as a direct consequence of the lowered levels of auxins in the petioles, the affected plants lose their leaves through abscission (P. M. Ray, 1958).

### Gibberellins

HISTORY. Japanese farmers have long been familiar with a disease of rice in which the affected plants were spindly and tall, a full 50 percent taller than the normal plants. Diseased plants were called *bakanae* ("foolish seedling"). The bakanae disease was traced to an ascomycetous fungus, *Gibberella fujikuroi,* and the unusual growth of the affected plants was found to be caused by substances called gibberellins secreted by the fungus. The case thus far seems not unusual: there are many examples of abnormal growth occurring in plants infected with various pathogens, for example, witch's broom, corn smut, and oak galls (McCalla et al., 1962). Physiologists are aware that the substances either produced or evoked by the pathogens are likely to be growth substances, yet strangely these cases have been only rarely exploited as systems for studying chemical control of morphogenesis.

The case of the gibberellins is extraordinary, not because it was carried successfully to the point of chemical identification, but because the work, pursued energetically by Japanese scientists and duly reported in the literature over an interval of 25 years, was completely ignored in the West until it was nearly completed. In 1955 the news of gibberellins burst on America and Europe as the most important development in the physiology of plant growth since the discovery of auxins!

The almost incredible history of the gibberellins has been told in detail by Stowe and Yamaki (1959).

CHEMISTRY. The structures of the gibberellins secreted by the *Gibberella* fungus were virtually secured by the painstaking work of Professor Sumiki and his group (cf. Kawarada et al., 1955). The authenticated structure of the fungal substances was finally established by Cross and his group at Imperial Chemicals

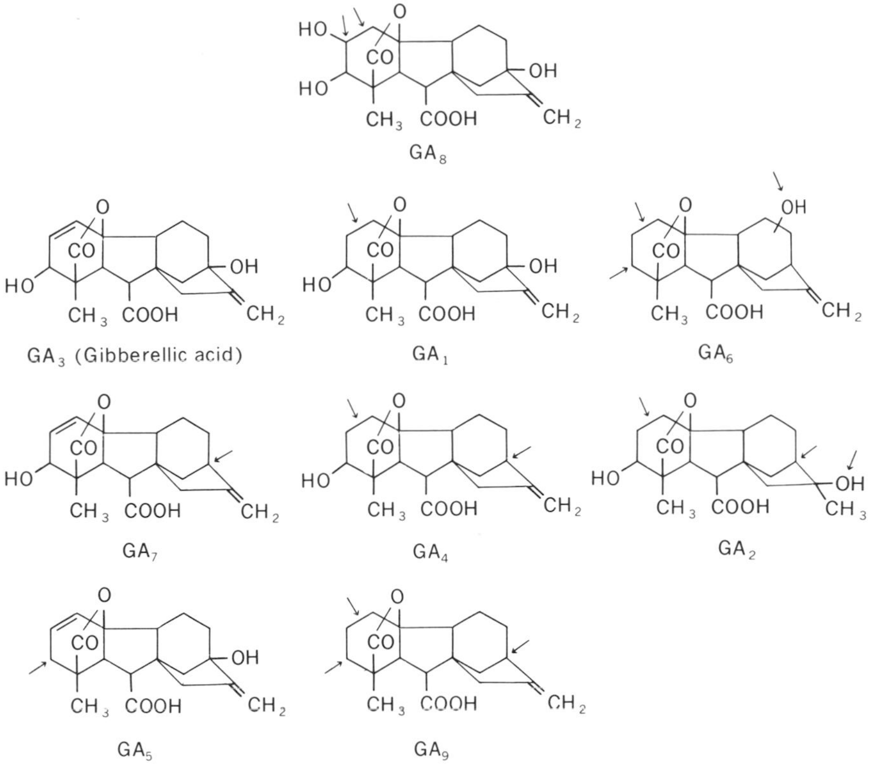

Fig. 5-25 Structure of gibberellins (Leopold, 1964). All the structures contain the same ring structure. The differences from gibberellic acid are shown by the arrows. The compounds are arranged with saturation increasing from left to right and polarity from bottom to top.

Industries (1956) (Fig. 5-25). Identical structures occur in higher plants (cf. Jones et al., 1963).

The gibberellin ring system is almost unique in the chemistry of natural products, especially the lactone arrangement. Because the double bonds are isolated, the extinction coefficients above 260 m$\mu$ are insignificant.

BIOSYNTHESIS. Gibberellins are synthesized in the plant in the mevalonic acid pathway along with terpenes, sterols, carotenoids, rubber, etc. The current views are based on isotopic labeling, enzyme activities, accumulation of and promotion by intermediates. The proposed biosynthetic pathway is shown in Fig. 5-26.

One of the most interesting features of the biosynthetic pathway is the strong inhibition that can be exerted by growth retardants, such as "AMO 1618" and phosphon, on the cyclization step leading to (—)-kaurene-19-ol (Ninnemann et al., 1964; Dennis et al., 1965). (See review of Cathey, 1964, on growth retardants.) These substances behave like "antigibberellins" in that treated

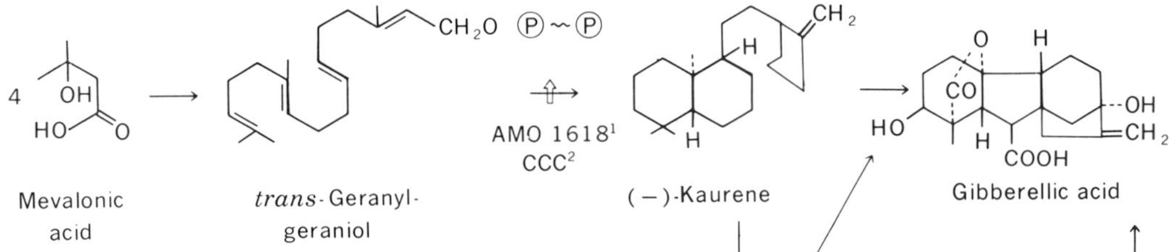

Mevalonic
acid

trans-Geranyl-
geraniol

AMO 1618[1]
CCC[2]

(−)-Kaurene

Gibberellic acid

CH₂OH

(−)-Kaurene-
19-ol

CHO

(−)-Kaurene-
19 al

COOH

(−)-Kaurene-
19-oic acid

[1] AMO 1618 is a synthetic growth retardant.

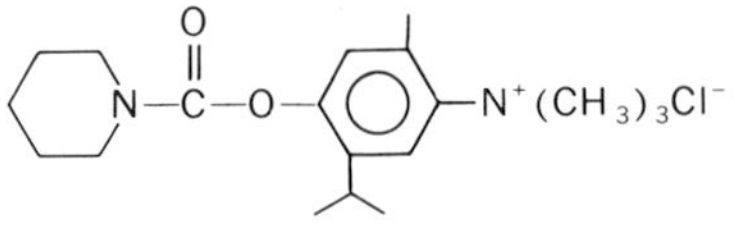

[2] CCC is another growth retardant,
2-chloroethyltrimethyl ammonium chloride.

$(CH_3)_3N^+CH_2-CH_2Cl \cdot Cl^-$

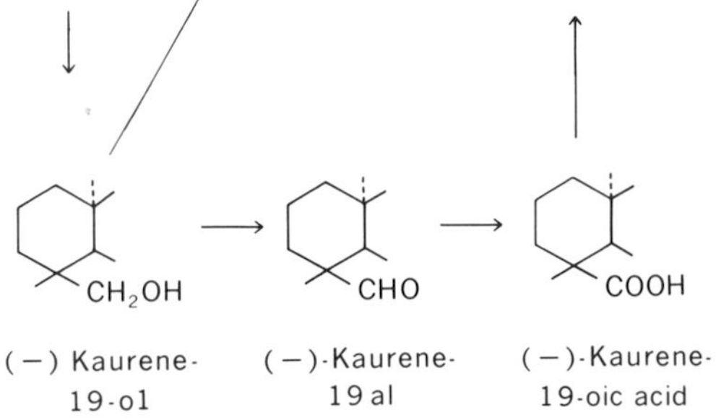

Fig. 5-26 The biosynthesis of gibberellins (Dennis and West, 1967; Cross et al., 1968). This scheme was inferred from the specific labeling of (−)-kaurene, gibberellic acid, and other gibberellins from labeled mevalonic acid lactone and (−)-kaurene, and from the demonstration of enzyme activities.

plants have short, thick internodes, as do genetic dwarfs. The effects of the growth retardants are reversed by the addition of gibberellins. That antiretardants actually decrease gibberellins in higher plants has been shown by extraction and bioassay (Zeevaart, 1966).

BIOLOGICAL ACTIVITY.    In common with other growth substances, gibberellins produce a variety of seemingly unrelated physiological responses.

Immature stem tissues of many plants respond to small quantities of gibberellins by greatly increased stem elongation (Fig. 5-27). These responses

*Fig. 5-27*  Response of rosettes to gibberellins (cf. Stowe and Yamaki, 1959; photograph, by E. C. Wassink and G. M. Curry, kindly supplied by B. B. Stowe). Rosettes, which are a form of plant with compressed stem internodes, typically respond to gibberellins by "bolting," that is, by elongating their stems without a corresponding increase in the number or size of the leaves. SD and LD mean short days and long days, respectively.

are especially striking in the case of rosettes which are induced to become tall and spindly, even to the point of assuming the habit of a vine (cf. Stowe and Yamaki, 1959). Whereas auxins are especially active in promoting growth of etiolated shoots, gibberellins promote growth of both etiolated and green tissue. Gibberellins have no effect on oat coleoptile growth. This is fortunate, since it provides a clear distinction between auxin and gibberellin activity.

The absolute sensitivity of plants to gibberellins is even greater than their responses to auxins. Amounts of the order of 5 $\mu$g per plant are often sufficient to evoke measurable stem elongation. This sensitivity and the correspondingly low concentration of gibberellin in plant tissues impose severe analytical problems. As with auxins, the analyst must rely on bioassays and enormous quantities of plant tissues must be processed to obtain enough for physical characterization.

Some but not all genetic dwarfs respond to gibberellins by growing to normal size (Fig. 5-27). Normal plants respond only slightly. This pattern of responses would occur if the dwarfs were defective in their ability to synthesize gibberellins. Analytical confirmation of this hypothesis is not yet available. Other dwarfs appear to have an impaired response to gibberellins. The specific responses of dwarf lines of corn and peas have made them favorite objects for bioassay of gibberellins.

A flowering response by certain plants to gibberellins is at least as dramatic as the growth response. Long-day plants are typically rosetted until induced to flower. Other rosetted plants are biennials that bolt and flower only after a cold treatment.

Long-day flowering plants and biennials are specifically induced to flower by gibberellins (Fig. 5-28) (Phinney and West, 1960).

*Fig. 5-28*  Flowering of long-day plants with gibberellin (Lang, *Proc. Natl. Acad. Sci.,* **43**:709, 1957; photograph kindly supplied by Anton Lang). The long-day plant *Silene armeria* was grown in short days (9 hr) and given increasing doses of gibberellin, left to right, 0, 2, 5, 10, 20, and 50 $\mu$g plant$^{-1}$ day$^{-1}$.

With some plants, bolting is easier to induce than flowering. Differences in the structures and or localization of naturally occurring gibberellins, as contrasted with applied chemicals, might account for the differences in physiological responses. It does appear nonetheless that shoot elongation and the formation of floral primordia are associated but distinct processes.

The overall pattern is consistent with gibberellins being long-day and biennial flowering hormones. Evidence in support of this is that the concentrations of extractable gibberellins are found to increase substantially with the initiation of flowering (Harada and Nitsch, 1967).

OTHER MORPHOGENETIC RESPONSES.    Gibberellins will permit seeds and other structures to break dormancy (Phinney and West, 1960), convert all-female cucumbers to male and all-male castor bean to female (Shifriss and George, 1964), increase leaf cell expansion (Stowe and Yamaki, 1959) and fruit growth in grapes (Weaver, 1961), and promote phloem differentiation (DeMaggio, 1966).

AMYLASE SECRETION.    When cereal grains germinate, the scutellum dissolves the endosperm causing a synthesis of amylase in amylolytic granules and in this way mobilizes the starch reserves of the seed. No such mobilization normally occurs in detached endosperms. The effect of the embryo can be duplicated by the application of gibberellins. The application of this principle through the use of gibberellins in accelerating the production of malt from barley has proved a boon to brewers. This phenomenon will be considered in a case study below (Sec. 5.7).

ACTIVITY *in vitro*.    Several investigators have reported small but repeatable promotion by gibberellins of RNA formation in isolated nuclei (Roychoudhury and Sen, 1965; Johri and Varner, 1968). The RNA increase is in high molecular weight and nonribosomal species; one naturally thinks that gibberellins may be promoting mRNA formation.

RELATIONS AMONG GIBBERELLINS AND WITH OTHER GROWTH REGULATORS. The various gibberellins are by no means equivalent in the various physiological responses of plants (Thimann, 1963) (Table 5-4). Taking $GA_7$ as a standard of comparison, $GA_5$ for example is equally active with maize Dwarf-3 and -5 but only 10 percent as active with Dwarf-1 and completely inactive in breaking the dormancy of lettuce seeds.

Various hypotheses have been constructed to account for gibberellin action through affecting auxin action. The disparity of response to these two classes of growth substances should effectively rule out this notion.

MODE OF ACTION.    Although gibberellins promote nucleic acid synthesis *in vivo* (Nitsan and Lang, 1966) and the *de novo* formation of enzymes (cf. Sec. 5.7), neither of these appears to be essential to biological responses. Haber et al. (1969) elicited a normal response to GA from "gamma plantlets" (page 298) where no nuclear nucleic acid synthesis was possible, and Pollard (1969)

Table 5-4   Biological Activities of the Gibberellins*

Activities expressed as percentages of activity with $GA_7$.

| | Dwarf Pea Stem | Dwarf Maize Leafsheath | | | Lettuce Hypocotyl | Cucumber Hypocotyl | Lettuce Seed Germination in Dark | Formation of Staminate Flowers on Gynoecious Cucumbers | Parthenocarpic Growth of Tomatoes |
|---|---|---|---|---|---|---|---|---|---|
| | | Dwarf-1 | -3 | -5 | | | | | |
| $GA_1$ | 100 | 100 | 33 | 20 | 2 | 2 | 3 | 8 | 50 |
| $GA_2$ | 33 | 10 | 5 | 10 | 2 | 2 | 0 | 40 | 11 |
| $GA_3$ | 330 | 100 | 50 | 100 | 50 | 2 | 10 | 27 | 33 |
| $GA_4$ | 16 | 100 | 50 | 33 | 16 | 100 | 100 | 80 | 100 |
| $GA_5$ | 33 | 10 | 100 | 100 | 2 | 0.2 | 3 | 2 | 330 |
| $GA_6$ | 33 | 10 | 5 | 20 | 0.5 | 0.2 | 0 | 3 | 25 |
| $GA_7$ | 100 | 100 | 100 | 100 | 100 | 100 | 100 | 100 | 100 |
| $GA_8$ | 3 | 0.3 | 0.5 | 1 | 0 | 0.02 | 0 | <2 | 3 |
| $GA_9$ | 0 | 0.3 | 100 | 33 | 16 | 100 | 0 | 40 | 20 |

* Compiled by Thimann, 1963.

observed enzyme secretion from GA-treated barley endosperm long before any *de novo* enzyme synthesis could be detected. As with the other growth substances, the molecular basis of gibberellin activity is unknown.

### Cytokinins

The arts of plant tissue culture have often been directed at the problems of morphogenesis. Folke Skoog and his group observed that cultures of tobacco stem pith cells and callus cultures were capable of developing occasional buds. In order to study the physiology of bud formation under controlled conditions, they tested additives to the tissue culture media that would increase the number of buds.

They discovered ultimately that a substance present in heated RNA would induce cell divisions in tobacco pith cells and increase bud formation in callus cultures. The substance, 6-furfuryladenine, was called *kinetin* and the class of factors inducing cell division was subsequently called cytokinins (cf. review of Skoog et al., 1965).

CHEMISTRY.   Cytokinins are a class of 6-substituted purines which promote cell division in certain types of tissue preparations. The general formula is shown in Fig. 5-29. The only requirement for R in synthetic cytokinins is moderate fat solubility; $C_5$ and larger radicals are active. Although it is now evident that many plant tissue extracts have cytokinin activity, only one, zeatin, has been chemically identified (Lethem and Miller, 1965). Triacanthine is a natural 1-substituted purine with a structure similar to synthetic cytokinins, but it is almost completely inactive until it is rearranged to a 6-substituted molecule (Leonard et al., 1966).

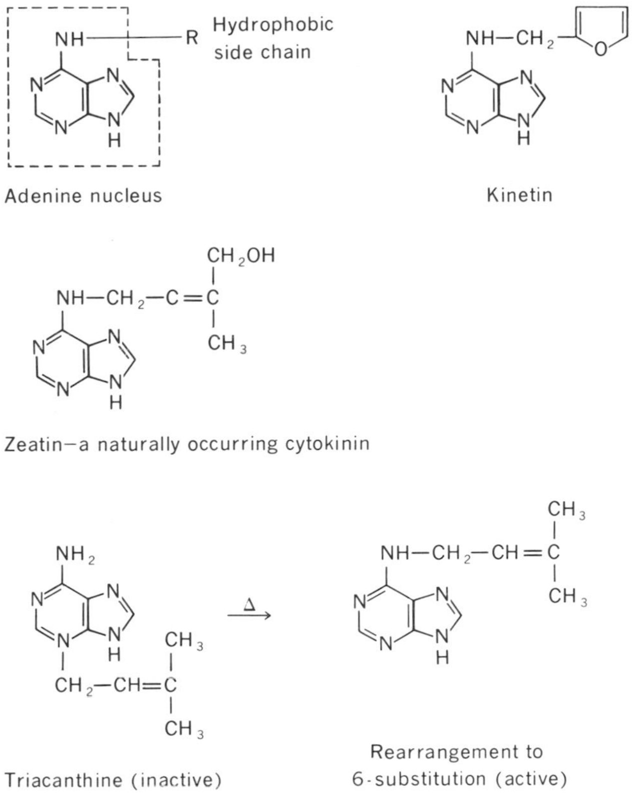

Adenine nucleus

Kinetin

Zeatin—a naturally occurring cytokinin

Triacanthine (inactive)

Rearrangement to
6-substitution (active)

*Fig. 5-29* Structure of cytokinins. To be an active cytokinin, a molecule must possess an adenine nucleus plus a hydrophobic side chain in the 6-position.

Nothing has been established concerning the biosynthesis and very little concerning the fate of cytokinins in tissues. There is good evidence that zeatin is incorporated into nucleotides (Miller, 1965). A small fraction of the label from radioactive kinetin is found in the soluble RNA fraction of plants, and especially in tRNA (Skoog et al., 1966; R. H. Hall et al., 1967), but a small change in the structure of a cytokinin (9-methylation of 6-benzylaminopurine) abolished incorporation into tRNA without changing the effectiveness of the molecule as a growth substance (Kende and Tavares, 1968).

Cytokinins are transported basipetally in stem sections, but little other than passive movements appear to occur within leaves or intact plants (McCready, 1966).

BIOLOGICAL ACTIVITY.   The most striking effects of cytokinins are in promoting cell division in cells which would not otherwise divide (Fig. 5-30). Such effects were first observed with tobacco callus and tobacco pith cultures. Similar responses were observed with carrot root parenchyma cultures, and with soybean cotyledon callus (cf. review of Miller, 1961). Plants as far down in the phylogenetic scale as mosses respond (Bopp and Brandes, 1964). It is interesting that growth of the red alga *Porphyra tenera* is stimulated by kinetin, but in this case the plant merely uses the benzimidazole moiety for the synthesis of vitamin $B_{12}$ (Iwasaki, 1965).

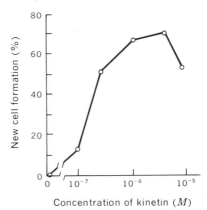

Concentration of kinetin ($M$)

*Fig. 5-30* Stimulation of cell division in tobacco pith by cytokinin (Skoog and Miller, 1957). Explants of tobacco pith were treated for 7 days in sterile culture with different concentrations of kinetin. Auxin was also present.

In some cases cytokinins also induce a massive initiation and development of buds, but always in tissue which produces some buds spontaneously. Such tissues include tobacco callus, *Isatis tinctoria* root cultures, *Convolvulus arvensis* roots, *Saintapaulia ionantha* (African violet) leaves, Begonia leaves, *Tortella caespitosa* (a moss), and even in *Ulva lactuca* (sea lettuce).

A number of other effects of cytokinins have been observed including stimulation of tyramine methylpherase activity (Steinhart et al., 1964), reversal of sex (Negi and Olmo, 1966), inhibition and promotion of root growth, promotion of leaf expansion, breaking of seed dormancy, formation of gametophores, and delay or reversal of senescence (Miller, 1961). Cytokinins are also essential growth factors for certain endosperm tissue cultures.

There is therefore an exceedingly wide variety of physiological responses to cytokinins. This variety of responses contributes to the great difficulty in deciding what cytokinins as well as other growth regulators actually do.

One of the most exciting responses of plants to cytokinins is the stimulation of protein and RNA synthesis in detached leaves. Treatment of a region of a detached leaf with cytokinin reverses the typical course of chlorophyll and protein degradation that occurs in excised leaves (Fig. 5-31) (Richmond and Lang, 1957; Mothes, 1960; Osborne, 1962). Amino acids, phosphate, and other metabolites flow to cytokinin-treated regions, and the movement shows the pro-

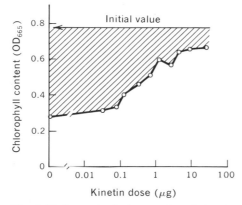

Fig. 5-31 Reversal of pigment degradation in intact leaves (Osborne and McCalla, 1961). *Xanthium* leaf discs were allowed to float for 48 hr on solutions containing different amounts of kinetin. Without any addition, the leaf disc in this interval will lose chlorophyll, protein, RNA, etc. Cytokinins reverse this process. In this experiment chlorophyll content was measured by extraction and photometry.

perties expected of mass flow through the phloem (cf. Sec. 3.9; Müller and Leopold, 1966). Cytokinins, therefore, create metabolic sinks.[1]

### Abscisins

When a fruit ripens or just prior to the loss of leaves, a peculiar, narrow zone of distinctive cells forms across the petiole near the stem. Cell division occurs, along with accumulation of various organic materials. Just distal to the *abscission zone,* the tissue becomes especially soft and the leaf or fruit breaks away.

The control of abscission has been of substantial interest to physiologists, especially in view of its enormous agricultural importance, as in the harvesting of cotton and sugar cane. Auxins have been shown to influence abscission both positively and negatively (cf. reviews of Carns, 1966; Leopold, 1964; and Jacobs, 1962). In addition a number of plant diffusates and squeezates were known to inhibit or promote abscission.

One of these, a diffusate from the base of young cotton fruit, ultimately yielded two substances, named abscisin I and abscisic acid (Addicott and Lyons, 1969), which act strongly to hasten abscission. The latter was found by Ohkuma et al. (1965) to be the unsaturated terpene acid shown in Fig. 5-32. As one

---

[1] Kinetin can protect a plant against powdery mildew (Dekker, 1963), possibly by stimulating the removal of extracellular soluble organic matter that could otherwise serve as substrates for the fungus. Another practical use of cytokinins is in increasing the shelf life of cut flowers and certain vegetables.

might predict from its structure, abscisic acid is synthesized via the mevalonate pathway (Robinson and Ryback, 1969).

The identical molecule was independently isolated and characterized by Cornforth et al. (1965) on the basis of the induction of bud dormancy.

At present there is no information on the structure-activity range for abscisins, although several variants on abscisic acid are active.

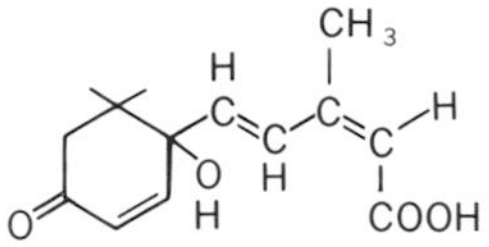

*Fig. 5-32* Structure of abscisic acid (Ohkuma et al., 1965). Abscisic acid promotes abscission and dormancy. It has also been called *dormin.*

Abscisins may turn out to be the biological analogs of the antiparticles of physics. Abscisic acid suppresses germination and extension growth and hastens the onset of senescence and dormancy; in doing so, it inhibits the actions of auxin, gibberellins, and cytokinins (cf. Addicott and Lyons, 1969; Galston, 1969), although in no case does the action appear to be competitive. Abscisic acid is not without positive action: *Pharbitis,* strawberry, and black currant, all normally short-day plants, will flower under long days when treated with abscisic acid (El-Antably et al., 1967), but these plants will also flower in long day when vegetative growth is inhibited; once again, abscisic acid may act by inhibiting.

## Ethylene

Kidd and West (cf. Burg, 1962) as early as 1933 proposed that ethylene gas could be produced by plants and in low concentrations promoted the ripening of plant fruits. Ethylene thus was accused of hormone-like behavior.

When infrared photometry and gas chromatography made feasible the measurements of trace quantities of ethylene, the charge was renewed (cf. Burg, 1962; Burg and Burg, 1966). This time it could be shown that the concentrations and sequence of events had sufficient correspondence that we may reasonably regard ethylene as a ripening hormone, at least for many fruits. For example (Burg and Burg, 1962), the concentration of ethylene within green Gros Michel bananas is 0.1 ppm. It rises to 0.3 ppm the day before the onset of ripening. The same concentration of ethylene is just sufficient to stimulate artificial ripening of bananas.[1]

---

[1] Whether or not we regard ethylene as a hormone is unimportant; bananas do, and the response is sufficiently important that The United Fruit Company has installed devices to absorb ethylene from the atmosphere in the holds of banana boats.

Ethylene also affects morphogenesis in at least two ways: it maintains the characteristic apical hook of etiolated pea shoots (Kang et al., 1967) and interacts with auxin in promoting transverse as opposed to longitudinal growth of cells (Burg and Burg, 1966).

## Phytochrome

Phytochrome was considered in the chapter on light (Sec. 2.8) as a substance which could control a variety of physiological processes. Phytochrome is a protein and there is no evidence for the movement of phytochrome from one cell to another, although its *influence* is clearly systemic. We should probably not think of it as a hormone, certainly not as *the* flowering hormone (cf. suggested case study on floral induction). We have no specific information on the mechanism of its action, but we can speculate that it might be a repressor.

## Rules of Evidence for the Mechanisms of Action of Growth Substances

Much of the literature on growth substances has been aimed at the question of the mechanisms of action, and has missed the mark. Let us see if we can frame rigorous questions and identify lines of evidence that will provide answers.

1. What growth substances exist in the test tissue? Although we might wish to understand the action of a synthetic growth substance, such as 2,4-D or kinetin, it seems crucial that we identify the endogenous growth substances. Different kinds of gibberellins differ in promoting different processes. The action of an applied growth substance might be to antagonize or potentiate an endogenous one. For this latter reason, it seems especially important to determine if antigrowth substances such as abscisins are present.
2. A quantitative corollary to the requirement of identity is that we know how much of the different growth substances are present. For example, it was thought at first that IAN was itself an auxin, until Thimann (1964) found that it is active only upon being converted to IAA.
3. Evidence on the possible action of a growth substance at a site in the information net is acceptable only if the site is isolated from other possible control points. This is a demanding, but I think necessary restriction. Because feedback controls appear to be widespread in nature, we cannot conclude automatically that a stimulation of, say, transcription is not an indirect effect of end-product inhibition.

Success in identifying sites of control in bacterial systems has often occurred through the release from control by other elements in the system by mutations, e.g., "relaxed" mutants whose RNA synthesis is not repressed under conditions of zero growth. Controls may also be "isolated" in time by observing events over very short intervals following a metabolic event; e.g., an induction by detecting a new species of RNA before the new protein is found. For example, estrogen can be shown to stimulate nuclear RNA synthesis within 2 min after its addition to animal tissues (cf. Hamilton, 1968). This interval is about equal to that required by animal cells for the complete formation of a macro-molecule. In contrast, ecdysone stimulates new RNA synthesis in intact chromosomes only after a lag period of 1 hr (Karlson, 1965).

The take-home lesson in studying the primary action of plant growth substances is that we must concern ourselves with events *before* the growth substance has exerted its full or steady-state influence. A coordinate part of our search for understanding growth substances will be evidence on their action *in vitro*. Thus far none of the *in vitro* effects of hormones have been proved to be related to their *in vivo* effects. Action on organelles and other subcellular systems provides an intermediate state between physiological and biochemical levels of investigation, and in this connection we should examine carefully reports of effects of growth substances on isolated nuclei (cf. Roychoudhury and Sen, 1965; Johri and Varner, 1968).

## Case Studies

Holding in mind our understanding of biological control systems and of the rules of evidence for growth substances outlined above, we can ask what kinds of information we shall require to characterize the physiology of growth, differentiation, and development. (Since it is my prejudice that a characterization at the molecular level would be necessary and sufficient, I should require molecular answers.)

• We must identify which growth substances, if any, are involved.
• We must identify what informational pathways are involved.
• We should characterize the phenomenon at the structural and especially the ultrastructural level. Only in this way can we determine with any precision just what changes are occurring and where.
• We must identify how informational systems lead to the structural changes observed.

Although this may seem like a harsh list of requirements in comparison to the information we can supply in the case studies below, we must also recognize that our innocence of the mechanisms of plant growth, differentiation, and development is virtually unsullied by proofs based on definitive evidence.

## 5.6  CASE STUDY: AUXIN-INDUCED CELL ELONGATION

As noted above, Went proved in 1928 that a substance, called auxin, was present in the tips of oat coleoptiles and could move into subjacent tissue and promote cell extension. We noted further that the substance, ultimately identified as indole-3-acetic acid (IAA), promoted cell extension in a variety of stem tissues and that a variety of synthetic molecules, including 2,4-D, elicited similar responses. These then constitute the phenomena we wish to study.

## Identity of the Growth Substance as IAA

Went showed (cf. Went and Thimann, 1937) that growth-promoting activity could be collected in agar blocks by placing excised coleoptile tips on them for an interval of several hours or by extraction with ether. This procedure proved in a simple and elegant way that the growth-promoting activity was a positive substance, not a release from inhibition. Evidence that auxin was produced in one place (coleoptile tips or stem nodes) and had its action elsewhere (the subjacent tissue) unmistakably placed auxins in the distinctive category of hormones.

Because of the extremely small amounts involved, the identification of IAA as the most common natural auxin of plants was not certain until it was possible to compare the migration of the activities of plant diffusates with that of authentic IAA in partition and gas chromatography (cf. Grunwald et al., 1967).

Although IAA is usually the auxin present in diffusates and extracts, we must also ask if this is necessarily the active form of auxin. Pea stem sections and many other elongating tissues contain active peroxidases (often known as "IAA oxidases"), and it was once proposed that an IAA oxidation product is the active auxin (Galston et al., 1964). In contrast, $C^{11}$-labeled IAA added to oat coleoptiles can be quantitatively recovered after an interval of growth by extraction with an aqueous solution of urea (Winter and Thimann, 1966). Thus the hypothesis of an IAA oxidation product appears untenable for the oat system.

The experiments of Winter and Thimann do not exclude the possibility that the active form of IAA might be an addition product, such as with inositol (Labarca et al., 1965), which is reversibly split by urea, although one could then expect to recover the ureide instead of free IAA. Nor is it excluded that oxidation products might play a role in other systems. For example, there is preliminary evidence that γ-methylene oxindole (Fig. 5-24), a peroxidase product of IAA (Still et al., 1965), is an allosteric effector (Tuli and Moyed, 1966).

## Transport

One of Went's initial observations was that auxin is transported only basipetally in coleoptile sections. This has become the general observation with stems.

The polarity of IAA transport, its specificity for IAA, and its speed (1 cm hr$^{-1}$) led to the early recognition that a very special transport process was involved (cf. Goldsmith, 1967a).

The phenomenon of polar transport is crucial to the understanding of tropisms and correlative growth, but is only indirectly related to the problem of how auxin promotes cell elongation. Suggestions were presented at the end of Chap. 3 for the development of auxin transport as a case study (see also review of McCready, 1966).

### Essential Features of Auxin-induced Growth

An enormous variety of physiological processes normally accompanies the growth of stems. In order to isolate variables, one can try to isolate the extension process itself. Auxin-induced growth of isolated coleoptile or stem sections occurs almost as rapidly as in the intact plant. The necessary conditions are that the tissue have a respirable substrate and be able to carry out oxidative phosphorylation.

If we are interested in the primary effect of auxin in cell expansion, we must discover how quickly it acts. By watching an oat coleoptile through a microscope, Kohler (1956) saw that the growth rate reached its maximum rate in about 10 min after the addition of auxin (Fig. 5-33). Ray and Ruesink

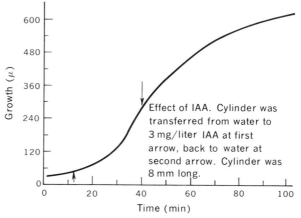

Effect of IAA. Cylinder was transferred from water to 3 mg/liter IAA at first arrow, back to water at second arrow. Cylinder was 8 mm long.

*Fig. 5-33*   Time course of response of coleoptile elongation to auxin (Ray and Ruesink, 1962). By following the growth of oat coleoptile sections under carefully controlled conditions over short intervals of time, it is possible to detect growth responses in less than 10 min after the addition of auxin.

(1962) similarly found a half-time of 12.6 min at 23° and 27.8 min at 13° and that these kinetics were essentially independent of the concentration of IAA. Even shorter times (ca. 5 min) are observed with the methyl ester of IAA (Evans and Ray, 1969).

Since a diffusion-limited process would be concentration-dependent, we can infer from Ray and Ruesink's data that some metabolic process probably intervenes between the primary action of auxin and cell expansion. Ten or twenty minutes is of course quite long enough for the synthesis of macromolecules. At the same time, we should discount any supposedly direct effects which require longer than 10 or 20 min; such slower effects could be indirect consequences of growth or requirements of the growth process.

At first, investigators focused on the generalization that auxin stimulates nearly all the synthetic processes in the cell. This did not reveal much: auxins might affect everything through some master reaction such as oxidative phosphorylation, but it might just as well affect directly only a single process, such as the formation of a membrane lipid, and influence the rest by release from end-product inhibition.

Later, investigators began looking at the cell wall, or rather took a second look. In 1934, Heyn (cf. Went and Thimann, 1937) had found that auxin increased the apparent plasticity of cell walls, that is, the amount of permanent deformation that results from an applied tension. Was this cause or effect? Employing the strategy of isolation in time, Thimann (1954) exposed plant tissue to auxin but prevented growth by immersing the tissue in hypertonic mannitol. After removal of both auxin and mannitol, the cells expanded to a length greater than the initial but without any increase in the osmotic pressure.

Ray and Ruesink (1962) reasoned that a time change in wall plasticity would be equivalent to expansion by *steady-flow viscosity*, but that the data of Thimann and others did not exclude an alternative model, a *chemorheological process*. In the steady-flow viscosity model, the viscosity of the cell wall is imagined to be under metabolic control; in the chemorheological model, subunits of the wall are thought to slip past one another by the continuous synthesis and breaking of the chemical cross links. The distinction is a fine one, but the two models would be expected to behave differently in time and in response to temperature change. Ray and Ruesink found that when the temperature was changed from 12 to 23°, the growth rate of oat coleoptiles changed within 20 sec. This property and the calculated energy of activation of 20,000 cal mole$^{-1}$ were more consistent with the *chemorheological* model of bond breaking and remaking.

Thus auxin can apparently promote wall loosening and synthesis, presumably by bond breaking and remaking, under conditions that exclude indirect consequences of growth itself. But how? One hypothesis is that auxin affects the methylation of pectic materials (Fig. 5-34): they include pectins and pectic acids, heterogeneous chains of galacturonic acids intermixed with other sugars. The free carboxylic acids of one chain can link with others through divalent metal ions, and one can imagine that auxin could affect the degree of methylation and hence the cross-linking and strength of cell walls.

In fact, the transfer of methyl groups from methionine to pectin is promoted by auxin (Ordin et al., 1957), but it can be completely inhibited by ethionine without affecting growth (Cleland, 1963b). There are also reported

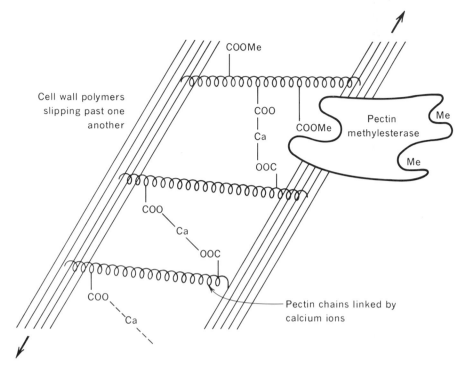

COOMe

Cell wall polymers
slipping past one
another

COO

Ca

COOMe

Pectin
methylesterase

Me

Me

OOC

COO

Ca

OOC

COO

Ca

Pectin chains linked by
calcium ions

*Fig. 5-34*   The pectin methylesterase hypothesis of auxin action (cf. Ordin et al., 1957).
According to this hypothesis, the plasticity of cell walls, which is required for cell enlargement, is controlled by the rate at which polymers in the wall can slip past one
another. In this hypothesis chain pectins, cemented by divalent ions such as calcium, are
thought to lock the wall polymers together. It was proposed that pectinmethylesterase,
somehow promoted by auxin released the pectins by esterifying the sites for calcium
binding. Similar hypotheses have substituted other polysaccharidases (cf. Galston, 1969).

correlations between pectin methylesterase activity and growth rates (Bryan
and Newcomb, 1954; Yoda, 1958), but there is no clear evidence that any
of these effects of auxin are causally related to cell enlargement. Wada et al.
(1968) reported that cell elongation was accelerated by externally applied $\beta$-1,3-
glucanase, but since the cell wall loosening process is reversed after the removal
of auxin (Cleland, 1968), auxin action cannot be simply explained by increased
activities of polysaccharide hydrolases. Cellulase ($\beta$-1,4-glucanase) is another
candidate for cell wall loosening, but none of the hydrolytic enzymes identified
increases nearly rapidly enough to account for auxin-induced growth (Datko
and Maclachlan, 1968).

Still another hypothesis is that auxin acts as a derepressor, that is, by
permitting the transcription of genes. There is certainly sufficient time (Fig.
5-33) for this to happen, but let us look at the evidence: Noodén and Thimann
(1963), Key and Ingle (1964), Morré (1965), and Cleland (1965) showed

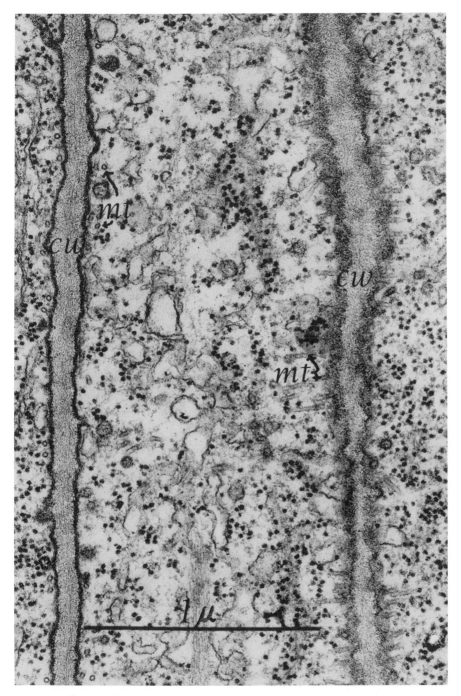

*Fig. 5-35* Ultrastructure of an elongating plant cell [electron photomicrograph of bean root tip (*Phaseolus vulgaris*) kindly provided by Eldon Newcomb]. Note cell wall (*cw*), microtubules (*mt*) oriented transverse to the wall, and longitudinal microfilaments.

that actinomycin D and chloramphenicol inhibit auxin-induced growth. Actino-mycin D is known to inhibit transcription, but although it begins to inhibit RNA synthesis in the plant tissues within 30 min and abolish RNA synthesis in 15 hr, growth is unaffected for 1.5 hr, and finally it inhibits no more than 70 percent. Chloramphenicol is now recognized as an inhibitor of trans-lation specifically in chloroplasts and mitochondria, but its effect on auxin-in-duced growth is delayed for about 5 hr. Additionally, the cell wall loosening effect of auxin is completely unaffected by actinomycin D over this same interval (Cleland, 1965).

Although these data clearly show that transcription and translation are required for the *continuation* of cell expansion, they do not constitute sufficient evidence that auxin acts directly in this region of metabolism.

If auxin were to act on transcription, one would predict that auxin-treated tissues should produce new or more kinds of mRNA, compared with the non-growing controls. Key and Shannon (1964) observed increased incorporation of $C^{14}$-precursors into species of RNA with many of the properties of mRNA but the auxin effects occurred only after about 3 hr. In contrast, estrogen, which probably does act at or near the level of transcription in animal tissues, can induce RNA synthesis within 2 min. If the formation of new species of mRNA had been a direct effect of auxin, we should expect an acceleration of incorporation within 10 to 20 min. The data then are consistent only with an *indirect* derepression by auxin of RNA and protein synthesis.[1]

In conclusion, we know enough about auxin-induced stem elongation to be able to predict the behavior of the metabolic system under the direct or at least early influence of auxin, but it has not yet been identified: It should respond maximally within 10 to 20 min after the addition of auxin, it should bring about a loosening and synthesis of cell wall materials, and over the longer run it should derepress and require certain kinds of mRNA and protein synthesis.

### Ultrastructural Basis of Growth

Systems of microtubules (Fig. 5-35) have been seen adjacent to and parallel to cell wall synthesis (Newcomb, 1969) and specifically in the oat coleoptile (Cronshaw and Bouck, 1965), but whether these structures represent the physical sites of action of auxin is also unknown.

### 5.7   CASE STUDY: CONTROL BY GIBBERELLINS OF MOBILIZATION OF ENDOSPERM RESERVES

#### The Phenomenon

As in the case of castor bean (Sec. 1.4), cereal embryos draw upon the food reserves of the endosperm during germination and seedling development. Haber-landt (1890) and Brown and Escombe (1898) perceived that both embryo

---

[1] *Note added in proof:* But see Y. Masuda and S. Kamisaka, 1969: Rapid stimulation of RNA biosynthesis by auxin, *Plant Cell Physiol.*, **10**:79–86.

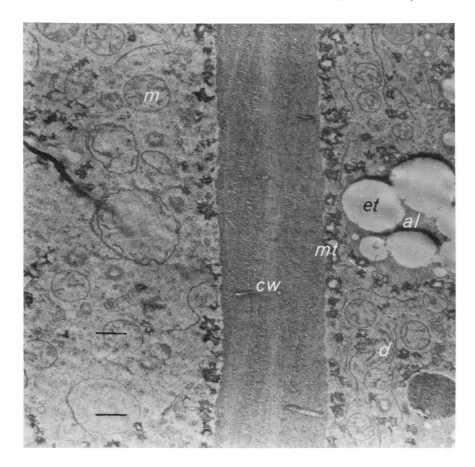

(a)

*Fig. 5-36* Ultrastructural changes in the aleurone layer of barley under the influence of gibberellin (Paleg and Hyde, 1964). Barley aleurone was excised and either incubated in water or treated with GA₃. Aleurone grains (*al*) contain electron-dense spheres (*ed*) and electron-transparent spheres (*et*). The photomicrographs also show mitochondria (*m*), dictyosomes (*d*), spherosomes or microtubules (*mt*), and cell walls (*cw*). The scale line represents 1μ. (*a*) Control tissue after 8 hr. (*b*) GA₃-treated tissue after 8 hr. The aleurone grain membranes are enlarged. The microtubules are less regular and enlarged. (*c*) Control tissue after 24 hr. The general appearance is similar to that at 8 hr. (*d*) GA₃-treated tissue after 24 hr. Aleurone grain membranes have been greatly extended and are indistinguishable from the endoplasmic reticulum. The electron-transparent inclusions have disappeared.

and endosperm somehow conspire to produce hydrolysis of the carbohydrates and proteins of the endosperm. In 1958, Yomo found that barley endosperms separated from the embryo remained inert, but suspended in the same flask with isolated embryos, they would develop amylase activity. That a hormonal connection existed between embryo and endosperm was then established when Yomo showed that purified embryo extracts, probably containing gibberellin,

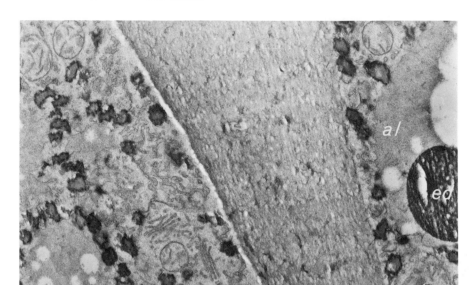

(*b*)

*Fig. 5-36*   (Continued)

could evoke amylase activity in isolated barley endosperm. Authentic gibberellins produced the same effects. Paleg (1960) independently reached the same conclusions on similar evidence.

### Identity of the Hormone

Nearly all known gibberellins produce the same response in isolated barley endosperms. GA, $GA_3$, and $GA_4$ are active at 1 mg liter$^{-1}$ (Paleg et al., 1964), and other gibberellins (Mori et al., 1965) are active at 10 to 100 mg liter$^{-1}$ (cf. Paleg, 1965). Unambiguous identification of the natural hormone as a gibberellin requires that the active principle in barley endosperm be completely separated and identified. Jones et al. (1963) have identified gibberellic acid ($GA_3$) as the only detectable gibberellin in immature barley ears.

The action of gibberellins in stimulating starch hydrolysis is mimicked by helminthosporol, another fungal product (Mori et al., 1965).

### Cellular and Subcellular localization

Brown and Escombe inferred in 1898 that the aleurone skin of the grain was the likely site of action causing hydrolysis of the interior of the endosperm. MacLeod and Millar (1962) and others (cf. Paleg, 1965) showed that only the aleurone was capable of secreting hydrolases in response to gibberellin.

The changes in the aleurone induced by gibberellin have also been seen at the ultrastructural level (Paleg and Hyde, 1964). The aleurone grains start off as discrete membrane-bounded organelles. Within 24 hr after treatment with $GA_3$, they expand and fuse until they form an apparent continuum with the rest of the cytoplasm. The membranes become layered in profile and become

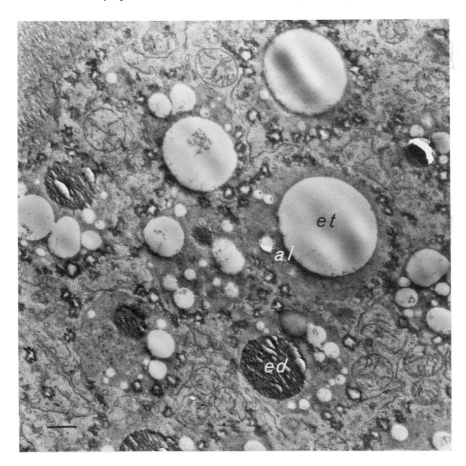

(c)

*Fig. 5-36*   (Continued)

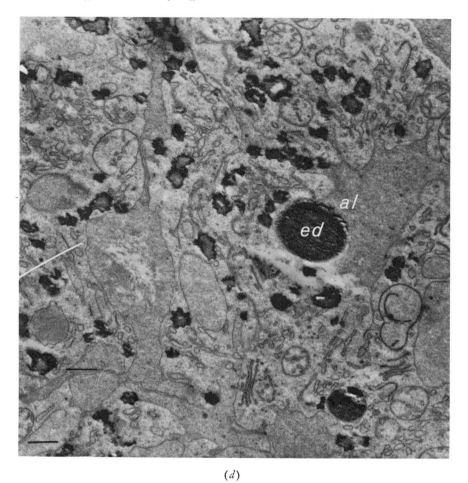

(*d*)

*Fig. 5-36* (Continued)

indistinguishable from the endoplasmic reticulum. The very first change to be seen occurs in 8 hr: 0.2-$\mu$ dense bodies, which Paleg and Hyde identify as spherosomes, begin to expand, become less dense at their centers and ultimately disappear (Fig. 5-36*a* to *d*).

Coincident with the expansion and fusion of the aleurone grains, the cell walls in the aleurone show etching and perforations and the outer layers of the starchy endosperm begin to dissolve. As more and more amylases and proteases are released the entire endosperm is gradually converted into a rich soup of dextrins, oligosaccharides, peptides, and amino acids.

### Information Systems

We noted in the beginning that gibberellin increases amylase and protease activities in the endosperm. This could be due to activation of pre-existing enzymes or *de novo* synthesis of new enzymes. There is reason to think that β-amylase activity is also increased by activation (cf. review of Paleg, 1965), but most of the amylolytic activity is due to α-amylase synthesis, not activation. Varner (1964) and Varner and Chandra (1964) fed C¹⁴-labeled phenylalanine to barley endosperms under sterile conditions and obtained incorporation into α-amylase molecules. No such incorporation occurred without gibberellin or

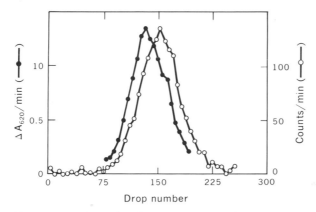

*Fig 5-37*   Evidence for *de novo* synthesis of α-amylase in response to gibberellin (Filner and Varner, 1967). A small amount of purified α-amylase labeled with O¹⁶ and H³ (●) was sedimented to equilibrium with a much larger amount of enzyme obtained from plants treated with gibberellin in the presence of $H_2O^{18}$ (○). Because the curve for the enzyme activity is shifted to higher densities from that of the normal or O¹⁶-enzyme, one can conclude that it was formed from amino acids enriched in O¹⁸ and therefore represents *de novo* synthesis.

when actinomycin D and gibberellin were added simultaneously. They concluded therefore that gibberellin causes the α-amylase to be synthesized *de novo*.[1]

Can we conclude therefore that gibberellin acts at the level of transcription? Although the data are consistent with this view, other alternatives have not been excluded. α-Amylase synthesis gets off to a slow start after gibberellin treatment; 10 to 12 hr elapse before the rates of enzyme synthesis become maximal. This delay could mean merely that mRNA for α-amylase is synthesized slowly, but it could also mean that other processes intervene: sequential inductions, derepressions, or release from end-product inhibition. To prove the proposed mechanism of action, it would be necessary to prove that formation of mRNA for α-amylase commences immediately upon addition of gibberellin and that over these initial hours the level of RNA is in fact limiting to α-amylase formation.

[1] That new enzyme molecules were formed was shown unambiguously by the method of density labeling (Filner and Varner, 1967) (see Fig. 5-37).

## 5.8   CASE STUDY: CHLOROPLAST DEVELOPMENT[1]

### The Phenomenon

Chloroplasts enjoy a kind of autonomy in the cells of plants. They have independent pathways for the synthesis of certain constituents, including ATP. Certain
genes affecting chloroplasts fail to segregate with chromosomal genes (whence
the terms "extrachromosomal" or "cytoplasmic" genes) and may well be in
the chloroplasts themselves (Sager, 1965). Finally there are inhibitors that
affect processes in chloroplasts without affecting the homologous processes elsewhere in the cell. One of them, streptomycin, can wipe out chloroplasts in
a cell line. Since plants are otherwise unaffected, we may say that the drug
"cures" plants of their chloroplasts.

The plastids of seed plants generally remain undeveloped until exposed

[1] References: Schiff and Zeldin, 1968; Smillie and Scott, 1969.

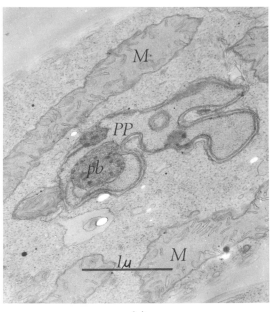

(*a*)

*Fig. 5-38*   Ultrastructure of the transformation of proplastids to chloroplasts in *Euglena gracilis* (plates kindly
supplied by Jerome A. Schiff). (*a*) Section of etiolated
cell. Note proplastid (*PP*) containing prolamellar bodies
(*pb*) and ribosomes, mitochondria (*M*), and numerous
cytoplasmic ribosomes. (*b*) Section of fully greened cell.
Note chloroplast (*C*) containing numerous lamellae and
ribosomes and associated paramylon granules (*PA*), the
pellicle (*PE*) system including transverse microtubules,
mitochondria (*M*), a dictyosome (*D*), endoplasmic reticulum (*ER*), and numerous cytoplasmic ribosomes.

to light; then there is a rapid conversion of pre-existing protochlorophyll to chlorophyll (Sec. 2.4) followed by a slow synthesis of much larger amounts of chlorophyll (cf. Fig. 2-10), roughly coordinate with the development of the proplastid into the highly differentiated lamellar structure of the mature chloroplast (Fig. 5-38) (cf. Bogorad, 1967; Anderson and Boardman, 1964).

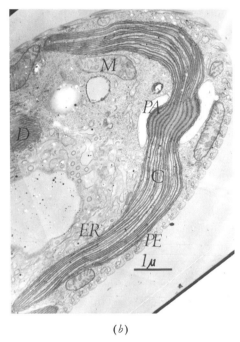

(b)

*Fig. 5-38*   (Continued)

The lamellae appear to be organized from folding of the internal membrane of the plastid. The initial structure is a flattened sac, called a disc or a thylakoid (cf. Heslop-Harrison, 1963; Wehrmeyer, 1964). Subsequent lamellae result from further proliferation and folding of the internal membrane.

Thus, with the flick of a light switch, this system is under the control of the experimenter. Various biological test systems have been used. Etiolated seedlings, especially of beans (cf. Boardman and Anderson, 1964), is one such system and has the advantage that intact plastids can be readily isolated (Boardman and Wildman, 1962).

Lower plants are typically green when grown in the light or the dark, but a few algae (*Euglena, Ochromonas,* certain species and mutants of *Chlorella* and *Chlamydomonas*) show light-dependent plastid development. Because of the advantages of microbial growth, these forms have been favored in studying plastid control (cf. Ben-Shaul et al., 1964), but they have the disadvantage of great difficulties in the separation of pure, intact plastids. For example, the $S$-$\rho$ coordinates (cf. Fig. 1-2) of *Euglena* proplastids are identical with those of *Euglena* mitochondria (Knight and Price, 1968).

As noted earlier (Sec. 5.1) the composition and properties of chloroplasts and mitochondria are sufficiently similar to those of bacteria to encourage the notion that they started their association with plants as ancient symbionts (cf. discussion of Ris and Plaut, 1962). Since most of our understanding of biological information systems comes from the study of bacteria, it may be enlightening to look for information systems that might operate in chloroplast development.

### The Model

Let us consider the hypothesis of essentially unidirectional information flow with the entire sequence operating in the plastid (Fig. 5-39). (We know imme-

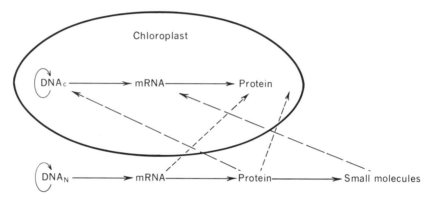

*Fig. 5-39*   Model for information flow in a plastid. The chloroplast appears to contain a complete informational system, including replication, transcription, and translation. But, since the chloroplast is also partially under the control of the nucleus, there must also be mechanisms for the passage of certain regulatory molecules into the chloroplast (dashed lines). If, as seems very likely in the case of mitochondria, the chloroplast should have only part of the information required to specify itself, there should also be some means for the entry of informational and/or functional molecules (shown by dotted lines).

diately that the model is inadequate because the chloroplasts are at best *semi*-autonomous; there is also information flow to and from the rest of the cell.) A more general, but a great deal more complicated model would show two parallel information flows, with possible interactions at almost any step. Let us see how far we can go in ignoring the rest of the cell.

DNA.   There were a variety of early indications that chloroplasts might contain DNA, but the data were indirect or ambiguous. Proof that chloroplasts contain DNA came from two lines: Electron photomicrographs of some chloroplasts show regions containing fibers about 24 Å in diameter and similar in appearance to the nucleoids of bacteria (Ris and Plaut, 1962) (Fig. 5-40). These fibers

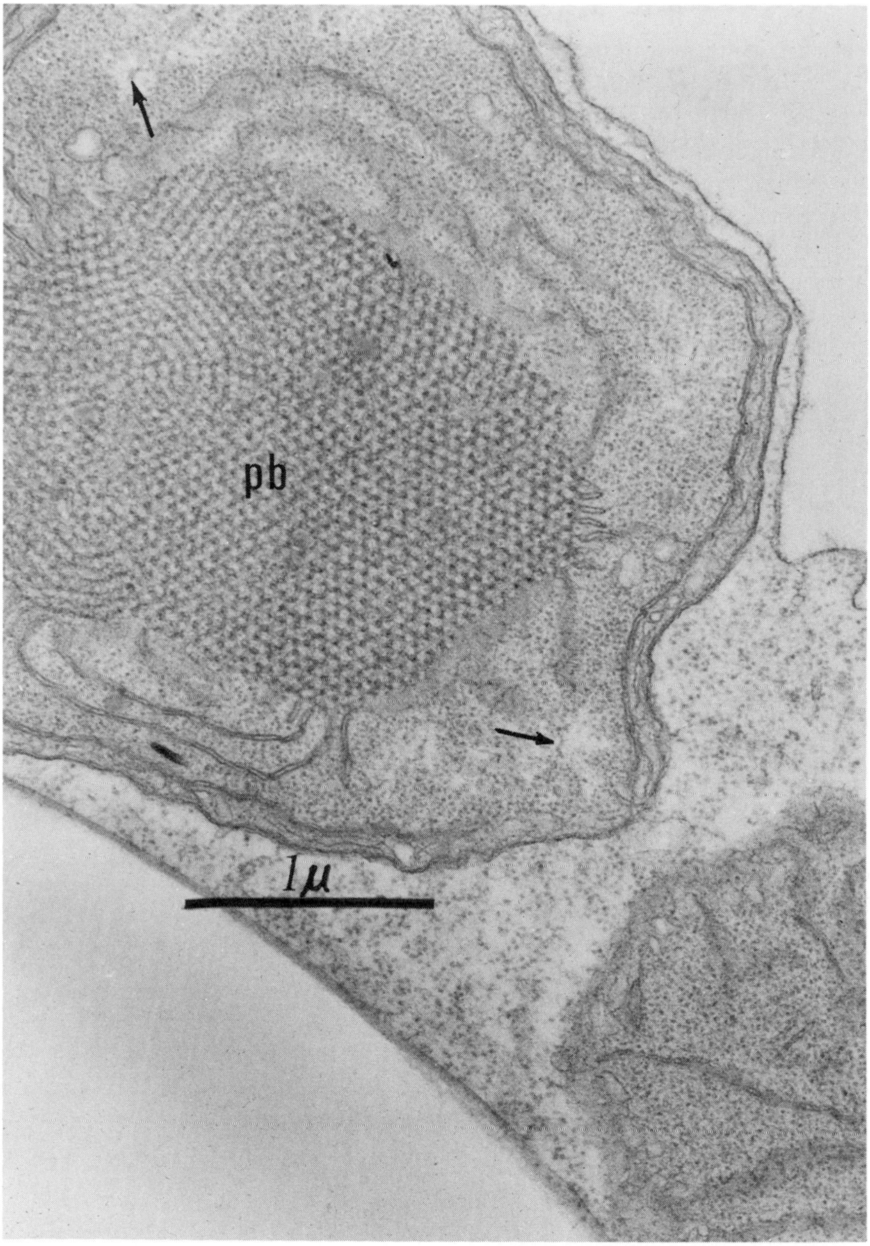

*Fig. 5-40* Ultrastructural evidence of DNA in proplastid (electron photomicrograph of maize proplastid kindly provided by Lawrence Bogorad). In proplastids and developing chloroplasts, interlamellar clear areas (arrows) are often seen to contain a tangle of very thin fibers (40 Å diameter) about the size expected of naked DNA. The fibers disappear after treatment of the sections with DNase. Note also highly ordered array of tubules in prolamellar body.

disappear when the sections are treated with DNase. Second, preparations of chloroplasts from enucleated *Acetabularia* contain DNA (Gibor and Izawa, 1963; Baltus and Brachet, 1963), and DNA of purified chloroplasts from this and other plants can be distinguished from the rest of the DNA of the cell by differences in its equilibrium density, its nucleotide composition, and its melting profile (Kirk, 1963; Sager and Ishida, 1963; Chun et al., 1963; Brawerman and Eisenstadt, 1964a; Edelman et al., 1964).

The equilibrium densities of DNA have proved to be the property most amenable to high resolution. Plant DNA typically shows a main band and one or more satellites (Fig. 5-41). One of these satellites is associated with the chloroplasts and another, when visible, with mitochondria.

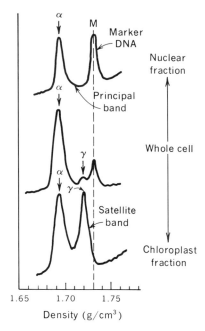

*Fig. 5-41* Equilibrium density profile of DNA from spinach (Chun et al., 1963). Profiles are densitometer tracings of UV absorption photographs of DNA in cesium chloride density gradient. DNA was centrifuged to equilibrium in an analytical ultracentrifuge.

The fact that chloroplast DNA after heat denaturation can be renatured into a molecule of the same density (Tewari and Wildman, 1966; Chiang and Sueoka, 1967) probably means that the molecules are in the form of closed, double-stranded circles, as is the case for most mitochondrial DNAs.

Chloroplast DNA in *Chlamydomonas* is replicated in a semiconservative (that is, conventional) fashion about midway in the cell division cycle, in contrast to nuclear DNA which doubles just before cell division (Fig. 5-42). The evidence for this is (1) the increase in DNA occurring in the middle of the light cycle and (2) the observation that chloroplast DNA from cells grown on $N^{15}H_4^+$ was fully converted to hybrid $N^{15}/N^{14}$ DNA about 4 hr after transfer to a medium containing $N^{14}H_4^+$ while nuclear DNA was not transformed until the middle of the dark period.

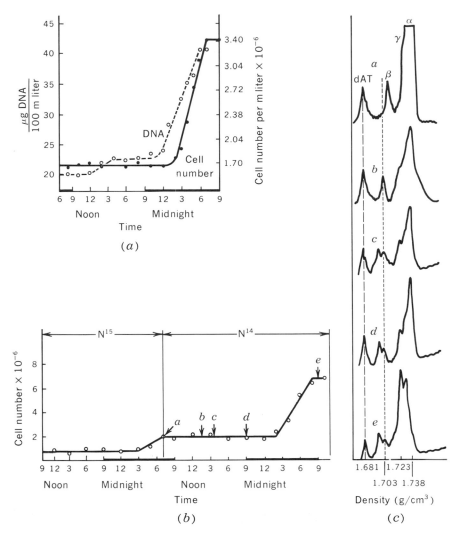

*Fig. 5-42* Time course of nuclear and chloroplast DNA synthesis in *Chlamydomonas* (Chiang and Sueoka, 1967). (*a*) Total DNA and cell number of *Chlamydomonas* in synchronous culture. (*b*) Schedule of growth on $N^{15}$ and transfer to $N^{14}$. Letters refer to times when samples were taken for isopycnic sedimentation of DNA. (*c*) Absorbancy profiles of DNA after isopycnic sedimentation. Letters correspond to times that samples were taken (Fig. 5-42*b*).

Differential rates of incorporation of precursors into main band and satellite DNA in tobacco was taken as evidence of independent replication of chloroplast DNA (Green and Gordon, 1966).

Chloroplast DNA in *Chlamydomonas* constitutes about 15 percent of the total DNA of the cells. This value appears to be rather high compared with the values of 5 percent or less found for most phototrophic tissues. Mitochondrial DNA by comparison rarely exceeds 1 percent of the total. Exceptions are yeast (up to 15 percent) and egg cells (nearly all).

Estimates of the amount of DNA per chloroplast range from $10^{-14}$ to $10^{-16}$ g. A median figure of $10^{-15}$ g would correspond to $2.4 \times 10^6$ nucleotides and, assuming no redundancy, it would be sufficient to code for about 70 proteins of 20,000 MW each (Edelman et al., 1966). Somewhat similar calculations reduce from the findings of Ray and Hanawalt (1965) that the particle weight of chloroplast DNA of *Euglena* is $20 \times 10^6$ to $40 \times 10^6$ daltons. This quantity of information, assuming no redundancy, makes chloroplasts appear highly intelligent (or at least well-informed) compared with the rather limited wit of mitochondria and of some of the moronic satellite viruses that can code only for their own coat protein. However, the fact that immature plastids have much less DNA than mature chloroplasts do makes it likely that mature chloroplasts are probably polyploid and that the information they contain is less than that calculated above.

A further complication is the finding of Woodcock and Bogorad (personal communication) that only one in three chloroplasts in *Acetabularia* have *any* DNA; it follows that the DNA in these chloroplasts would be three times the average value.

That isolated chloroplasts can synthesize DNA was shown by Spencer and Whitfeld (1967).

RNA AND RIBOSOMES.  One would like to show that chloroplast DNA can be transcribed.

After dark-grown *Euglena* are illuminated, small differences can be detected in the nucleotide composition of the total cell RNA (Brawerman and Chargaff, 1959; Smillie, 1963). These differences are due to the appearance of chloroplast ribosomes which differ appreciably from cytoplasmic ribosomes in their base ratios as well as in their size (Brawerman and Eisenstadt, 1964) (cf. Sec. 5.1). Dyer and Leech (1967) report two chromatographically separable heterogeneous bands of RNA from intact chloroplasts of broad bean leaves which can be distinguished from corresponding bands from the remainder of the leaves. One of these bands shows tRNA activity.

Chloroplasts from broad beans (Kirk, 1964) and *Acetabularia* (Schweiger and Berger, 1964) can synthesize RNA under conditions that resemble DNA-dependent synthesis. In maize the formation of a chloroplast-specific RNA polymerase and of rRNA occurs as an early event in greening (Bogorad, 1967, and personal communication).

As we noted above (Sec. 5.1), chloroplast ribosomes are all 70s, whereas those of the cytoplasm are 80s. This peculiarity not only excludes the possibility that the ribosomes in chloroplast preparations represent contaminating cytoplasmic ribosomes but is strongly consistent with a special origin of these ribosomes and of the other kinds of RNA within the chloroplast. The simplest mechanism for the existence of chloroplast-specific RNA would be through the transcription of chloroplast-specific DNA.

One kind of evidence that would be sufficient proof that chloroplast DNA is transcribed would be evidence that RNA complementary to chloroplast DNA was formed *in vitro* and *in vivo*. The latter has been shown by Scott and Smillie (1967) and Tewari and Wildman (1968).

PROTEIN SYNTHESIS.    A second kind of evidence that would be sufficient proof of the genetic autonomy of chloroplasts would be data showing the formation of a specific protein coded for by chloroplast DNA.

Efforts have been directed first at evidence that isolated chloroplasts can carry out protein synthesis. Bové and Raacke (1959) demonstrated amino acid activation and Dyer and Leech (1967) showed the presence of tRNA species in chloroplasts. There was a great deal of confusion at one time over whether isolated organelles, mitochondria or chloroplasts, could in fact synthesize proteins. Although $C^{14}$-labeled amino acids were taken into chloroplast preparations, it was difficult to exclude contamination by bacteria. It was especially helpful, therefore, when workers were able to demonstrate that the substantial rates of protein synthesis in chloroplasts were uncorrelated with the numbers of contaminating bacteria (Eisenstadt and Brawerman, 1964; Spencer, 1965; Goffeau and Brachet, 1965; Bamji and Jagendorf, 1966; Boardman et al., 1966).

Chen and Wildman (1967) detected polyribosomes in tobacco chloroplasts that were synthesizing protein. M. F. Clark et al. (1964) found that single ribosomes of cabbage leaf chloroplasts were converted to polysomes when protein synthesis in the intact leaf was turned on by light.

One might expect that completely intact chloroplasts might require fewer externally supplied intermediates than are required by isolated ribosomes. In fact, Goffeau and Brachet (1965) showed that chloroplasts from *Actabularia* incorporated at high rates without any addition to the suspension. Also, Davies and Cocking (1966) reported similar independence by plastids from sterile tomato fruit tissue.

Unfortunately, we do not know *which* proteins are synthesized in chloroplasts. No net synthesis has been demonstrated in isolated chloroplasts and the identities of the labeled proteins have not been identified with certainty.

In summary, therefore, we see that all the necessary informational systems are present in chloroplasts, but we have yet to trace the flow of any identified information from DNA to protein. Nor do we know in what way the informational systems of the rest of the cell interact with the chloroplasts.

CONTROL SYSTEMS. A great deal of information has been accumulated on the environmental conditions affecting plastid development. For example, red light acting through the phytochrome system can initiate formation of NADP-triose phosphate dehydrogenase without synthesis of chlorophyll. The fact that enucleated *Actabularia* also responds to light in this way (Ziegler et al., 1967) is strong evidence that the control and indeed the complete specification of this enzyme is contained in the chloroplast. Red light also potentiates plastids so that subsequent exposure to light or shorter wavelengths will cause chlorophyll formation at an accelerated rate (cf. Margulies, 1965). It is also known that the chloroplast replication system is differentially damaged by high temperatures.

Chloroplast development is also controlled by metabolites. Glucose in particular, but almost any respirable carbon source tends to inhibit chloroplast development (cf. Aoki et al., 1965).

Chloramphenicol, which we saw specifically inhibits protein synthesis in organelles, inhibits chloroplast development. Also streptomycin, which inhibits DNA replication in certain kinds of bacteria (Gorini and Beckwith, 1966), abolishes chloroplast replication.

With the exception of the effects of the antibiotics, none of these phenomena is understood at the level of informational systems.

Taken together, there is growing certainty that the informational systems of chloroplasts operate with a surprising degree of autonomy. What we lack is specific information on the points in the informational chain where controls operate and how these are translated into the morphological transformation of proplastids into mature chloroplasts.

## SUGGESTED CASE STUDIES

### Control of Morphogenesis in *Acetabularia*

*Acetabularia* is a large (3 to 5 cm), unicellular marine alga. It consists of a stalk surmounted by a cap or umbrella. Through most of its life cycle, the single nucleus remains at the basal end of the stalk. If the stalk is cut, the nucleated half will completely regenerate a normal cap.

Because of its large size and relative ease of manipulation, *Acetabularia* provides a virtually unique test organism for studying the morphogenesis of a cell. It has proved especially fruitful for studying the intracellular localization and movement of morphogenetic messages which may be forms of or contain mRNA.

*References.* Werz, 1964; Zetsche, 1964, 1965; Schweiger and Berger, 1964.

### Cyclitols as Growth Substances

Certain tissue cultures are known to respond to specific inositols; scyllo-inositol is one of the active substances from coconut milk. These cyclitols have also been implicated in the esterification of IAA. In addition, there is strong evidence for them as intermediates in polysaccharide synthesis. Thus, a number of the criteria are at hand for the identification of these molecules as plant growth substances. Are they also hormones and do they play regulatory functions in the normal plant?

*References.* Steinhart et al., 1962; Pollard et al., 1959.

### Induction of Floral Primordia

A cyclic sequence of differentiation of buds and leaves occurs as an unvarying geometric progression in the apical meristems of vegetative plants. Then, through whatever agency—a change in photoperiod, or the increased level of some hormone—the cycle of differentiation suddenly shifts to the production of ovaries, stamens, and other floral parts. These events, often under the precise control of the experimenter, offer exceptional opportunities for the student of plant physiology to ask questions about tissue and organ differentiation.

*References.* Cherry and van Huystee, 1965; Clowes, 1965; Evans, 1964; Bonner and Zeevart, 1962; Hamner and Bonner, 1938; Takimoto and Hamner, 1964.

### Leaf Abscission

When plants are about to enter a dormant phase or in certain cases of senescence, the plant sheds its leaves. The process and various details of it is of obvious survival value to the plant.

The mechanism of abscission has been investigated principally at the physiological level and the accumulated information, although somewhat overwhelming in quantity, offers a nice opportunity for weighing evidence for hypotheses.

*References.* Schwertner and Morgan, 1966; Carns, 1966; Jacobs et al., 1962; Osborne, 1962.

## Infection of Plant Cells by RNA Viruses

Ever since the remarkably detailed elucidation of phage infection of bacteria, the study of virus infection of plant cells has rather suddenly assumed prominence. For one thing, more is known of the chemistry of tobacco mosaic virus than any other virus—plant, animal, or bacterial. Plants may provide much simpler systems than animals for studying virus entry, replication, and parasitism of a eucaryote.

In any event, substantial progress has been made in tracing the movement of virus particles into the nuclei of plant cells, their transformation there into the double-stranded replicative form of the RNA, and the subsequent replication of nucleic acid and translation of messages for coat proteins and other components.

*References.* Clark et al., 1965; Reddi, 1964, 1966; Hirai and Hirai, 1964; Ralph et al., 1965; Smith and Schlegel, 1964; Reichmann, 1964; Shipp and Haselkorn, 1964 Spiegelman et al., 1965.

## Dormancy in Seed Germination

The germination of plant seeds contain a number of remarkable inventions for adapting plants to a variety of environments. Some of them involve maintaining dormancy until certain conditions in the environment have been met. For example, certain desert annuals germinate only after several centimeters of water has flowed past them, that is, until the soil has received enough water to suffice for a full plant generation. Dormancy of necessity involves blocks in  information transfer and/or metabolism. We can ask where these blocks are, how they operate, and how they are relieved.

*References.* Went, 1955; Ikuma and Thimann, 1964; Tuan and Bonner, 1964; Marcus and Feeley, 1964; Mayer and Poljakoff-Mayber, 1963.

# REFERENCES

Addicott, F. T., and J. L. Lyons, 1969. Physiology of abscisic acid and related substances, *Ann. Rev. Plant Physiol.,* **20**:139–164.

Alfridi, M. M. R. K., and E. J. Hewitt, 1964. The inducible formation and stability of nitrate reductase in higher plants: 1, *J Exp. Botany,* **15**:251–271.

Allfrey, V. G., and A. E. Mirsky, 1962. Evidence for the complete DNA-dependence of RNA synthesis in isolated thymus nuclei, *Proc. Nat. Acad. Sci.,* **48**:1590–1596.

Anderson, J. M., and N. K. Boardman, 1964. Studies on the greening of dark-grown bean plants: II. Development of photochemical activity, *Aust. J. Biol. Sci.,* **17**:93–101.

Aoki, S., M. Matuska, and E. Hase, 1965. De- and re-generation of chloroplasts in the cells of *Chlorella protothecoides:* V. Degeneration of chloroplasts induced by different carbon sources, and effects of some antimetabolites upon the process induced by glucose, *Plant and Cell Physiol.,* **6**:487–497.

Asahi, T., Y. Honda, and I. Uritani, 1966. Increase of mitochondrial content in sweet potato root tissue after wounding, *Arch. Biochem. Biophys.,* **113**:498–500.

Atkinson, D. E., 1966. Regulation of enzyme activity, *Ann. Rev. Biochem.,* **35**:85–124.

Avers, C. J., 1967. Heterogeneous length distribution of circular DNA filaments from yeast mitochondria, *Proc. Nat. Acad. Sci.,* **58**:620–627.

———, F. E. Billheimer, H.-P. Hoffmann, and R. M. Pauli, 1968. Circularity of yeast mitochondrial DNA, *Proc. Nat. Acad. Sci.,* **61**:90–97.

Baltus, E., and J. Brachet, 1963. Presence of deoxyribonucleic acid in the chloroplasts of *Acetabularia mediterranea, Biochim. Biophys. Acta,* **76**:490–492.

Bamji, M. S., and A. T. Jagendorf, 1966. Amino acid incorporation by wheat chloroplasts, *Plant Physiol.,* **41**:764–770.

Barnett, W. E., D. H. Brown, and J. L. Epler, 1967. Mitochondrial-specific aminoacyl-RNA synthetases, *Proc. Nat. Acad. Sci.,* **57**:1775–1781.

Basilio, Carlos, M. Bravo, and J. E. Allende, 1966. Ribonucleic acid code words in wheat germ, *J. Biol. Chem.,* **241**:1917–1919.

Beckman, L., J. G. Scandalios, and J. L. Brewbaker, 1964. Catalase hybrid enzymes in maize, *Science,* **146**:1174–1175

Beevers, L., and R. H. Hageman, 1969. Nitrate reduction in higher plants, *Ann. Rev. Plant Physiol.,* **20**:495–522.

Benjamin, W., O. A. Levander, A. Gellhorn, and R. H. DeBellis, 1966. An RNA-histone complex in mammalian cells: The isolation and characterization of a new RNA species, *Proc. Nat. Acad. Sci.,* **55**:858–865.

Ben-Shaul, Y., J. A. Schiff, and H. T. Epstein, 1964. Studies of chloroplast development in *Euglena:* VII. Fine structure of the developing plastid, *Plant Physiol.,* **39**:231–240.

Berlowitz, L., and M. L. Birnstiel, 1967. Histones in the wild-type and the anucleoplate mutant of *Xenopus laevis, Science,* **156**:78–80.

Bhattacharya, P. K., J. M. Naylor, and M. Shaw, 1965. Nucleic acid and protein changes in wheat leaf nuclei during rust infection, *Science,* **150**:1605–1607.

Boardman, N. K., and S. G. Wildman, 1962. Identification of proplastids by fluorescence microscopy and their isolation and purification, *Biochim. Biophys. Acta,* **59**:222–224.

———— and J. M. Anderson, 1964. Studies on the greening of dark-grown bean plants: I. Formation of chloroplasts from proplastids, *Aust. J. Biol. Sci.,* **17**:86–92.

————, R. I. B. Francki, and S. G. Wildman, 1966. Protein synthesis by cell-free extracts of tobacco leaves: III. Comparison of the physical properties and protein synthesizing activities of 70S chloroplast and 80S cytoplasmic ribosomes, *J. Mol. Biol.,* **17**:470–489.

Bogorad, L., 1967. The role of cytoplasmic units, *Developmental Biology,* Supplement I, 1–31.

Bonner, J., 1965. "The Molecular Biology of Development," pp. 1–155, Oxford University Press, London.

———— and J. A. D. Zeevaart, 1962. Ribonucleic acid synthesis in the bud essential component of floral induction in *Xanthium, Plant Physiol.,* **37**:43–49.

———— and P. Ts'o. (eds.), 1964. "The Nucleohistones," 398 pp., Holden-Day, San Francisco.

Bonnett, H. T, and J. G. Torrey, 1964. Movement of IAA-C$^{14}$ in segments of *Convolvulus* roots, *Plant Physiol.,* **39**:lxvi.

Bopp, M., and H. Brandes, 1964. Versuche zur Analyse der Protonema-Entwicklung der Laubmoose: Über den Zusammenhang zwischen Protonema-Differenzierung und Kinetin-Wirkung bei der Bilding von Moosknospen, *Planta,* **62**:116–136.

Bové, J., and I. D. Raacke, 1959. Amino acid activating enzymes in isolated chloroplasts from spinach leaves, *Arch. Biochem. Biophys.,* **85**:521–531.

Brawerman, G., 1963. The isolation of a specific species of ribosomes associated with chloroplast development in *Euglena gracilis, Biochim. Biophys. Acta,* **72**:317–331.

———— and E. Chargaff, 1959. Relation of ribonucleic acid to the photosynthetic apparatus in *Euglena gracilis, Biochim. Biophys. Acta,* **31**:172–177.

———— and J. M. Eisenstadt, 1964a. Deoxyribonucleic acid from the chloroplasts of *Euglena gracilis, Biochim. Biophys. Acta,* **91**:477–485.

———— and ————, 1964b. Template and ribosomal ribonucleic acids associated with the chloroplasts and the cytoplasm of *Euglena gracilis, J. Mol. Biol.,* **10**:403–411.

Briggs, W. R., 1963. Mediation of phototropic responses of corn coleoptiles by lateral transport of auxin, *Plant Physiol.*, **38**:237–247.

Brown, H. T., and F. Escombe, 1898. On the depletion of the endosperm of *Hordeum vulgare* during germination, *Proc. Roy. Soc. (London)*, **63**:3–25.

Bryan, W. H., and E. H. Newcomb, 1954. Stimulation of pectin methylesterase activity of cultured tobacco pith by indoleacetic acid, *Physiol. Plantarum,* **7**:290–297.

Burdett, A. N., and P. F. Wareing, 1966. The effect of kinetin on the incorporation of labeled orotate into various fractions of ribonucleic acid of excised radish leaf discs, *Planta,* **71**:20–26.

Burg, S. P., 1962. The physiology of ethylene formation, *Ann. Rev. Plant Physiol.*, **13**:265–302.

———— and E. A. Burg, 1962. Role of ethylene in fruit ripening, *Plant Physiol.,* **37**:179–189.

———— and ————, 1966. The interaction between auxin and ethylene and its role in plant growth, *Proc. Nat. Acad. Sci.,* **55**:262–268.

Burton, D., and P. K. Stumpf, 1966. Fat metabolism in higher plants: XXXII. Control of plant acetyl-CoA carboxylase activity, *Arch. Biochem. Biophys.*, **117**:604–614.

Cairns, J., 1963. The form and duplication of DNA, *Endeavour,* **XXII**:141–143.

Carns, H. R., 1966. Abcission and its control, *Ann. Rev. Plant Physiol.*, **17**:295–314.

Carpenter, W. J. G., and J. H. Cherry, 1966a. Effects of protein inhibitors and auxin on nucleic acid metabolism in peanut cotyledons, *Plant Physiol.*, **41**:919–922.

———— and ————, 1966b. Effects of benzyladenine on accumulation of [32]P into nucleic acids of peanut cotyledons, *Biochem. Biophys. Acta,* **114**:640–642.

Cathey, H. M., 1964. Physiology of growth retarding chemicals, *Ann. Rev. Plant Physiol.*, **15**:271–302.

Chen, J. L., and S. G. Wildman, 1967. Functional chloroplast polyribosomes from tobacco leaves, *Science,* **155**:1271–1273.

Cherry, J. H., 1964. Association of rapidly metabolized DNA and RNA, *Science,* **146**:1066–1069.

———— and R. B. van Huystee, 1965. Comparison of messenger RNA in photoperiodically induced and noninduced *Xanthium* buds, *Science,* **150**:1450–1453.

Chiang, K. S., and N. Sueoka, 1967. Replication of chloroplast DNA in *Chlamydomonas reinhardi* during vegetative cell cycle: Its mode and regulation, *Proc. Nat. Acad. Sci.,* **57**:1506–1513.

Chun, E. H. L., M. H. Vaughan, Jr., and A. Rich, 1963. The isolation and characterization of DNA associated with chloroplast preparations, *J. Mol. Biol.*, **7**:130–141.

Clark, J. M., Jr., A. Y. Chang, S. Spiegelman, and M. E. Reichmann, 1965. The *in vitro* translation of a monocistronic message, *Proc. Nat. Acad. Sci.*, **54**:1193–1197.

Clark, M. F., R. E. F. Mathews, and R. K. Ralph, 1964. Ribosomes and polyribosomes in *Brassica pekinensis, Biochim. Biophys. Acta,* **91**:289–304.

Cleland, Robert, 1963a. The occurrence of auxin-induced pectin methylation in plant tissues, *Plant Physiol.*, **38**:738–740.

———, 1963b. Independence of effects of auxin on cell wall methylation and elongation, *Plant Physiol.*, **38**:12–18.

———, 1965. Auxin-induced cell wall loosening in the presence of actinomycin D, *Plant Physiol.*, **40**:595–600.

———, 1968. Auxin and wall extensibility: Reversibility of auxin-induced wall-loosening process, *Science,* **160**:192–194.

Clever, V., 1964. Puffing in giant chromosomes of *Diptera* and the mechanisms of its control, in J. Bonner and P. Ts'o (eds.), "The Nucleohistones," pp. 314–317, Holden-Day, San Francisco.

Clowes, F. A. L., 1965. Synchronization in a meristem by 5-amino-uracil, *J. Exp. Botany,* **16**:581–586.

Cook, J. R., and M. Carver, 1966. Partial photo-repression of the glyoxylate by-pass in *Euglena, Plant Cell Physiol.,* **7**:377–383

Corneo, G., C. Moore, D. R. Sanadi, L. I. Grossman, and J. Marmur, 1966. Mitochondrial DNA in yeast and some mammalian species, *Science,* **151**:687–689.

Cornforth, J. W., B. V. Milborrow, G. Ryback, and P. F. Wareing, 1965. Identity of sycamore "dormin" with abscisin II, *Nature,* **205**:1269–1270.

Crawley, J. C. W., 1965. The fine structure of isolated *Acetabularia* nuclei, *Planta,* **65**:205–217.

Cronshaw, J., and G. B. Bouck, 1965. The fine structure of differentiating xylem elements, *J. Cell Biol.*, **24**:415–431.

Cross, B. E., J. F. Grove, J. MacMillan, and T. P. C. Mulholland, 1956. Gibberellic acid. IV. Structures of gibberic and allogibberic acids and possible structures of gibberellic acid, *Chemistry & Industry,* 954–955.

———, K. Norton, and J. C. Stewart, 1968. The biosynthesis of gibberellins, III, *J. Chem. Soc.,* 1054–1056.

Datko, A. H., and G. A. Maclachan, 1968. Indoleacetic acid and synthesis of glucanases and pectic enzymes, *Plant Physiol.,* **43**:735–742.

Davies, J. W., and E. C. Cocking, 1966. Protein synthesis studies with a cell-free plastid preparation from tomato fruit locule tissue, *Biochem. J.,* **101**:28P.

Dekker, J., 1963. Effect of kinetin on powdery mildew, *Nature,* **197**:1027–1028.

DeMaggio, A. E., 1966. Phloem differentiation: induced stimulation by gibberellic acid, *Science,* **152**:370–372.

Dennis, D. T., C. D. Upper, and C. A. West, 1965. An enzymic site of gibberellin biosynthesis by Amo 1618 and other plant growth retardants, *Plant Physiol.*, **40**:948–952.

———— and C. A. West, 1967. Biosynthesis of gibberellins: III. The conversion of (—)-kaurene to (—)-kauren-19-oic acid in endosperm of *Echinocystis macrocarpa* Greene, *J. Biol. Chem.*, **242**:3293–3300.

Devlin, R. M., and W. T. Jackson, 1961. Effect of *p*-chlorophenoxyisobutyric acid on rate of elongation of root hairs of *Argrostic alba* L., *Physiol. Plantarum,* **14**:40–48.

Dyer, T. A., and R. M. Leech, 1967. Two distinct low molecular weight ribonucleic acids localized within the chloroplasts of *Vicia faba* leaves, *Biochemical J.*, **102**:6.

Edelman, M., C. A. Cowan, H. T. Epstein, and J. A. Schiff, 1964. Studies of chloroplast development in *Euglena:* VII. Chloroplast-associated DNA, *Proc. Nat. Acad. Sci.*, **52**:1214–1219.

————, H. T. Epstein, and J. A. Schiff, 1966. Isolation and characterization of DNA from the mitochondrial fraction of *Euglena, J. Mol. Biol.*, **17**:463–469.

Eidlic, L., and F. C. Neidhardt, 1965. Role of valyl-sRNA synthetase in enzyme repression, *Proc. Nat. Acad. Sci.*, **53**:539–543.

Eisenstadt, J. M., and G. Brawerman, 1964. The protein-synthesizing systems from the cytoplasm and the chloroplasts of *Euglena gracilis, J. Mol. Biol.*, **10**:392–402.

El-Antably, H. M. M., P. F. Wareing, and J. Hillman, 1967. Some physiological responses to D,L abscisin (dormin), *Planta,* **73**:74–90.

Evans, L. T., 1964. Inflorescence initiation in *Lolium temulentrum* L.: VI. Effects of some inhibitors of nucleic acid, protein, and steroid biosynthesis, *Aust. J. Bio. Sci.*, **17**:24–35.

Evans, M., and P. M. Ray, 1969. *J. Gen. Physiol.*, **53**, in press.

Fairfield, S. A., C. J. Penington, and W. E. Barnett, 1969. Light-induced modifications in the tRNAs of *Euglena gracilis, Fed. Proc.*, **28**:350.

Fan, D. P., A. Higa, and C. Levinthal, 1964. Messenger RNA decay and protection, *J. Mol. Biol.*, **8**:210–222.

Filner, P., and J. Varner, 1967. A test for *de novo* synthesis of enzymes: Density labeling with $H_2O^{18}$ of barley $\alpha$-amylase induced by gibberellic acid, *Proc. Nat. Acad. Sci.*, **58**:1520–1526.

————, J. L. Wray, and J. E. Varner, 1969. Enzyme induction in higher plants, *Science,* **165**:358–367.

Fitch, W. M., and E. Margoliash, 1967. Construction of phylogenetic trees, *Science,* **155**:279–284.

Fosket, D. E., and J. G. Torrey, 1969. Hormonal control of cell proliferation and xylem differentiation in cultured tissues of *Glycine max* var. *Biloxi, Plant Physiol.*, **44**:871–880.

Fowden, L., 1958. New amino acids of plants, *Biol. Rev.*, **33**:393–441.

Frey-Wyssling, A., and K. Mühlethaler, 1965. "Ultrastructural Plant Cytology," 377 pp., Elsevier Publishing Co., Amsterdam.

Fujimoto, Y., 1963. Kinetics of microbial growth and substrate consumption, *J. Theoret. Biol.*, **5**:171–191.

Galston, A. W., 1969. Hormonal regulation in higher plants, *Science*, **163**:1288–1297.

———, P. Jackson, R. Kaur-Sawhney, N. P. Kefford, and W. J. Meudt, 1964. *Colloq. Intern. Centre. Natl. Rech. Sci. (Paris)*, **123**:251–264.

Garner, W. W., and H. A. Allard, 1920. Effect of length of day on plant growth, *J. Agr. Res.*, **18**:553–606.

Garren, L. D., R. R. Howell, and G. M. Tomkins, 1964. Mammalian enzyme induction by hydrocortisone: The possible role of RNA, *J. Mol. Biol.*, **9**:100–108.

Gautheret, R. J., 1955. The nutrition of plant tissue cultures, *Ann. Rev. Plant Physiol.*, **6**:433–484.

Gerhart, J. C., and A. B. Pardee, 1962. The enzymology of control by feedback inhibition, *J. Biol. Chem.*, **237**:891–896.

——— and H. K. Schachman, 1965. Distinct subunits for the regulation and catalytic activity of asparate transcarbamylase, *Biochemistry*, **4**:1054–1062.

Gibor, A., and M. Izawa, 1963. The DNA content of the chloroplasts of *Acetabularia*, *Proc. Nat. Acad. Sci.*, **50**:1164–1169.

——— and S. Granick, 1964. Plastid and mitochondria: Inheritable systems, *Science*, **145**:890–897.

Gilbert, W., and B. Müller-Hill, 1966. Isolation of the LAC repressor, *Proc. Nat. Acad. Sci.*, **56**:1891–1898.

Glasziou, K. T., and J. C. Waldron, 1964. The regulation of invertase synthesis in sugar-cane: Effects of sugars, sugar derivatives, and polyhydric alcohols, *Aust. J. Biol. Sci.*, **17**:609–618.

Goffeau, A., and J. Brachet, 1965. Deoxyribonucleic acid-dependent incorporation of amino acids into the proteins of chloroplasts isolated from anucleate *Acetabularia* fragments, *Biochim. Biophys. Acta*, **95**:302–313.

Goldsmith, M. H. M., 1966a. Maintenance of polarity of auxin movement by basipetal transport, *Plant Physiol.*, **41**:749–754.

———, 1966b. Movement of indoleacetic acid in coleoptiles of *Avena sativa* L: II. Suspension of polarity by total inhibition of the basipetal transport, *Plant Physiol.*, **41**:15–27.

———, 1967a. Separation of transit of auxin from uptake: Average velocity and reversible inhibition by anaerobic conditions, *Science*, **156**:661–663.

———, 1967b. Movement of pulses of labeled auxin in corn coleoptiles, *Plant Physiol.*, **42**:258–263.

——— and M. B. Wilkins, 1964. Movement of auxin in coleoptiles of *Zea mays* L. during geotropic stimulation, *Plant Physiol.*, **39**:151–162.

Goodfriend, T. L., and N. O. Kaplan, 1964. Effects of hormone administration on lactic dehydrogenase, *J. Biol. Chem.*, **239**:130–135.

Gorini, L., and V. R. Beckwith, 1966. Suppression, *Ann. Rev. Microbiol.*, **20**:401–422.

Graham, J. S. D, R. K. Morton, and J. K. Raison, 1964. The *in vivo* uptake and incorporation of radioisotopes into proteins of wheat endosperm, *Aust. J. Biol. Sci.*, **17**:102–114.

Green, B. R., and M. P. Gordon, 1966. Replication of chloroplast DNA of tobacco, *Science*, **152**:1071–1074.

Gregolin, C., E. Ryder, R. C. Warner, A. K. Kleinschmidt, and M. D. Lane, 1966. Liver acetyl CoA carboxylase: The dissociation-reassociation process and its relation to catalytic activity, *Proc. Nat. Acad. Sci.*, **56**:1751–1758.

Grunwald, C., M. Vendrell, and B. B. Stowe, 1967. Gas and other chromatographic separations of indolic methyl esters, *Anal. Biochem.*, **20**:484–494.

Haber, A. H., 1962. Nonessentiality of concurrent cell divisons for degree of polarization of leaf growth: I. Studies with radiation-induced mitotic inhibition, *Am. J. Botany*, **49**:583–589.

———, D. E. Foard, and S. W. Perdue, 1969. Actions of gibberellic and abscisic acids on lettuce seed germination without actions on nuclear DNA synthesis, *Plant Physiol.*, **44**:463–467.

Haberlandt, G., 1890. Die Kleberschicht des Gras-Endosperm als Diastase ausscheidendes Drüsengewebe, *Bei. Den. Botan. Ges.*, **8**:40–48.

Hall, R. H., L. Sconka, H. David, and B. McLennan, 1967. Cytokinins in the soluble RNA of plant tissues, *Science*, **156**:69–71.

Hall, T. C., and E. C. Cocking, 1966. Studies on protein synthesis in tomato cotyledons and leaves: II. Intermediate stages of protein synthesis, *Plant Cell Physiol.*, **7**:343–356.

Halperin, W., 1966. Single cells, coconut milk, and embryogenesis *in vitro*, *Science*, **153**:1287–1288.

Hamilton, T. H., 1968. Control by estrogen of genetic transcription and translation, *Science*, **161**:649–661.

Hamner, K. C., and J. Bonner, 1938. Photoperiodism in relation to hormones as factors in floral initiation, *Botan. Gaz.*, **100**:388–431.

Harada, H., and J. P. Nitsch, 1967. Isolation of gibberellins $A_1$, $A_3$, $A_9$ and of a fourth growth substance from *Althaea rosea* Cav, *Phytochemistry*, **6**:1695–1703.

Haselkorn, R., and V. A. Fried, 1964. Cell-free protein synthesis: Messenger competition for ribosomes, *Proc. Nat. Acad. Sci.*, **51**:1001–1007.

Hellebust, J. A., and R. G. S Bidwell, 1964. Protein turnover in attached wheat and tobacco leaves, *Can. J. Botany*, **42**:1–12.

Heller, J., and E. L. Smith, 1965. The amino acid sequence of cytochrome *c* of *Neurospora crassa*, *Proc. Nat. Acad. Sci.*, **54**:1621–1625.

Heslop-Harrison, J., 1963. Structure and morphogenesis of lamellar systems in grana-containing chloroplasts: I. Membrane structure and lamellar architecture, *Planta,* **60:**243–260.

Hewitt, E. J., 1963. The essential nutrient elements, in F. C. Steward (ed.), "Plant Physiology," vol. III, pp. 137–360, Academic Press, New York.

Hirai, T., and A. Hirai, 1964. Tobacco mosaic virus: Cytological evidence of the synthesis in the nucleus, *Science,* **145:**589–591.

Hock, B., and H. Beevers, 1966. Development and decline of the glyoxalate cycle enzymes in watermelon seedlings, *Z. Pflanzenphysiologie,* **55:**405–414.

Holm-Hansen, O., 1969. Algae: Amounts of DNA and organic carbon in single cells, *Science,* **163:**87–88.

Hoober, J. K., and G. Blobel, 1969. Characterization of the chloroplastic and cytoplasmic ribosomes of *Chlamydomonas reinhardi, J. Mol. Biol.,* **41:**121–138.

Hopkinson, J. M., 1964. Studies on the expansion of the leaf surface. IV. The carbon and phosphorus economy of a leaf, *J. Exp. Botany,* **15:**125–137.

Housley, S., 1961. Kinetics of auxin-induced growth, in W. Ruhland (ed.), "Encyclopedia of Plant Physiology," vol. XIV, pp. 1007–1043, Springer-Verlag, Berlin.

Hu, A. S. L., R. M. Bock, and H. O. Halvorson, 1962. Separation of labeled from unlabeled proteins by equilibrium density gradient centrifugation, *Anal. Biochem.,* **4:**489–504.

Huang, R. C. C., J. Bonner, and K. Murray, 1964. Physical and biological properties of soluble nucleohistones, *J. Mol. Biol.,* **8:**54–64.

———— and ————, 1965. Histone-bound RNA, a component of native nucleohistone, *Proc. Nat. Acad. Sci.,* **54:**960–967.

Huxley, J. S., 1932. "Problems of Relative Growth," Dial Press, New York.

Ikuma, H., and K. V. Thimann, 1964. Analysis of germination processes of lettuce seed by means of temperature and anaerobiosis, *Plant Physiol.,* **39:**756–767.

Ingram, V. M., 1965. "The Biosynthesis of Macromolecules," 223 pp., Benjamin Press, New York.

Iwasaki, H., 1965. Nutritional studies of the edible seaweed *Porphyra tenera:* I. The influence of different $B_{12}$ analogues, plant hormones, purines and pyrimidines on the growth of *Conchocelis, Plant Cell Physiol.,* **6:**325–336.

Jacobs, W. P., 1962. Longevity of plant organs: Internal factors controlling abcission, *Ann. Rev. Plant Physiol.,* **13:**403–436.

————, J. A. Shield, Jr., and D. J. Osborne, 1962. Senescence factor and abcision of Coleus leaves, *Plant Physiol.,* **37:**104–106.

Johri, M. M., and J. E. Varner, 1968. Enhancement of RNA synthesis in isolated pea nuclei by gibberellic acid, *Proc. Nat. Acad. Sci.,* **59:**269–276.

Jones, D. F., J. MacMillan, and M. Radley, 1963. Plant hormones: III. Identification of gibberellic acid in immature barley and immature grass, *Phytochemistry*, **2**:307–314.

Kang, B. G., C. S. Yocum, S. P. Burg, and P. M. Ray, 1967. Ethylene and carbon dioxide: Mediation of hypocotyl hook-opening response, *Science*, **156**:958–959.

Karlson, P. (ed.), 1965. "Mechanism of Hormone Action," Academic Press, New York.

Kawarada, A., H. Kitamurg, Y. Seta, N. Takahashi, M. Takai, S. Tamura, and Y. Sumiki 1955. Biochemical studies on bakanae fungus: XXXV. The relation among gibberellins. $A_1$, $A_2$, and gibberellic acid, *Bull. Agr. Chem. Soc. Japan*, **19**:278–281.

Kende, H., and J. E. Tavares, 1968. On the significance of cytokinin incorporation into RNA, *Plant Physiol.*, **43**:1244–1248.

Kenney, F. T., and W. L. Albritton, 1965. Repression of enzyme synthesis at the translational level and its hormonal control, *Proc. Nat. Acad. Sci.*, **54**:1693–1698.

Key, J. L., and J. Ingle, 1964. Requirement for the synthesis of DNA-like RNA for growth of excised plant tissue, *Proc. Nat. Acad. Sci.*, **52**:1382–1388.

———— and J. C. Shannon, 1964. Enhancement by auxin of ribonucleic acid synthesis in excised soybean hypocotyl tissue, *Plant Physiol.*, **39**:360–364.

Khorana, H. G., 1965. Polynucleotide synthesis and the genetic code, *Fed. Proc.*, **24**:1473–1487.

Kirk, J. T. O., 1963. The deoxyribonucleic acid of broad bean chloroplasts, *Biochim. Biophys. Acta*, **76**:417–424.

————, 1964. DNA-dependent RNA synthesis in chloroplast preparations, *Biochim. Biophys. Res. Comm.*, **14**:393–397.

Kleinsmith, L. J., V. G. Allfrey, and A. E. Mirsky, 1966. Phosphorylation of nuclear protein early in the course of gene activation in lymphocytes, *Science*, **154**:780–781.

Knight, G. J., and C. A. Price, 1968. Measurements of $S$-$\rho$ coordinates during the development of *Euglena* chloroplasts, *Biochim. Biophys. Acta*, **158**:283–285.

Kohler, D., 1956. Über die Beziehungen zwischen der Länge von Haferkoleoptilen und der Wachstumgeschwindigkeit ihrer isolierten Ausschnitte, *Plante*, **47**:159–164.

Korner, A., 1964. Regulation of the rate of synthesis of messenger ribonucleic acid by growth hormone, *Biochem J.*, **92**:449–456.

Koshland, D. E., and K. E. Neet, 1968. The catalytic and regulatory properties of enzymes, *Ann. Rev. Biochem.*, **37**:359–410.

Kuenzler, E. J., 1965. Glucose-6-phosphate utilization by marine algae, *J. Phycology*, **1**:156–164.

———— and B. H. Ketchum, 1962. Rate of phosphorus uptake by *Phaeodactylum tricornutum, Biol. Bull.,* **123**:134–135.

———— and J. P. Parras, 1965. Phosphatases of marine algae, *Biol. Bull.,* **128**:271–284.

Labarca, C., P. B. C. Nicholls, and R. S. Bandurski, 1965. A partial characterization of indoleacetylinositols from *Zea mays, Biochem. Biophys. Res. Comm.,* **20**:641–646.

————, R. S. Bandurski, and P. B. C. Nicholls, 1966. Separation and characterization of four low-molecular-weight bound auxins, *Plant Research,* 38–40.

Lang, A., 1957. The effect of gibberellin on flower formation, *Proc. Nat. Acad. Sci.,* **43**:709–717.

Larner, J., M. Rosell-Preez, D. Friedman, and J. Craig, 1963. Hormonal control of UDPG-glycogen transglucosylase, *Biochem. J.,* **89**:36.

Leonard, N. J., S. Achmatowixz, R. N. Loeppky, K. L. Carraway, W. A. H. Grimm, A. Szweykowska, H. Q. Hamzi, and F. Skoog, 1966. Development of cytokinin activity by rearrangement of 1-substituted adenines to 6-substituted aminopurines: Inactivation by $N^6$, 1-cyclization, *Proc. Nat. Acad. Sci.,* **56**:709–716.

Leopold, A. C., 1964. "Plant Growth and Development," 466 pp., McGraw-Hill Book Company, New York.

———— and M. Kawase, 1964. Benzyladenine effects on bean leaf growth and senescence, *Am. J. Botany,* **51**:294–298.

———— and O. F. Hall, 1966. Mathematical model of polar auxin transport, *Plant Physiol.,* **41**:1476–1480.

Lethem, D. S., and C. O. Miller, 1965. Identity of kinetin-like factors from *Zea mays, Plant and Cell Physiol.,* **6**:355–359.

Levine, R. P., 1962. "Genetics," 180 pp., Holt, Reinhart and Winston, New York.

Lockhart, J. A., 1965a. The analysis of interactions of physical and chemical factors on plant growth, *Ann. Rev. Plant Physiol.,* **16**:37–52.

————, 1965b. An analysis of irreversible plant cell elongation, *J. Theoret. Biol.,* **8**:264–275.

Lyman, H., 1967. Specific inhibition of chloroplast replication in *Euglena gracilis* by nalidixic acid, *J. Cell Biol.,* **35**:726–730.

Lyttleton, J. W., 1960. Nucleoproteins of white clover, *Biochem. J.,* **74**:82–90.

————, 1962. Isolation of ribosomes from chloroplasts, *Exp. Cell Res.,* **26**:312–317.

McCalla, D. R., M. K. Genthe, and W. Hovanitz, 1962. Chemical nature of an insect gall growth factor, *Plant Physiol.,* **37**:98–103.

McConkey, E. H., and J. W. Hopkins, 1964. The relationship of the nucleolus to the synthesis of ribosomal RNA in HeLa cells, *Proc. Nat. Acad. Sci.,* **51**:1197–1204.

McCready, C. C., 1966. Translocation of growth regulators, *Ann. Rev. Plant Physiol.,* **17**:283–294.

McRae, D. H., and J. Bonner, 1953. Chemical structure and antiauxin activity, *Physiol. Plantarum,* **6**:485–510.

Mahadevan, S., 1963. Conversion of 3-indoleacetaldoxime to 3-indoleacetonitrile by plants, *Arch. Biochem. Biophys.,* **100**:557–558.

———— and K. V. Thimann, 1964. Nitrilase: II. Substrate specificity and possible mode of action, *Arch. Biochem. Biophys.,* **107**:62–68.

Mans, R., and G. Novelli, 1964. Stabilization of the maize seedling amino acid incorporating system, *Biochim. Biophys. Acta,* **80**:127–136.

Marcus, A., and J. Feeley, 1964. Activation of protein synthesis in the imbibition phase of seed germination, *Proc. Nat. Acad. Sci.,* **51**:1075–1079.

———— and ————, 1966. Ribosome activation and polysome formation *in vitro:* Requirement for ATP, *Proc. Nat. Acad. Sci.,* **56**:1770–1777.

Margulies, M. M., 1965. Relationship between red light mediated glyceraldehyde-3-phosphate dehydrogenase formation and light dependent development of photosynthesis, *Plant Physiol.,* **40**:57–61.

Mayer, A. M., and A. Poljakoff-Mayber, 1963. "The Germination of Seeds," 236 pp., Pergamon Press, New York.

Medawar, P. B., 1945. In W. E. Le Gros Clark and P. B. Medawar (eds.), "Essays on Growth and Form: Presented to D'Arcy Wentworth Thompson," pp. 157–187, Oxford Press, London.

Mendiola, L. R., A. P. Hirvonen, A. Kovacs, and C. A. Price, 1969. Zonal sedimentation of chloroplast ribosomes from *Euglena gracilis, Fed. Proc.,* **28**:879.

Miller, C. O., 1961. Kinetin and related compounds in plant growth, *Ann. Rev. Plant Physiol.,* **12**:395–408.

————, 1965. Evidence for the natural occurrence of zeatin and derivatives: Compounds from maize which promote cell division, *Proc. Nat. Acad. Sci.,* **54**:1052–1058.

Milthorpe, F. L., and P. Newton, 1963. Studies on the expansion of the leaf surface: III. The influence of radiation on cell division and leaf expansion, *J. Exp. Botany,* **14**:483–495.

Minamikawa, T., T. Akazawa, and I. Uritani, 1963. Analytical study of umbelliferone and scopoletin synthesis in sweet potato roots infected by *Ceratocystis fimbriata, Plant Physiol.,* **38**:493–497.

Monod, J., 1942. "La croissance des cultures bacteriennes," Hermann et cic., Paris.

————, J. P. Changeux, and F. Jacob, 1963. Allosteric proteins and cellular control systems, *J. Mol. Biol.,* **6**:306–329.

————, J. Wyman, and J. P. Changeux, 1965. On the nature of allosteric transitions: A plausible model, *J. Mol. Biol.,* **12**:88–118.

Morgan, A. R., R. D. Wells, and H. G. Khorana, 1966. Studies on polynucleotides: LIX. Further condon assignments from amino acid incorporations directed by ribopolynucleotides containing repeating trinucleotide sequences, *Proc. Nat. Acad. Sci.,* **56**:1899–1906.

Mori, S., K. Kumazawa, and S. Mitsui, 1965. Stimulation of release of reducing sugars from the endosperms of rice seeds by helminthosporol, *Plant Cell Physiol.*, **6:**571–574.

Morré, D. J., 1965. Changes in tissue deformability accompanying actinomycin D inhibition of plant growth and ribonucleic acid synthesis, *Plant Physiol.*, **40:**615–619.

Morton, K., B. A. Palk, and J. K. Raison, 1964. Intracellular components associated with protein synthesis in developing wheat endosperm, *Biochem. J.*, **91:**522–528.

Mothes, K., 1960. Über das Altern der Blätter und die Möglichkeit ihrer Wiederverjungung, *Naturwiss.*, **47:**337–351.

Moudrianakis, E. N., and M. Beer, 1965. Base sequence determination in nucleic acids with the electron microscope: III. Chemistry and microscopy of guanine-labelled DNA, *Proc. Nat. Acad. Sci.*, **53:**564–571.

Müller, K., and A. C. Leopold, 1966. The mechanism of kinetin-induced transport in corn leaves, *Planta*, **68:**186–205.

Naqvi, S. M., and S. A. Gordon, 1967. Auxin transport in *Zea mays* coleoptiles: II. Influence of light on the transport of indoleacetic acid-2-$^{14}$C, *Plant Physiol.*, **42:**138–143.

Nation, J. L., and F. A. Robinson, 1966. Gibberellic acid: Effects of feeding in an artificial diet for honeybees, *Science*, **152:**1765–1766.

Negi, S. S., and H. P. Olmo, 1966. Sex conversion in a male *Vitis vinifera* L. by a kinin, *Science*, **152:**1624–1625.

Neumann, J., and M. E. Jones, 1964. End-product inhibition of aspartate transcarbamylase in various species, *Arch. Biochem. Biophys.*, **104:**438–447.

Newcomb, E. H., 1969. Plant microtubules, *Ann. Rev. Plant Physiol.*, **20:**253–288.

Nickerson, W. J., and S. Bartnicki-Garcia, 1964. Biochemical aspects of morphogenesis in algae and fungi, *Ann. Rev. Plant Physiol.*, **15:**327–344.

Ninnemann, H., J. A. D. Zeevaart, H. Kende, and A. Lang, 1964. The plant-growth retardant CCC as inhibitor of gibberellin biosynthesis in *Fusarium moniliforme* (*Gibberella fujikurvi*), *Planta*. **61:**229–235.

Nitsan, J., and A. Lang, 1966. DNA synthesis in the elongating nondividing cells of the lentil epicotyl and its promotion by gibberellin, *Plant Physiol.*, **41:**965–970.

Nitsch, J. P., and C. Nitsch, 1969. Haploid plants from pollen grains, *Science*, **163:**85–87.

Noodén, L. O., and K. V. Thimann, 1963. Evidence for a requirement for protein synthesis for auxin-induced cell enlargement, *Proc. Nat. Acad. Sci.*, **50:**194–200.

Ohkuma, K., F. T. Addicott, O. E. Smith, and W. E. Thiessen, 1965. Structure of abcissin II, *Tetrahedron Letters*, **29:**2529–2535.

Ordin, L., R. Cleland, and J. Bonner, 1957. Methyl esterification of cell wall constituents under the influence of auxin, *Plant Physiol.*, **32**:216–220.

Osborne, D. J., 1962. Effect of kinetin on protein and nucleic acid metabolism in *Xanthium* leaves during senescence, *Plant Physiol.*, **37**:595–602.

——— and D. R. McCalla, 1961. Rapid bioassay for kinetin and kinins using senescing leaf tissue, *Plant Physiol.*, **36**:219–221.

Paleg, L. G., 1960. Physiological effects of gibberellic acid: II, *Plant Physiol.*, **35**:902–906.

———, 1965. Physiological effects of gibberellins, *Ann. Rev. Plant Physiol.*, **16**:291–322.

———, D. Aspinall, B. Coombe, and P. Nicholls, 1964. Physiological effects of gibberellic acid: VI. Other gibberellins in three test systems, *Plant Physiol.*, **39**:286–290.

Paleg, L., and B. Hyde, 1964. Physiological effects of gibberellic acid: Electron microscopy of barley aleurone cells, *Plant Physiol.*, **39**:673–680.

Parsons, P., and M. V. Simpson, 1967. Biosynthesis of DNA by isolated mitochondria: Incorporation of thymidine triphosphate-2-$C^{14}$, *Science,* **155**:91–93.

Phinney, B. O., and C. A. West, 1960. Gibberellins as native plant growth regulators, *Ann. Rev. Plant Physiol.*, **11**:411–436.

Pickard, B. G., and K. V. Thimann, 1964. Transport and distribution of auxin during tropistic response: II. The lateral migration of auxin in phototropism of coleoptiles, *Plant Physiol.*, **39**:341–349.

Pirie, N. W., 1966. Leaf protein as a human food, *Science,* **152**:1701–1705.

Pitot, H. C., C. Peraino, C. Lamra, and A. L. Kennan, 1965. Template stability of some enzymes in rat liver and hepatoma, *Proc. Nat. Acad. Sci.,* **54**:845–851.

Pogo, G. T., V. G. Allfrey, and A . E. Mirsky, 1966. RNA synthesis and histone acetylation during the course of gene activation in lymphocytes, *Proc. Nat. Acad. Sci.,* **55**:805–812.

Pollard, C. J., 1969. The action of gibberellic acid in barley aleurone layers, *Fed. Proc.,* **28**:414.

Pollard, J., E. Shantz, and F. Steward, 1959. The growth-promoting activity of coconut milk: The nature of the nonionic component, *Plant Physiol.*, **34**:vii.

Pollock, M. R., 1959. Induced formation of enzymes, in P. D. Boyer, H. Lardy, and K. Mybäck (eds.), "The Enzymes," vol. I, 2d ed., pp. 619–680, Academic Press, New York.

Poole, R. J., and K. V. Thimann, 1964. Uptake of indole-3-acetic acid and indole-3-acetonitrile by *Avena* coleoptile sections, *Plant Physiol.*, **39**:98–103.

Preiss, J., H. P. Ghosh, and J. Wittkop, 1967. Regulation of the biosynthesis in spinach leaf chloroplasts, in T. Goodman (ed.), "Biochemistry of Chloroplasts," vol. II, pp. 131–153, Academic Press, New York.

Price, C. A., 1962. Repression of phosphatase synthesis in *Euglena gracilis, Science,* **135**:46.

Raacke, I. D., 1959. Protein synthesis with ribonucleoprotein particles from pea seedlings, *Biochim. Biophys. Acta,* **34**:1–9.

Ralph, R. K., R. E. F. Matthews, A. I. Matus, and H. G. Mandel, 1965. Isolation and properties of double-stranded viral RNA from virus-infected plants, *J. Mol. Biol.,* **11**:202–212.

Rawson, J. R., and E. Stutz, 1968. Characterization of *Euglena* cytoplasmic ribosomes and ribosomal RNA by zone velocity sedimentation in sucrose gradients, *J. Mol. Biol.,* **33**:309–314.

Ray, D. S., and P. C. Hanawalt, 1965. Satellite DNA components in *Euglena gracilis* cells lacking chloroplasts, *J Mol. Biol.,* **11**:760–768.

Ray, P. M., 1958. Destruction of auxin, *Ann. Rev. Plant Physiol.,* **9**:81–118.

―――― and A. W. Ruesink, 1962. Kinetic experiments on the nature of the growth mechanism in oat coleoptile cells, *Develop. Biol.,* **4**:377–397.

Reddi, K. K., 1964. Studies on the formation of tobacco mosaic virus ribonucleic acid: Presence of tobacco mosaic virus in the nucleus of the host cell, *Proc. Nat. Acad. Sci.,* **52**:397–401.

―――― , 1966. Studies on the formation of tobacco mosaic virus ribonucleic acid: VII. Fate of tobacco mosaic virus after entering the host cell, *Proc. Nat. Acad. Sci.,* **55**:593–598.

Reed, D. J., 1965. Tryptamine oxidation by extracts of pea seedlings: Effect of growth retardant $\beta$-hydroxyethylhydrazine, *Science,* **148**:1097–1099.

Reichmann, M. E., 1964. The satellite tobacco necrosis virus: A single protein and its genetic code, *Proc. Nat. Acad. Sci.,* **52**:1009–1017.

Richmond, A. E., and A. Lang, 1957. Effect of kinetin on protein content and survival of detached *Xanthium* leaves, *Science,* **125**:650–651.

Riddiford, L. M., and C. M. Williams, 1967. Volatile principle for oak leaves: Role in sex life of the *Polyphemus* moth, *Science,* **155**:589–590.

Ris, H., and W. Plaut, 1962. Ultrastructure of DNA-containing areas in the chloroplast of *Chlamydomonas, J. Cell Biol.,* **13**:383–391.

Robinson, D. R., and G. Ryback, 1969. Incorporation of tritium [(4R)-4-³H] mevalonate into abscisic acid, *Biochem. J.,* **113**:895–897.

Roychoudhury, R., and S. P. Sen, 1965. The effect of gibberellic acid on nucleic acid metabolism in coconut milk nuclei, *Plant Cell Physiol.,* **6**:761–765.

Sager, R., 1965. Genes outside the chromosomes, *Sci. Am.,* **212**:70–79.

―――― and M. R. Ishida, 1963. Chloroplast DNA in *Chlamydomonas, Proc. Nat. Acad. Sci.,* **50**:725–730.

―――― and Z. Ramanis, 1963. The particulate nature of nonchromosomal genes in *Chlamydomonas, Proc. Nat. Acad. Sci.,* **50**:260–268.

Sampson, M., A. Katoh, Y. Hotta, and H. Stern, 1963. Metabolically labile deoxyribonucleic acid, *Proc. Nat. Acad. Sci.,* **50**:459–463.

Schaffer, A. G., and M. Alexander, 1966. Morphogenic substance in legume nodule formation, *Science,* **152**:82–83.

Schiewer, U., and E. Libbert, 1965. Indolacetamid—ein Intermediat der Indolessigsaurebildung aus Indolacetonitril bei der Alge *Furcellaria, Planta,* **66**:377–380.

Schiff, J. A., and M. H. Zeldin, 1968. The developmental aspect of chloroplast continuity in *Euglena, J. Cell Physiol.,* **72**: Supplement 7, 103–128.

Schwartz, D., 1964. A second hybrid enzyme in maize, *Proc. Nat. Acad. Sci.,* **51**:602–605.

Schweiger, H. G., and S. Berger, 1964. DNA-dependent RNA synthesis in chloroplasts of *Acetabularia, Biochim. Biophys. Acta,* **87**:533–535.

Schwertner, H. A., and P. W. Morgan, 1966. Role of IAA-oxidase in abscission control in cotton, *Plant Physiol.,* **41**:1513–1519.

Scott, N. S., and R. M. Smillie, 1967. Evidence for the direction of chloroplast ribosomal RNA synthesis by chloroplast DNA, *Biochem. Biophys. Res. Comm.,* **28**:598–603.

Shifriss, O., and W. L. George, Jr., 1964. Sensitivity of female inbreds of *Cucumis sativus* to sex reversion by gibberellin, *Science,* **143**:1452–1453.

Shipp, W., and R. Haselkorn, 1964. Double-stranded RNA from tobacco leaves infected with TMV, *Proc. Nat. Acad. Sci.,* **52**:401–408.

Sinclair, R., and B. J. Stevens, 1966. Circular DNA filaments from mouse mitochondria, *Proc. Nat. Acad. Sci.,* **56**:508–514.

Sinnott, E. W., 1960. "Plant Morphogenesis," 550 pp., McGraw-Hill Book Company, New York.

Sissokian, N. J., I. Filippovich, E. N. Svetialo, and K. A. Aliyev, 1965. On the protein-synthesizing system of chloroplasts, *Biochim. Biophys. Acta,* **95**:474–485.

Skoog, F., and C. O. Miller, 1957. Chemical regulation of growth and organ formation in plant tissues cultivated *in vivo, Symp. Soc. Exp. Biol.,* **11**:118–131.

——, F. M. Strong, and C. O. Miller, 1965. Cytokinins, *Science,* **148**:532–533.

——, D. J. Armstrong, J. D. Cherayil, A. E. Hampel, and R. M. Bock, 1966. Cytokinin activity: Localization in transfer RNA preparations, *Science,* **154**:1354–1356.

Slama, K., and C. M. Williams, 1966. Juvenile hormone: V. Sensitivity of the bug *Pyrrhocoris apterus* to a hormonally-active factor in American paper pulp, *Biol. Bull.,* **130**:235–246.

Slonimski, P. P., 1953. "La formation des enzymes respiratoires chez la levure," 203 pp., Masson et Cie., Paris.

Smillie, R. M., 1963. Formation and function of soluble proteins in chloroplasts, *Can. J. Botany,* **41**:123–154.

———— and N. S. Scott, 1969. Organelle biosynthesis, *Prog. Molecular and Subcell. Biology,* **1**: in press.

Smith, E. L., R. J. De Lange, and J. Bonner, 1969. On the chemistry of histones, *Fed. Proc.,* **28**: in press.

Smith, S. H., and D. E. Schlegel, 1964. Incorporation of uridine-H³ into nuclei of virus-infected tobacco, *Science,* **145**:1058–1059.

Spencer, D., 1965. Protein synthesis by isolated spinach chloroplasts, *Arch. Biochem. Biophys.,* **111**:381–390.

———— and P. R. Whitfeld, 1966. The nature of the ribonucleic acid of isolated chloroplasts, *Arch. Biochem. Biophys.,* **117**:337–346.

———— and ————, 1967. DNA synthesis in isolated chloroplasts, *Biochem. Biophys. Res. Comm.,* **28**:538–542.

Spiegelman, S., I. Haruna, I. D. Holland, G. Beaudreau, and D. Millls, 1965. The synthesis of self-propagating and infectious nucleic acid with a purified enzyme, *Proc. Nat. Acad. Sci.,* **54**:927.

Stahmann, M. A., B. G. Clare, and W. Woodbury, 1966. Increased disease resistance and enzyme activity induced by ethylene and ethylene production by black rot infected sweet potato tissue, *Plant Physiol.,* **41**:1505–1512.

Steinhart, C. E., L. Anderson, and F. Skoog, 1962. Growth promoting effect of cyclitols on spruce tissue cultures, *Plant Physiol.,* **37**:60–65.

————, J. D. Mann, and S. H. Mudd, 1964. Alkaloids and plant metabolism: VII. The kinetin-produced elevation in tyramine methylpherase levels, *Plant Physiol.,* **39**:1030–1038.

Stern, A. I., J. A. Schiff, and H. T. Epstein, 1964. Studies of chloroplast development in *Euglena:* V. Pigment biosynthesis, photosynthetic oxygen evolution and carbon dioxide fixation during chloroplast development, *Plant Physiol.,* **39**:220–226.

Stern, H., 1963. Internal regulatory mechanisms in chromosome replication and segregation, *Fed. Proc.,* **22**:1097–1102.

Steward, F. C., M. O. Mapes, A. E. Kent, and R. D. Holsten, 1964. Growth and development of cultured plant cells, *Science,* **143**:20–27.

———— and R. G. S. Bidwell, 1966. Storage pools and turnover systems in growing and non-growing cells: Experiments with ¹⁴C-sucrose, ¹⁴-C-glutamine, and ¹⁴C-asparagine, *J. Exp. Botany,* **17**:726–741.

Still, C. C., C. C. Oliver, and H. S. Moyed, 1965. Inhibitory oxidation product of indole-3-acetic acid: Enzyme formation and detoxification by pea seedlings, *Science,* **149**:1249–1251.

Stowe, B. B., 1959. Occurrence and metabolism of simple indoles in plants, *Fortschr. Chem. org Naturstoff,* **17**:248–297.

———— and T. Yamaki, 1959. Gibberellins: Stimulants of plant growth, *Science,* **129**:807–816.

Strittmatter, C. F., 1957. Adaptive variation in the level of oxidative activity in *Saccharomyces cerevisiae, J. Gen. Microbiol.,* **16**:169–183.

Suyama, Y., and W. D. Bonner, Jr., 1966. DNA from plant mitochondria, *Plant Physiol.,* **41**:383–388.

Svetailo, E. N., I. I Fillippovich, and N. M. Sissakian, 1967. Differences in sedimentation properties of chloroplast and cytoplasmic ribosomes of pea seedlings, *J. Mol. Biol.,* **24**:405–415.

Sweeney, B. M., and J. W. Hastings, 1962. Rhythms, in R. A. Lewin (ed.), "Physiology and Biochemistry of Algae," pp. 687–700, Academic Press, New York.

Syrett, P. M., 1966. The kinetics of isocitratelyase formation in *Chlorella:* Evidence for the promotion of enzyme synthesis by photophosphorylation, *J. Exp. Botany,* **17**:641–654.

Takimoto, A., and K. C. Hamner, 1964. Effect of temperature and preconditioning on photoperiodic response of *Pharbitis nil, Plant Physiol.,* **39**:1024–1030.

Talal, N., and G. M. Tomkins, 1964. Allosteric properties of glutamate dehydrogenases from different sources, *Science,* **146**:1309–1311.

Tamiya, H., T. Sasa, T. Nihei, and S. Ishibashi, 1955. Effect of variation of day-length, day and night-temperatures, and intensity of daylight upon the growth of *Chlorella, J. Gen. Appl. Microbiol.,* **4**:298–307.

Taylor, M. M., and R. Storck, 1964. Uniqueness of bacterial ribosomes, *Proc. Nat. Acad. Sci.,* **52**:958–965.

Tewari, K. K., and S. G. Wildman, 1966. Chloroplast DNA from tobacco leaves, *Science,* **153**:1269–1271.

———— and ————, 1968. Function of chloroplast DNA with RNA from cytoplasmic (80S) and chloroplast (70S) ribosomes, *Proc. Nat. Acad. Sci.,* **59**:569–576.

Thatch, R. E., M. A. Cecere, T. A. Sundararajan, and P. Doty, 1965. The polarity of messenger translation in protein synthesis, *Proc. Nat. Acad. Sci.,* **54**:1167–1173.

Thimann, K. V., 1937. On the nature of inhibition caused by auxins, *Am. J. Botany,* **24**:407–412.

————, 1954. The physiology of growth in plant tissues, *Am. Scientists,* **42**:589–606.

————, 1963. Plant growth substances; past, present, and future. *Ann. Rev. Plant Physiol.,* **14**:1–18.

————, 1964. Nitrilase: I. Occurrence, preparation, and general properties of the enzyme, *Arch. Biochem. Biophys.,* **105**:133–141.

————, 1967. Phototropism, in M. Florkin and E. H. Stotz (eds.), "Comprehensive Biochemistry," vol. 27, pp. 1–29, Elsevier, Amsterdam.

———— and G. M. Curry, 1960. Phototropism and phototaxis, in M. Florkin and H. S. Mason (eds.), "Comparative Biochemistry," vol. 1, pp. 243–309, Academic Press, New York.

———— and ————, 1961. Phototropism, in W. D. McElroy and B. Glass (eds.), "Light and Life," pp. 646–670, Johns Hopkins Press, Baltimore.

Tomkins, G. M., K. L. Yielding, J. F. Curran, M. R. Summers, and M. W. Bitensky, 1965. The dependence of the substrate specificity on the conformation of crystalline glutamate dehydrogenase, *J. Biol. Chem.*, **240**:3793–3798.

Torrey, J. G., and R. S. Loomis, 1967. Auxin-cytokinin control of secondary vascular tissue formation in isolated roots of *Raphanus, Amer. J. Bot.*, **54**:1098–1106.

Trosko, J. E., and S. Wolff, 1965. Strandedness of *Vicia faba* chromosomes as revealed by enzyme digestion studies, *J. Cell Biol.*, **26**:125–135.

Tuan, D. Y. H., and J. Bonner, 1964. Dormancy associated with repression of genetic activity, *Plant Physiol.*, **39**:768–772.

Tuli, V., and H. S. Moyed, 1966. Desensitization of regulatory enzymes by a metabolite of plant auxin, *J. Biol. Chem.*, **241**:4564–4566.

van Overbeek, J., 1966. Plant hormones and regulators, *Science,* **152**:721–730.

————, H B. Tukey, F. W. West, and R. M. Muir, 1954. Nomenclature of chemical plant regulators, *Plant Physiol.*, **29**:307–308.

Varner, J. E., 1964. Gibberellic acid controlled synthesis of α-amylase in barley endosperm, *Plant Physiol,* **39**:413–415.

———— and A. R. Chandra, 1964. Hormonal control of enzyme synthesis in barley endosperm, *Proc. Nat. Acad. Sci.,* **52**:100–106.

Vogel, H. J., 1957. Repression and induction as control mechanisms of enzyme biogenesis: The "adaptive" formation of acetylornithinease, in W. D. McElroy and B. Glass (eds.), "A Symposium on the Chemical Basis of Heredity," pp. 276–289, Johns Hopkins Press, Baltimore.

————, 1961. Control by repression, in D. M. Bonner (ed.), "Control Mechanisms in Cellular Processes," pp. 23–65, Ronald Press, New York.

————, D. F. Bacon, and A. Baich, 1963. Induction of acetylorinthine α-transaminase during pathway-wide repression, in H. Vogel, V. Bryson, and J. O. Lampen (eds.), "Informational Macromolecules," pp. 293–300, Academic Press, New York.

———— and R. H. Vogel, 1967. Regulation of protein synthesis, *Ann. Rev. Biochem.*, **36**:519–538.

Wada, S., E. Tanimoto, and Y. Masuda, 1968. Cell elongation and metabolic turnover of the cell wall, as affected by auxins and cell wall-degrading enzymes, *Plant and Cell Physiol.*, **9**:369–376.

Wain, R. L., 1967. Some developments in research on auxins and kinins, *Ann. N.Y. Acad. Sci.,* **144**:223–234.

Watson, J. D., 1965. "The Molecular Biology of the Gene," 494 pp., W. A. Benjamin, New York.

———— and F. Crick, 1953. Molecular structure of nucleic acid, *Nature,* **171**:737–738.

Weaver, R. J., 1961. Growth of grapes in relation to gibberellin, *Adv. Chem.,* **28**:89–108.

Weber, D. J., and M. A. Stahmann, 1964. Ceratocystis infection in sweet potato: Its effect on proteins, isozymes, and acquired immunity, *Science,* **146**:929–931.

Wehrmeyer, W., 1964. Über Membranbildungsprozesse im Chloroplasten: Mitteilung zur Entstehung der Grana durch Membranüberschiebung, *Planta,* **63**:13–30.

Went, F. W., 1955. Ecology of desert plants, *Sci. Am.,* April, p. 68.

———— and K. V. Thimann, 1937. "Phytohormones," 294 pp., Macmillan, New York.

Werz, Gunther, 1964. Untersuchungen zur Feinstruktur des Zellkernes und des perinuclearen Plasmas von *Acetabularia, Planta,* **62**:255–271.

Whaley, W. G., 1961. Growth as a general process, in W. Ruhland (ed.), "Encyclopedia of Plant Physiology," vol. XIV, pp. 71–112, Springer-Verlag, Berlin.

Wilkins, M. B., and T. K. Scott, 1968. Auxin transport in roots, *Nature,* **219**:1388–1389.

Willetts, N. S., 1967. Intracellular protein breakdown in non-growing cells of *Escherichia coli, Biochem. J.,* **103**:453–461.

Wilson, A. C., and A. B. Pardee, 1964. Comparative aspects of metabolic control, in M. Florkin (ed.), "Comparative Biochemistry," vol. VI, pp. 73–118, Academic Press, New York.

Winter, A., 1966. A hypothetical route for the biogenesis of IAA, *Planta,* **71**:229–239.

———— and K. V. Thimann, 1966. Bound indoleacetic acid in *Avena* coleoptiles, *Plant Physiol.,* **41**:335–342.

Yoda, S., 1958. Auxin action and pectic enzyme, *Botan. Mag. (Tokyo),* **71**:207–213.

Yomo, J., 1958. Barley malt: Sterilization of barley seeds and the formation of amylase by separated embryos and endosperms, *Hakko Kyokaishi,* **16**:444–448.

Zeevaart, J. A. D., 1966. Reduction of the gibberellin content of *Pharbitis* seeds by CCC and after-effects in the progeny, *Plant Physiol.,* **41**:856–862.

Zenk, M. H., and H. Scherf, 1964. Verbreitung der *d*-Tryptophan-Konjugations-Mechanismen im Pflanzenreich, *Planta,* **62:**350–354.

Zetsche, K., 1964. Der Einfluss von Actinomycin D auf die Abgabe morphogenetischer Substanzen aus dem Zellkern von *Acetabularia mediterranea, Naturwissenschaften,* **1:**18–19.

————, 1965. Anreicherung von morphogenetischen Substanzen in Lichtpflanzen von *Acetabularia mediterranea* unter dem Einfluss von Puromycin, *Planta,* **64:**119–128.

Ziegler, H., I. Ziegler, and K. Beth, 1967. The light-induced increase in the activity of NADP+-dependent glyceraldehyde-3-phosphate-dehydrogenase: VI. The influence of the nucleus on the effect, *Planta,* **72:**247–251.

Zucker, M., 1969. Induction of phenylalanine ammonia-lyase in *Xanthium* leaf disks, *Plant Physiol.,* **44:**912–922.

Zuckerkandl, E., and L. Pauling, 1965. Molecules as documents of evolutionary history, *J. Theoret. Biol.,* **8:**357–366.

# Appendix a
# Metabolic
# Pathways

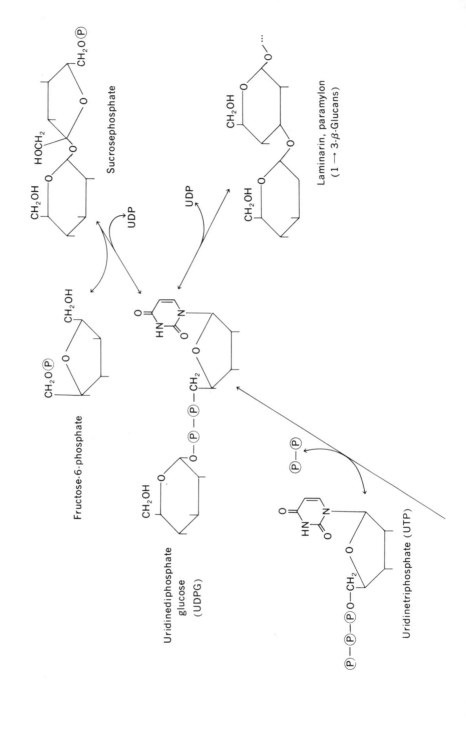

Sucrosephosphate

Laminarin, paramylon
$(1 \rightarrow 3\text{-}\beta\text{-Glucans})$

UDP

UDP

Fructose-6-phosphate

Uridinediphosphate
glucose
(UDPG)

Uridinetriphosphate (UTP)

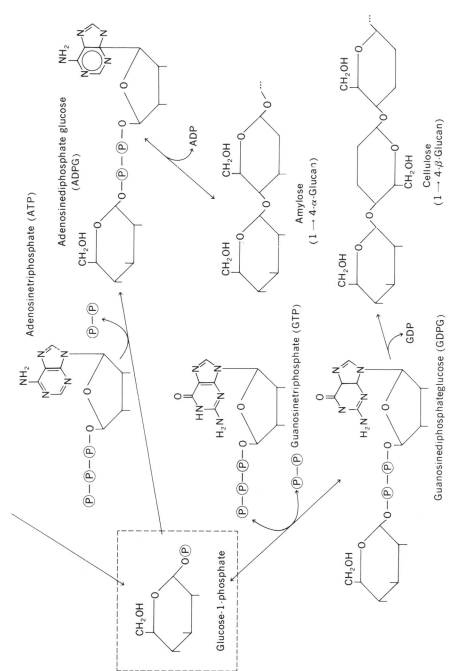

Scheme 1   Glycan synthesis

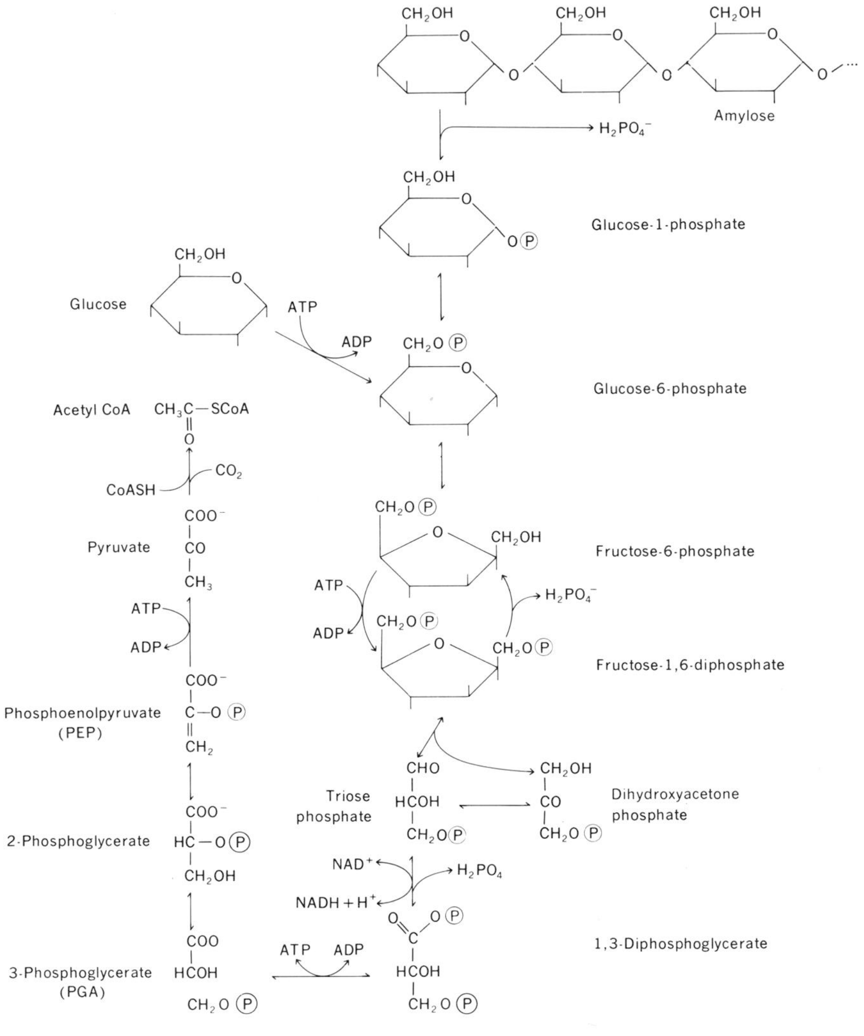

*Scheme 2*   Embden-Meyerhof-Parnas  pathway

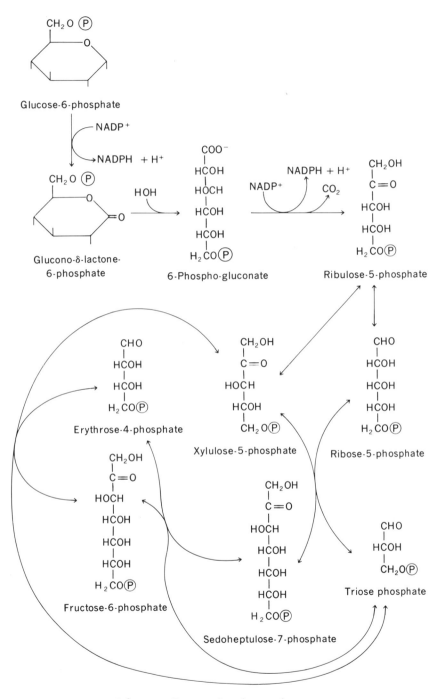

*Scheme 3* Pentose phosphate pathway

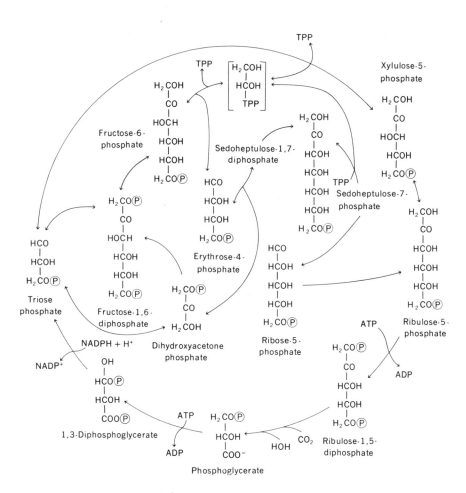

*Scheme 4*   Calvin-Benson  cycle

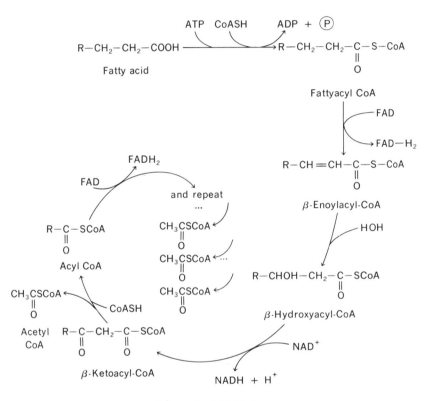

*Scheme 5*    β-Oxidation

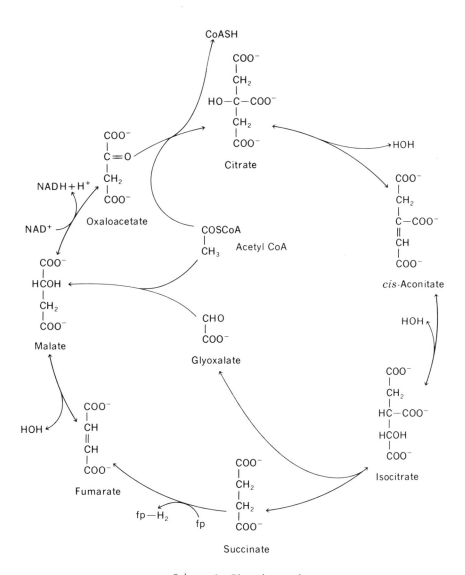

*Scheme 6*   Glyoxalate cycle

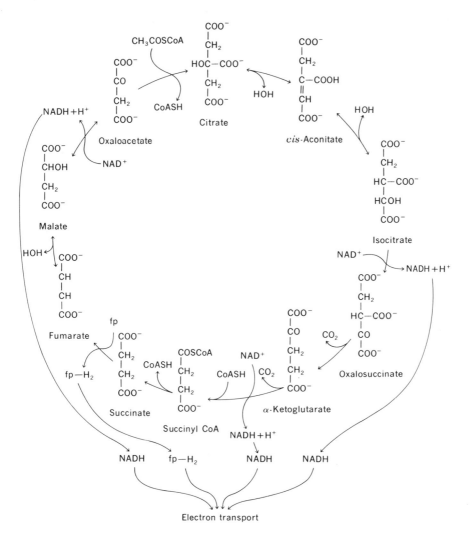

*Scheme* 7   Krebs  cycle

# Appendix b
# Equations

## 1. MICHAELIS-MENTEN EQUATIONS (BRIGGS AND HALDANE FORMULATION)

Consider the reaction of enzyme with substrate:

$$E_f + S \underset{k_2}{\overset{k_1}{\rightleftharpoons}} \overline{ES} \overset{k_3}{\to} E_f + P$$

where

$E_f \equiv$ free enzyme
$S \equiv$ substrate
$\overline{ES} \equiv$ enzyme-substrate complex
$P \equiv$ product
$k_i =$ arbitrary rate constants

At steady state, the rate of formation of $\overline{ES}$ must equal its rate of decomposition:

$$\left(\frac{d\,\overline{ES}}{dt}\right)_{forward} = -\left(\frac{d\,\overline{ES}}{dt}\right)_{backward}$$

Hence,

$$k_1\,E_f S = -(k_2 + k_3)\,\overline{ES}$$

The rate of substrate disappearance $v$ equals the forward reaction:

$$v = k_3\,\overline{ES}$$

At the limit of $S \to \infty$, all of the enzyme will tend to become $\overline{ES}$:

$$V_{max} = k_3\,E_t$$

where

$$E_t = E_f + \overline{ES}$$

Combining and substituting,

$$v = \frac{k_3\,E_t\,S}{(k_2 + k_3)/k_1 + S} \quad \text{and}$$

$$v = \frac{V_{max}\,S}{(k_2 + k_3)/k_1 + S} \qquad \text{Eq. 1-3}a$$

The Michaelis constant

$$K_M \equiv \frac{k_2 + k_3}{k_1}$$

Hence

$$v = \frac{V_{max}\,S}{K_M + S} \qquad \text{Eq. 1-3}$$

## 2. EFFECT OF TEMPERATURE ON EQUILIBRIA: THE VAN'T HOFF EQUATION

$$\frac{d \ln K}{dT} = \frac{\Delta H}{RT^2}$$

where

$K =$ equilibrium constant
$T =$ absolute temperature
$\Delta H =$ enthalpy
$R =$ gas constant

In the integrated form,

$$\log \frac{K_1}{K_2} = \frac{\Delta H}{2.303R} \left( \frac{1}{T_1} - \frac{1}{T_2} \right)$$

## 3. EFFECT OF TEMPERATURE ON REACTION RATES: THE ARRHENIUS EQUATION

$$k = A\epsilon^{-E/RT}$$

where

$k =$ reaction rate constant
$A =$ arbitrary constant, but one which can have physical significance
$E =$ activation energy
$R =$ gas constant
$T =$ absolute temperature

In the integrated form,

$$\log \frac{k_1}{k_2} = - \frac{E}{2.303R} \left( \frac{1}{T_1} - \frac{1}{T_2} \right)$$

## 4. FIRST-ORDER KINETICS

When the disappearance of a chemical species $N$ is proportional to the concentration of that species, as in radioactive decay, the rate will be

$$\frac{-dN}{dt} = kN$$

where $k$ is the reaction rate constant. Integrating,

$$\int_{N_1}^{N_2} \frac{dN}{N} = -k \int_0^t dt$$

$$\ln \frac{N_2}{N_1} = -kt \qquad \text{or} \qquad \frac{N_2}{N_1} = \epsilon^{-kt}$$

## 5.  BINARY FISSION KINETICS

If the rate of increase of a species $N$ is proportional to the concentration of $N$, the reaction velocity will be

$$\frac{dN}{dt} = kN$$

where $k$ is the reaction rate constant. Integrating,

$$\int_{N_1}^{N_2} \frac{dN}{N} = k \int_0^t dt$$

$$\ln \frac{N_2}{N_1} = kt \qquad \text{or} \qquad \frac{N_2}{N_1} = \epsilon^{kt}$$

Note that these equations differ in sign from the first-order kinetics in 4.

## 6.  KINETICS OF APPEARANCE OF LABEL IN TRANSFORMATION OR TRANSPORT PROCESSES[1]

In each case a negligible quantity of radioactive species $(A*)$ is introduced at zero time. The appearance of label in B or C is calculated for different kinetic models (from Francis et al., 1959). $A_0^*$ refers to specific activity at zero time.

Irreversible binary process:

$$A \xrightarrow{k} B$$

$$A* = A_0^* \, \epsilon^{-kt}$$

$$B* = A_0^* \, (1 - \epsilon^{-kt})$$

or

$$\ln \left( \frac{A_0^* - B^*}{A_0^*} \right) = -kt$$

Reversible binary process:

$$A \underset{k_2}{\overset{k_1}{\rightleftharpoons}} B$$

$$A* = \frac{k_2}{k_1 + k_2} A_0^* \left[ \left( 1 + \frac{k_1}{k_2} \right) \epsilon^{-(k_1 + k_2)t} \right]$$

$$B* = \frac{k_1}{k_1 + k_2} A_0^* \left[ 1 - e^{-(k_1 + k_2)t} \right]$$

[1] *Reference:* G. E. Francis, W. Mulligan, and A. Wormall, 1959. "Isotopic Tracers," 2d ed., Athlone Press, London, 524 pp.

Irreversible catenary process (cf. Fig. 1-4) :

$$A \xrightarrow{k_1} B \xrightarrow{k_2} C$$

$$\frac{A^*}{A} = \epsilon^{-k_1 t}$$

$$\frac{B^*}{B} = \frac{(2k_2 - k_1)\, \epsilon^{-k_1 t} + k_1(1 - \epsilon^{-k_2 t}) - k_2}{k_2 - k_1}$$

$$\frac{C^*}{C} = \frac{k_2(1 - \epsilon^{-k_1 t}) - k_1(1 - \epsilon^{-k_2 t})}{k_2 - k_1}$$

# Appendix c
# Physical
# Constants
# and
# Relations

$$1 \text{ atmosphere} = 760 \text{ mm Hg}$$
$$= 1.0332 \text{ Kg cm}^{-2}$$
$$= 1.0133 \text{ bars}$$
$$\text{Avogadro's number} = 6.023 \times 10^{23} \text{ molecules mole}^{-1}$$
$$1 \text{ curie} = 3.7 \times 10^{10} \text{ disintegrations sec}^{-1}$$
$$1 \text{ day} = 1.44 \times 10^3 \text{ min} = 8.64 \times 10^4 \text{ sec}$$
$$1 \text{ year} = 5.26 \times 10^5 \text{ min} = 3.16 \times 10^7 \text{ sec}$$
$$c \text{ (velocity of light)} = 3.00 \times 10^{10} \text{ cm sec}^{-1}$$
$$1 \text{ dalton} = 1 \text{ atomic weight unit (e.g., 1 molecule of}$$
$$10^5 \text{ molecular weight weighs } 10^5 \text{ daltons)}$$
$$= 1.66 \times 10^{-24} \text{ g}$$
$$h \text{ (Planck's constant)} = 6.6237 \times 10^{-27} \text{ erg sec}$$
$$1 \text{ poise} = 1.0 \text{ dyne sec cm}^{-2}$$
$$R \text{ (gas constant)} = 82.06 \text{ cc atm deg}^{-1} \text{ mole}^{-1}$$
$$= 8.3144 \times 10^7 \text{ ergs deg}^{-1} \text{ mole}^{-1}$$
$$= 8.3144 \text{ joules deg}^{-1} \text{ mole}^{-1}$$
$$= 1.013 \times 10^6 \text{ dynes cm}^{-2} \text{ deg}^{-1} \text{ mole}^{-1}$$
$$= 1.987 \text{ cal g}^{-1} \text{ mole}^{-1} \text{ deg}^{-1}$$
$$1 \text{ Svedberg (sedimentation unit)} = 10^{-13} \text{ sec}$$

## PREFIXES USED IN METRIC SYSTEM

| | | |
|---|---|---|
| p | pico | $10^{-12}$ |
| n | nano | $10^{-9}$ |
| $\mu$ | micro | $10^{-6}$ |
| m | milli | $10^{-3}$ |
| c | centi | $10^{-2}$ |
| d | deci | $10^{-1}$ |
| | unit | 1 |
| | deka | 10 |
| | hecto | $10^2$ |
| k | kilo | $10^3$ |
| | mega | $10^6$ |
| | giga | $10^9$ |
| | tera | $10^{12}$ |

## SOME UNITS OF WEIGHTS AND MEASURES

$$1 \text{ milligram (mg)} = 10^{-3} \text{ gram (g)}$$
$$1 \text{ microgram } (\mu g, \gamma, \text{mcg}) = 10^{-6} \text{ gram (g)}$$
$$1 \text{ milliliter (ml, cc)} = 1 \text{ cubic centimeter} = 10^{-3} \text{ liter (l)}$$
$$1 \text{ microliter } (\mu l, \lambda) = 10^{-6} \text{ liter (l)}$$
$$1 \text{ micron } (\mu) = 10^{-3} \text{ millimeter (mm)} = 10^{-6} \text{ meter (m)}$$
$$1 \text{ nanometer (nm)} = 1 \text{ millimicron (m}\mu) = 10^{-9} \text{ meter (m)}$$
$$1 \text{ angstrom (Å)} = 0.1 \text{ nanometer (nm)} = 10^{-10} \text{ meter (m)}$$

## SOLUBILITY OF CARBON DIOXIDE AND OXYGEN IN WATER

Milliliters of gas dissolved in 1 ml of water at 1 atm partial pressure and at indicated temperature:

|                | 0°      | 10°     | 15°     | 20°     | 25°     | 30°     |
|----------------|---------|---------|---------|---------|---------|---------|
| Carbon dioxide | 1.713   | 1.194   | 1.019   | 0.878   | 0.759   | 0.665   |
| Oxygen         | 0.04872 | 0.03793 | 0.03441 | 0.03091 | 0.02822 | 0.02612 |

# Appendix d
# Properties of Some Biologically Important Isotopes

| Isotope | Half-life | Specific Activity of Carrier-free Isotope, curies·gram atom$^{-1}$ | Radiation | |
|---|---|---|---|---|
| | | | Kind | Maximum Energy, mev |
| Hydrogen-3(Tritium) | 12.3y | $2.90 \times 10^4$ | $\beta^-$ | 0.018 |
| Carbon-14 | 5600y | 63.8 | $\beta^-$ | 0.155 |
| Calcium-45 | 180d | $7.25 \times 10^5$ | $\beta^-$ | 0.255 |
| Iron-55 | 2.91y | $1.23 \times 10^5$ | $\kappa$ | 0.006 |
| Iron-59 | 46.3d | $2.82 \times 10^6$ | $\gamma, \beta^-$ | 1.29, 0.462 |
| Manganese-54 | 310d | $4.21 \times 10^5$ | $\kappa$ | 0.835 |
| Zinc-65 | 250d | $5.22 \times 10^5$ | $\gamma$ | 1.11 |
| Phosphorus-32 | 14.3d | $9.12 \times 10^6$ | $\beta^-$ | 1.71 |
| Phosphorus-33 | 25d | $5.22 \times 10^6$ | $\beta^-$ | 0.248 |
| Sulfur-35 | 88d | $1.48 \times 10^6$ | $\beta^-$ | 0.167 |

*Note:* specific activity of carrier-free isotope $= 1.3043 \times 10^8$ curies (half-life in days)$^{-1}$

# Index